Energy for the 21st Century

NEW HORIZONS IN ENVIRONMENTAL AND ENERGY LAW

Environmental law – including the pressing considerations of energy law and climate change – is an increasingly important area of legal research and practice. Given the growing interdependence of global society and the significant steps being made towards environmental protection and energy efficiency, there are few people untouched by environmental and energy lawmaking processes.

At the same time, environmental and energy law is at a crossroads. The command and control methodology that evolved in the 1960s and 1970s for air, land and water protection may have reached the limit of its environmental protection achievements. New life needs to be injected into our environmental protection regimes – perhaps through the concept of sustainability in its environmental, economic and social forms. The same goes for energy policy and law, where liberalisation, environmental protection and security of supply are at the centre of attention. This important series seeks to press forward the boundaries of environmental and energy law through innovative research into environmental and energy law, doctrine and case law. Adopting a wide interpretation of environmental and energy law, it includes contributions from both leading and emerging international scholars.

Titles in the series include:

Conservation, Biodiversity and International Law
Alexander Gillespie

Comparative Ocean Governance
Place-Based Protections in an Era of Climate Change
Robin Kundis Craig

Environmental Governance of the Great Seas
Law and Effect
Joseph F.C. DiMento and Alexis Jaclyn Hickman

Conservation on the High Seas
Harmonizing International Regimes for the Sustainable Use of Living Resources
Simone Borg

Global Environmental Governance
Law and Regulation for the 21st Century
Louis Kotzé

Climate Law in EU Member States
Towards National Legislation for Climate Protection
Edited by Marjan Peeters, Mark Stallworthy and Javier de Cendra de Larragán

International Law and Freshwater
The Multiple Challenges
Edited by Laurence Boisson de Chazournes, Christina Leb and Mara Tignino

Genetic Resources, Equity and International Law
Camena Guneratne

Energy for the 21st Century
Opportunities and Challenges for Liquefied Natural Gas (LNG)
Susan L. Sakmar

Energy for the 21st Century

Opportunities and Challenges for Liquefied Natural Gas (LNG)

Susan L. Sakmar

University of Houston Law Center and University of San Francisco School of Law, USA

NEW HORIZONS IN ENVIRONMENTAL AND ENERGY LAW

Edward Elgar

Cheltenham, UK • Northampton, MA, USA

Published by
Edward Elgar Publishing Limited
The Lypiatts
15 Lansdown Road
Cheltenham
Glos GL50 2JA
UK

Edward Elgar Publishing, Inc.
William Pratt House
9 Dewey Court
Northampton
Massachusetts 01060
USA

Reprinted with amendments 2014

A catalogue record for this book
is available from the British Library

Library of Congress Control Number: 2012951741

This book is available electronically in the ElgarOnline.com
Law Subject Collection, E-ISBN 978 1 78100 588 0

ISBN 978 1 84980 421 9

Typeset by Servis Filmsetting Ltd, Stockport, Cheshire
Printed and bound in Great Britain by the CPI Group (UK Ltd)

Contents

Figures

Tables

Appendices

Preface and acknowledgments

This book was written out of a desire to learn as much as possible about one of the fastest-growing segments of the energy industry – liquefied natural gas (LNG). When I began my research, massive LNG export projects were underway in the tiny country of Qatar, the US was running out of natural gas and was widely expected to become one of the world's largest importers of LNG, the global financial crisis had not yet hit markets, and shale gas was not on anyone's radar. What a difference a few years make!

Qatar is now the world's largest LNG exporter, having reached its production capacity goal of 77 million tonnes per annum (MTPA). The US is no longer going to be the world's largest LNG importer but instead may become one of the world's largest LNG exporters thanks to the shale gas revolution that has swept through that country. While the financial crisis seems to ebb and flow, the interest in natural gas and LNG remains persistent with more and more countries turning to LNG to meet growing energy demand with cleaner burning fuels.

While it is impossible to be an expert on all of the issues involving the complex LNG industry, my goal in writing this book was to identify as many as possible of the key opportunities and challenges for LNG in the coming years with the hope that the book would serve as a useful reference for others interested in learning more about the dynamic LNG industry.

I am very grateful to the many experts in the LNG industry that lent me not only their wisdom and knowledge but also their enthusiastic support for the book. This includes Mr Warren True, LNG Editor, *Oil & Gas Journal*, Mr Albert Nahas, VP International Government Affairs, Cheniere Energy Inc., Mr Steven Miles, Partner, Baker Botts, Mr David Wochner, Partner, Sutherland, Mr Jay Copan, Executive Director, LNG17 and Senior Advisor, American Gas Association, and many others whom I have met over the years at various LNG related events.

I also thank the numerous organizations that gave me permission to use their materials for this book, including the International Energy Agency (IEA), EA Gibson, Zeus Intelligence, ExxonMobil, Shell, Chevron, BP, Charles River Associates, Poten & Partners Inc., and others.

I would especially like to thank The CWC Group and, in particular, Tracy Clark, Jenny O'Mahony, and Tanya Crossick for allowing me to attend their excellent energy conferences including CWC's World LNG Summit, CWC's World LNG Series, and CWC's World Shale Gas. I also thank Gavin Sutcliffe from DMG Events for allowing me access to the annual Gastech Conference and Exhibition, which is a highlight in the global gas industry.

I also thank the University of Houston Law Center and the law firm of Andrews Kurth for inviting me to serve as a Visiting Law Professor, Andrews Kurth Energy Law Scholar for 2012–2013 and for providing the opportunity to teach a law course focused on Global Gas Markets. I especially thank Professor Jacqueline L. Weaver, A.A. White Professor of Law, University of Houston Law Center, for her support and interest in the book and my academic pursuits in general. I also extend my appreciation to Professor Ronald D. Ripple, Director, Centre for Research in Energy and Minerals Economics at Curtin Business School, Perth, Australia for his insights and assistance in reviewing the book.

I also extend my appreciation to the University of San Francisco School of Law and Dean Jeffrey S. Brand and Associate Dean Ronald Micon for supporting my academic pursuits in general and for allowing me the privilege to teach at USF as an adjunct law professor for many years.

I also extend my sincere appreciation to my editors and colleagues at Edward Elgar including Tara Gorvine, Rebecca Hastie, Alison Hornbeck, Sheila Milne, and others for their patience and support while I kept extending deadlines to write about key emerging issues such as shale gas and North American LNG exports.

Most importantly, I thank my friends and family for their support and encouragement and for listening to more than they ever wanted to know about natural gas markets and LNG. And finally, I dedicate this book to my two sons, Owen and Dallas, who patiently waited for me to finish what they refer to as "the most anticipated LNG book of the decade!"

Susan L. Sakmar
Visiting Assistant Professor
Andrews Kurth Energy Law Scholar
University of Houston Law Center

Abbreviations

API	American Petroleum Institute
ASEAN	Association of Southeast Asian Nations
BREE	Bureau of Resources and Energy Economics (Australia)
CAAGR	compound average annual growth rate
CAFE	corporate average fuel economy (standards in the US)
CAPEX	capital expenditures
CBM	coal bed methane
CCS	carbon capture and storage
CCT	clean coal technologies
CH_4	methane
CHP	combined heat and power (generation)
CNG	compressed natural gas
CO	carbon monoxide
CO_2	carbon dioxide
CO_2-eq	carbon-dioxide equivalent
COP	Conference of the Parties
CSG	coal seam gas
CTG	coal-to-gas
CTL	coal-to-liquids
DOE	Department of Energy (United States)
DOI	Department of the Interior (United States)
DOT	Department of Transportation (United States)
E&P	exploration and production
EC	European Commission
EGR	enhanced gas recovery
EIA	Energy Information Administration (United States)
EITs	economies in transition: Russia, Former Soviet Union (FSU) and East & Central Europe (ECE)
EOR	enhanced oil recovery
EPA	Environmental Protection Agency (United States)
ERU	emission reduction units
EU	European Union

EU ETS	European Union Emissions Trading System
EUA	European Union allowances
Eurostat	Eurostat is the statistical office of the European Union
EV	electric vehicle
FDI	foreign direct investment
FEED	front-end engineering and design
FERC	Federal Energy Regulatory Commission (United States)
FID	final investment decision
FIT	feed-in tariff
FOB	free-on-board
FSU	Former Soviet Union
FTA	free trade agreement
FYP	Five-Year Plan (China)
G8	group of eight industrialized nations – Canada, France, Germany, Italy, Japan, Russia, United Kingdom and the United States
G20	group of 20 finance ministers and central bank governors. The group is made up of all G8 members, plus Argentina, Australia, Brazil, China, India, Indonesia, Mexico, Saudi Arabia, South Africa, South Korea and Turkey. The European Union is the twentieth member
GCCSI	Global Carbon Capture and Storage Institute
GDP	gross domestic product
GHG	greenhouse gases
GIIGNL	Groupe Internationale des Importateures de Gaz Nature 1 Liquéfié (The International Group of Liquefied Natural Gas Importers)
GTL	gas-to-liquids
H_2	hydrogen
HDV	heavy-duty vehicle
HH	Henry Hub
IAEA	International Atomic Energy Agency
IEA	International Energy Agency
IGU	International Gas Union
IOC	international oil company
IPCC	Intergovernmental Panel on Climate Change
IPO	initial public offering
IPR	intellectual property rights
IRENA	The International Renewable Energy Agency

IRR	internal rates of return
JCC	Japan Customs-cleared Crude
JODI	Joint Organisations Data Initiative (IEA database)
kWe	kilowatt electrical capacity
LDC	least-developed country
LDV	light-duty vehicle
LNG	liquefied natural gas
LPG	liquefied petroleum gas
M&A	mergers and acquisitions
MENA	Middle East and North Africa
NBP	national balancing point
NEM	National Energy Market
NGL	natural gas liquid
NGO	non-governmental organization
NGV	natural gas vehicle
NIMBY	not in my back yard
NOC	national oil company
NO_x	nitrogen oxides
NWS	North West Shelf (Australia)
NYMEX	New York Mercantile Exchange
OCGT	open-cycle gas turbine
OECD	Organisation for Economic Co-operation and Development
OPEC	Organization of the Petroleum Exporting Countries
PCC	pulverized coal combustion
PM	particulate matter
PSA	production-sharing agreement
R&D	research and development
RE	renewable energy (renewables)
RES	renewable energy sources
RET(s)	renewable energy technology(ies)
ROW	rest of the world
SNG	synthetic natural gas
SO2	sulfur dioxide
SWF	sovereign wealth funds
Synfuel	synthetic fuel
Syngas	synthetic gas
UAE	United Arab Emirates
UN	United Nations
UNDP	United Nations Development Programme

UNEP	United Nations Environment Programme
UNFCCC	United Nations Framework Convention on Climate Change
US AID	United States Agency for International Development
USGS	United States Geological Survey
VOC	volatile organic compound
WEC	World Energy Council
WEO	*World Energy Outlook* (IEA publication)
WTI	West Texas Intermediate (crude oil category)
WTO	World Trade Organization
WTW	well-to-wheel

ABBREVIATIONS FOR QUANTITIES

bbl	barrel
bbl/d	barrels per day
bcf or Bcf	billion cubic feet
bcf/d	billion cubic feet per day
bcm	billion cubic meters
bcm/y	billion cubic meters per year
bcma	billion cubic meters per annum
cm	cubic meter
Gj	gigajoule
mb/d	million barrels per day
mcm	thousand cubic meters
MMBtu	million British thermal units
mmcm or MMcm	million cubic meters
MMt	million tonnes
MMt/y	million tonnes per year
MMtpa	million tonnes per annum
Mt or MT	million tonnes
Mta or MTA	million tonnes per annum
Mtoe	million tonnes oil equivalent
Mtpa or MTPA	million tonnes per annum
Pj	petajoule
Tcf	trillion cubic feet
tcm	trillion cubic meters
Tj	terajoule

NATURAL GAS QUICK CONVERSIONS

This book will generally use the measurements provided in the cited source, which unfortunately may result in some inconsistency throughout. For the reader's convenience, the following is provided to make the necessary conversions:

1 million tonnes (liquid) = 48.7 Bcf (gas)
1 tonne (LNG) (liquid) = 53.57 MMBtu (gas)*
1 cf (gas) = 0.0283 cubic meters (gas)
1 cf (gas) = 0.000045 cubic meters (liquid)
1 Bcf (gas) = 45,000 m^3 (LNG)
1 cubic meter (m^3) = 35.315 cubic feet (cf)
Typical tanker holds 2.8 Bcf of gas = 126,000 cubic meters of liquid
1 million tonnes (LNG) = 48.7 Bcf**(gas) = 1.379 billion m^3/year (gas)
1 million tonnes per year (mtpa) (LNG) = 48.7 Bcf/year**(gas) = 1.379 billion m^3/year (gas)

* Assumes a natural gas heating value of 1,100 Btu/cf
** Assumes a specific gravity of LNG at .45

NATURAL GAS CONVERSIONS TABLE

Under approximations for gas conversions

From	To	mmcm	bcm	tcm	mmcf	bcf	Mt LNG	GJ	TJ	PJ	MMBtu
mmcm	Multi-ply by:	1	0.001	1.00×10^{-6}	35.31	3.53×10^{-2}	7.35E-04	38800	38.80	3.88×10^{-2}	36775
Bcm		1000	1	1.00×10^{-3}	35313	35.31	0.735	3.88×10^{7}	38800	38.80	3.68×10^{7}
Tcm		1.00×10^{6}	1000	1	3.53×10^{7}	35313	735	3.88×10^{10}	3.88×10^{7}	38800	3.68×10^{10}
mmcf		0.028	2.83×10^{-5}	2.83×10^{-8}	1	1.00×10^{-3}	2.08×10^{-5}	1099	1	1.10×10^{-3}	1041
Bcf		28.32	0.028	2.83×10^{-5}	1000	1	0.021	1.10×10^{6}	1099	1.099	1.04×10^{6}
Mt LNG		1361	1.361	1.36×10^{-3}	48045	48.04	1	5.28×10^{7}	52787	52.79	5.00×10^{7}
GJ		2.58×10^{-5}	2.58×10^{-8}	2.58×10^{-11}	9.10×10^{-4}	9.10×10^{-7}	1.89×10^{-8}	1	1.00×10^{-3}	1.00×10^{-6}	0.948
TJ		0.026	2.58×10^{-5}	2.58×10^{-8}	0.910	9.10×10^{-4}	1.89×10^{-5}	1000	1	1.00×10^{-3}	948
PJ		25.77	0.026	2.58×10^{-5}	910	0.910	0.019	1.00×10^{6}	1000	1	9.48×10^{5}
MMBtu		2.72×10^{-5}	2.72×10^{-8}	2.72×10^{-11}	9.60×10^{-4}	9.60×10^{-7}	2.00×10^{-8}	1.055	1.06×10^{-3}	1.06×10^{-6}	1

Notes:

1. To convert 10 Mt of LNG into million cubic meters, multiply by 1361
   10Mt LNG = 13610 million cubic meters of gas
2. 1 million cubic meters = 10^{6} × 1.0 cubic meter (m^3)
3. 1 billion cubic meters = 10^{9} × 1.0 cubic meter (m^3)
4. 1 trillion cubic meters = 10^{12} × 1.0 cubic meter (m^3)
5. 1 gigajoule = 10^{9} × 1.0 joule (J)
6. 1 terajoule = 10^{12} × 1.0 joule (J)
7. 1 petajoule = 10^{15} × 1.0 joule (J)
8. 1 British thermal unit = 1055 joules (J)
9. 1 tonne = 10^{3} × 1.0 kilogram (kg) = 2205 pounds (lbs)

Source: Australian Government, Bureau of Resources and Energy Economics (BREE), *Gas Market Report July 2012*, www.bree.gov.au

Introduction

Policy makers around the globe continue to grapple with issues related to energy security, energy affordability, and an expected increase in demand for all energy sources. At the same time, concerns about global climate change and reducing greenhouse gas emissions remain in focus as the world struggles to define the path to a sustainable energy future. Regarding it as an abundant, affordable, and clean-burning fuel, many countries around the world are increasingly looking to natural gas to play a key role in powering the future. The prospects for natural gas are so promising that the International Energy Agency (IEA) has suggested that the 21st century could be the "Golden Age of Gas" with demand for natural gas projected to increase by more than 50 percent from 2010 levels and account for over 25 percent of the world's energy supply mix by 2035.

Along with the increased demand for natural gas comes a corresponding increase in international trade in natural gas, with most of the increased trade expected to be in the form of liquefied natural gas or LNG. LNG is natural gas that has been cooled to approximately −161 °C, at which point it condenses to a liquid that can then be shipped via LNG tanker anywhere in the world. Since the majority of natural gas reserves are located away from key demand markets, LNG offers an important solution for the global gas markets in terms of moving natural gas to markets where it is most needed.

In recent years, the significant increase in interregional LNG trade has led many to question whether the gas markets were "globalizing" and whether LNG would someday trade as a global commodity. Though this issue will be discussed in more detail in Chapter 6, the general consensus that seems to have emerged is that while LNG markets are "globalizing" in terms of the increase in trade and the number of countries now involved in LNG trade, LNG is still not likely to become a global commodity anytime soon for lack of a single pricing structure. Nonetheless, there is widespread recognition that LNG is the "glue" linking global gas markets and, indeed, the Golden Age of Gas would not be possible without LNG.

Energy for the 21st Century explores the growing role of LNG as the "glue" linking global gas markets and identifies the key opportunities and challenges for the LNG industry in the context of a number of competing drivers, including economic development, energy security, and climate change. Going forward, the pace and scale of demand growth for all forms of energy, including natural gas and LNG, ultimately will rest on the climate and energy policies adopted by countries, the global economic recovery, and industry investment. Perhaps the most difficult to predict is global energy policy since decisions about energy policy are inextricably linked to economic, environmental and national security policy, and have significant consequences in all three areas.

Though the pace and scale of the global economic recovery remain uncertain as this book goes to print with energy policies in a state of flux in most regions around the world, this book takes the long-term view that, over time, demand for natural gas and LNG will continue to increase as more countries look to meet rising demand for energy with lower emission fuels. Accordingly, it seems likely that LNG will play an increasingly large role in the world's energy future.

While there are countless opportunities for LNG in markets around the world, there are also numerous challenges that must be confronted by the industry going forward. This book explores both the opportunities and the challenges for LNG in the current contextual reality wherein energy law and policy are increasingly intersecting with environmental law and geopolitics. The book navigates the myriad of legal, policy, and environmental issues facing the LNG industry and provides the reader with a thorough understanding of the critical issues:

- the role of natural gas and LNG in the 21st century (Chapter 1)
- the entire LNG value chain, including a discussion of the liquefaction process, LNG shipping, the regasification process, and the complicated world of natural gas and LNG measurements (Chapter 2)
- the evolution of LNG markets including the history of LNG and an overview of the three major LNG markets – the Asia-Pacific LNG market, the European LNG market, and the North American/Atlantic Basin LNG market (Chapter 3)
- a description and overview of key LNG supply projects around the world (Chapter 4)
- the primary markets driving LNG demand around the world (Chapter 5)
- the increased globalization of LNG markets and whether LNG could someday trade as a global commodity (Chapter 6)
- the numerous safety and environmental issues that have been raised in the context of constructing LNG projects as well as the environmental sustainability of LNG as a fuel for the future (Chapter 7)

- an overview of current LNG mega projects around the world (Chapter 8)
- a discussion of the new LNG players and LNG projects around the world (Chapter 9)
- an in-depth analysis of the shale gas revolution (Chapter 10)
- the potential impact of shale gas on global gas markets including the prospects for North American LNG exports (Chapter 11)
- an overview of some of the key emerging issues in the LNG industry including whether North America will become a major LNG exporter, the potential impact of the Panama Canal expansion project on LNG trade, the growing role of floating LNG (FLNG), the potential influence of the Gas Exporting Countries Forum (GECF) to act as a "Gas OPEC," and the emergence of LNG as a shipping and vehicle fuel to aid in emission reduction efforts around the world (Chapter 12).

Accessible and non-technical in nature, this timely book will serve as an essential reference for practitioners, government officials, energy professionals, academics and anyone interested in 21st-century energy solutions.

1. The role of natural gas and LNG in the 21st century

1.1 OVERVIEW

As the world entered the 21st century, policy makers around the globe were grappling with issues related to energy security, energy poverty, and an expected increase in future demand for all energy sources. At the same time, concerns about climate change and reducing greenhouse gas emissions emerged as primary issues to be addressed in the search for a sustainable energy future.

Regarding it as a clean-burning fuel, many business and policy leaders began to look to natural gas to meet growing energy demand using more environmentally sustainable fuels. As will be discussed in detail in Section 1.5, an important aspect of the increased role of natural gas is the growing importance of LNG, which offers a critical solution for global gas markets in terms of moving natural gas from where it is found to key demand markets. During the first decade of the 21st century, natural gas demand increased significantly, as did LNG's share in worldwide natural gas trade.[1]

For example, according to the United States Energy Information Administration (US EIA), global natural gas consumption doubled between 1980 and 2010, rising from 53 trillion cubic feet (Tcf) to 113 Tcf. North America consumed 29 Tcf of natural gas, which accounted for more than 25 percent of the world's natural gas consumption (although the region had the slowest regional growth rate of 29 percent). The Middle East had the highest growth rate, increasing more than ten-fold from 1.3 Tcf in 1980 to 13.2 Tcf in 2010. (See Figure 1.1.)

---

[1] As will be discussed in more detail in Chapter 3, there is often a tendency to discuss natural gas and LNG as two separate fuels but this is a bit misleading, so at the outset it is important to note that LNG is not a fuel but merely a means of delivering natural gas. Nonetheless, natural gas and LNG are sometimes referred to separately since much of the debate about the role of natural gas has been focused on natural gas in general without a distinction being made as to whether it is piped gas or LNG. This book starts with a discussion of natural gas in general and then focuses in particular on LNG.

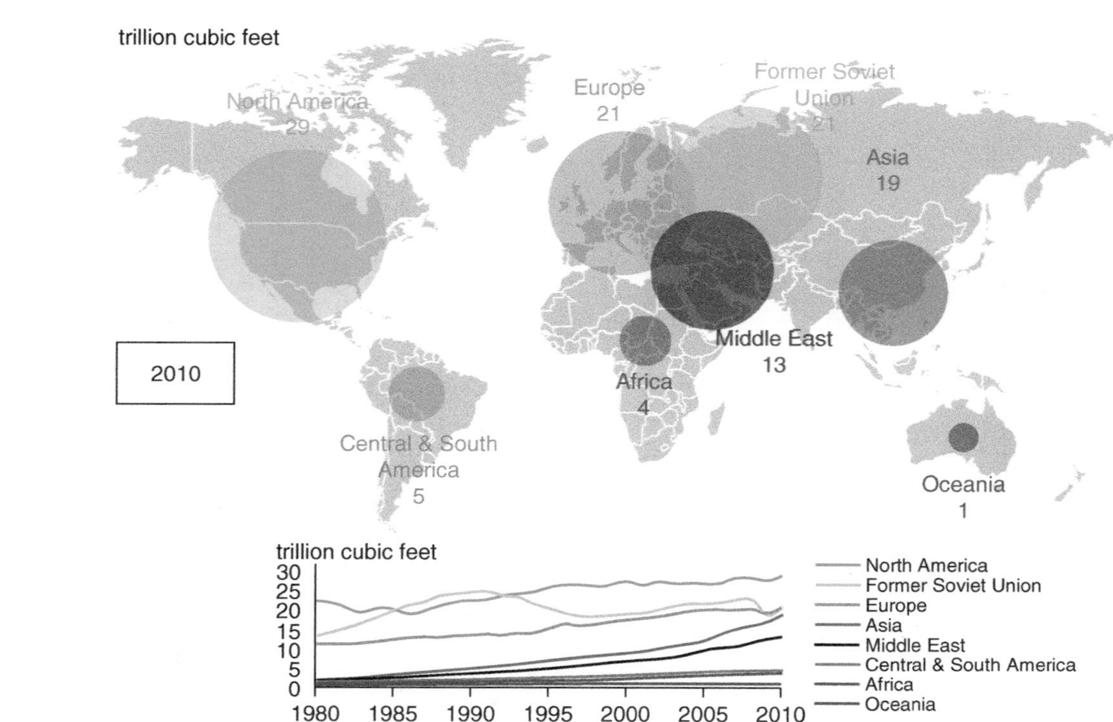

Source: US EIA, Today in Energy, April 12, 2012, http://www.eia.gov/todayinenergy/detail.cfm?id=5810

Figure 1.1 World natural gas consumption by region, 1980–2010

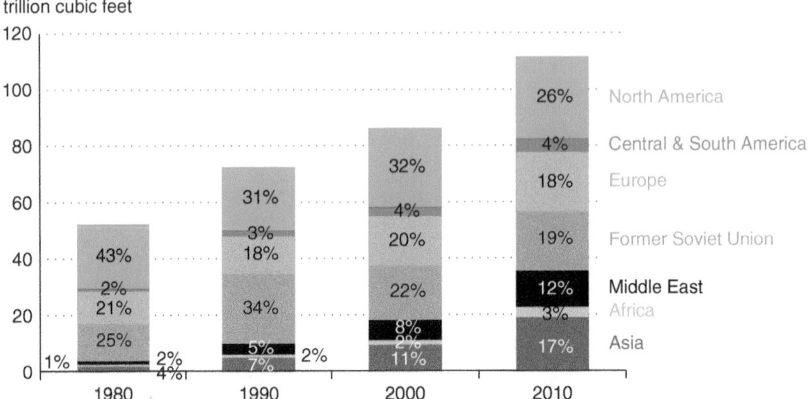

Note: Percents on graph represent that region's share of global dry natural gas consumption in that year. Percents do not sum to 100% for each year because the graph does not include Oceania, which only accounted for 1% of global consumption each year.

Source: US EIA, Today in Energy, April 12, 2012, http://www.eia.gov/todayinenergy/detail.cfm?id=5810

Figure 1.2 World natural gas consumption by region, 1980–2010

According to the US EIA, Asia had the second-highest growth rate, increasing more than eight-fold from 2.2 Tcf to 19.2 Tcf over the three decades. By 2010, Asian natural gas demand approached the level of Europe and the Former Soviet Union.[2] (See Figure 1.2.)

As noted by the US EIA, unlike in the crude oil markets, trends in regional natural gas consumption and production are more similar because of the historically limited role played by inter-continental movements of natural gas. In more recent years, however, inter-regional trade in natural gas via LNG has increased significantly with global LNG shipments increasing 58 percent from 6.7 Tcf to 10.5 Tcf between 2005 and 2010. Between 2009 and 2010, global LNG shipments were up 23 percent.[3]

Despite the growth in natural gas consumption, natural gas as a fuel for the future has received mixed reviews in the global marketplace, especially

---

[2] According to the US EIA, future growth in each region will depend on economic growth rates, natural gas production trends, differences in natural gas prices across regions, and future energy and environmental policies.

[3] US EIA, Today in Energy, April 12, 2012, http://www.eia.gov/todayinenergy/detail.cfm?id=5810, citing BP Statistical Review of World Energy 2011 (http://bp.com/statisticalreview).

as concerns about global climate change grew in the mid-2000s. Some environmental groups view natural gas as yet another fossil fuel with its own set of environmental and emissions considerations. Other groups and policy makers have taken the view that, at the least, natural gas could be a "bridge fuel" to a renewable energy future. Not surprisingly, the energy industry has embraced natural gas as a "foundation" fuel for the 21st century.

During the first decade of the 21st century, these divergent views tended to influence whether or not natural gas and LNG were perceived as a fuel for the future. As new technologies are developed and governments and industry seek new opportunities for natural gas and LNG, these debates are likely to continue throughout the 21st century. Despite the often divergent views about natural gas and LNG, however, as the world enters the second decade of the 21st century, natural gas and LNG seem poised to assume a far greater role in the energy supply mix for many reasons that will be discussed throughout this book.

1.2 THE DIVERGENT VIEWS ABOUT THE ROLE OF NATURAL GAS

As the world entered the 21st century, the role of natural gas in the energy supply mix was anything but clear. As concerns about climate change grew in the early to mid-2000s, there were a number of competing views regarding the role of natural gas coming from the industry, environmentalists, and a large group in the middle.

1.2.1 The Benefits of Natural Gas

Not surprisingly, the worldwide energy industry has embraced natural gas as a "foundation" fuel for the 21st century.[4] In support of its view, the natural gas industry has focused on the many benefits of natural gas and has set forth a coordinated view that highlights natural gas as a clean, affordable, reliable, efficient and abundant source of energy.[5]

---

[4] The American Gas Association (AGA) refers to natural gas as "America's foundation fuel." According to the AGA, "Natural gas is clean, domestic, abundant and efficient, making it the perfect foundation fuel to help strengthen America's economic recovery, meet our environmental challenges and improve our overall national security by reducing our dependence on foreign energy sources." AGA 2012 Policy Handbook, www.aga.org.

[5] International Gas Union (IGU), *Advocacy Messages for the Natural Gas*

Natural gas is clean Natural gas produces less emissions than any other fossil fuel and the most advanced combined-cycle gas turbine (CCGT) power plants emit almost 50 percent less CO_2 than coal-fired power plants.

Natural gas is affordable Natural gas power plants have a capital cost of less than half of the cost of coal, one-third of the cost of nuclear, and one-fifth of the cost of onshore wind.

Natural gas is reliable In contrast to renewable technologies that in some cases may take decades of research, natural gas is readily available now from a variety of sources. Natural gas is also a reliable back-up power source for intermittent energy sources such as wind and solar which facilitates the phase-in of renewables.

Natural gas is efficient Modern gas-fired power plants are 40 percent more efficient than coal-fired power plants and require less construction time than coal or nuclear power plants.

Natural gas is abundant Global production of natural gas is expected to increase in the coming decade with growing supplies coming from both conventional and unconventional resources such as shale gas. The impact of shale gas on global gas markets is discussed in detail in Chapters 10 and 11.

1.2.2 From Big Oil to Big Gas?

In addition to more focused efforts to highlight the benefits of natural gas to the public, many of the world's largest international oil companies (IOCs) are increasingly focusing their core businesses on natural gas.[6] For example, in December 2009 ExxonMobil (Exxon) announced plans to buy XTO Energy (XTO) in an all-stock transaction worth about $41 billion (including debt of $10 billion), which would make Exxon the world's largest natural gas company in terms of gas reserves.[7] Exxon's interest in

---

Sector, Nov. 2010, http://www.igu.org/gas-advocacy/Gas%20Advocacy%20IGU%20Presentation.pptx/view. Similarly, the Canadian Gas Association has advocated that "Natural gas is smart energy. Abundantly available, clean and affordable, natural gas is a safe and reliable energy choice for buildings and communities, power generation, and transportation." Canadian Gas Association, http://www.cga.ca/about-natural-gas/.

[6] *Beyond the Black Stuff, Big Oil is Being Forced to Rethink its Future*, The Economist, Feb. 4, 2010, http://www.economist.com/node/15473681.

[7] *ExxonMobil to Boost Unconventional Focus by Acquiring XTO*, Oil & Gas J.,

XTO was driven primarily by XTO's strong unconventional gas resource base and its technical expertise in extracting shale gas through hydraulic fracturing technology.[8]

The Exxon/XTO merger was seen by many in the oil and gas industry as a show of confidence in the future of shale gas and many praised the deal as a boost for natural gas to play a greater role in supplying the world with abundant, affordable, and cleaner-burning energy. At the same time, however, the merger also focused more scrutiny on the potential environmental impacts of US shale gas development.[9] Despite the intense scrutiny of the deal, including a US Congressional Hearing, the XOM/XTO merger closed on June 25, 2010.[10]

Royal Dutch Shell PLC (Shell) is also betting big on natural gas with plans to make gas roughly half of its total production by 2012.[11] Shell also believes that natural gas, and LNG especially, will play critical roles in meeting global energy demand to 2050, during which time the world must reduce greenhouse gas emissions by half.[12]

Another example of Big Oil's shift to Big Gas occurred in February 2009

Dec. 21, 2009, at 31; see *Natural Gas Helps Exxon and Shell Lift Profits*, N.Y. Times, July 30, 2010, at B4.

[8] Some industry experts have opined that Exxon's rush to natural gas is being driven largely by declining oil reserves and a shrinking access to oil fields around the world due to geopolitical reasons. Russell Gold and Angel Gonzalez, *Exxon Struggles to Find New Oil*, Wall St. J., Feb. 6, 2011.

[9] The ExxonMobil-XTO Merger: Impact on U.S. Energy Markets: Hearing Before the Subcomm. on Energy and Env't of the H. Comm. on Energy and Commerce, 111th Cong. 53 (2010), available at http://energycommerce.house.gov/Press_111/20100120/transcript_01202010_ee.pdf; Tom Doggett, *Exxon-XTO Merger Draws Scrutiny from Congress*, Reuters, Jan. 20, 2010, available at http://www.reuters.com/article/idUSTRE60J53920100120; see also Russell Gold, *Exxon Can Stop Deal if Drilling Method Is Restricted*, Wall St. J., Dec. 17, 2009, at B3.

[10] *ExxonMobil Announces Completion of All-Stock Transaction for XTO*, Business Wire, June 25, 2010, available at http://www.businesswire.com/portal/site/exxonmobil/index.jsp?ndmViewId=news_view&ndmConfigId=1001106&newsId=2010062; *XTO Energy Inc., Agreement and Plan of Merger*, Dated as of Dec. 13, 2009–Dec. 15, 2009, art. I, IX, available at http://www.faqs.org/sec-filings/091215/XTOENERGY-INC_8-K/dex21.htm.

[11] Nick Snow, *Half of Shell's Production Will be Gas by 2012, CEO Says*, Oil & Gas Journal, Oct. 9, 2009, http://www.pennenergy.com/index/petroleum/display/0480477872/articles/oil-gas-journal/drilling-production-2/2009/10/half-of_shell_s_production.html.

[12] Warren R. True, *Gastech: Shell Sees Critical Roles for Gas, LNG*, Oil & Gas Journal, Apr. 11, 2011, available at http://www.ogj.com/index/article-display/7479183575/articles/oil-gas-journal/volume-109/issue-15/general-interest/gastech-shell-sees-critical-roles-for-gas.html.

when Woodside Petroleum, a leading Australian oil and gas company, unveiled a new corporate logo designed to place a greater emphasis on the future of its liquefied natural gas business. It was just the fourth version of Woodside's logo in the 55-year history of the company, and the first substantial change in 32 years. According to the company, the changed logo, composed of three ellipses coming together to form a "W" and symbolize a flame, better acknowledges Woodside's emergence as a global leader in LNG and the expectation that natural gas will dominate Woodside's production portfolio going forward.[13]

The role of natural gas as an accessible, relatively inexpensive, environmentally friendly and widespread natural source of energy was outlined in a report issued in December 2010 by the European Gas Advocacy Forum. The Gas Advocacy Forum is an informal group of players from the European gas industry and includes Centrica, Eni, E.ON Ruhrgas, Gazprom Export, GDF SUEZ, Qatar Petroleum, Shell and Statoil. According to the report, Europe can reach its climate targets of reducing CO_2 emissions by 80 percent (compared to 1990 levels) by 2050 in a faster and more cost-efficient way if natural gas plays a significant part in the energy mix going forward. If Europe were to switch from coal to gas now, the reduction target can be met at a savings of 400–450 billion Euros if one compares it to the European Climate Foundation roadmap launched earlier this year. Additional cost savings for the period 2030–50 would most likely also be achieved because natural gas in power generation requires lower investments.[14]

The universal support for natural gas by major energy companies is significant since any transformation in the energy sector is almost impossible without such support. This is primarily because most energy companies, whether multinational or national (for example, controlled by the state) are vertically integrated. This means they actively participate along the entire supply chain, from locating the natural reserves to drilling and extracting the reserves, transporting the products around the world, and then refining

---

13 Woodside Media Release, *New Look for Woodside*, Feb. 4, 2009, http://www.woodside.com.au/Investors-Media/Announcements/Documents/04.02.2009%20Media%20Release%20-%20New%20look%20for%20Woodside.pdf.

14 Jorn Madslien and Damian Kahya, *Coal-to-Gas Power Shift* "*to Cut Energy Costs*," BBC News, Dec. 10, 2010, http://www.bbc.co.uk/news/business-11961564; *Gas Paves the Way for European Energy Savings*, Natural Gas for Europe, http://naturalgasforeurope.com/gas-paves-the-way-for-european-energy-savings.htm; *The Future Role of Natural Gas*, A Position Paper by the European Gas Advocacy Forum, http://www.gazpromexport.com/content/file/egaf/Making_the_green_journey_work_web_version.pdf.

and distributing the final products to end users.[15] So, for example, it would be extremely difficult, absent perhaps significant government intervention, to significantly expand the use of natural gas vehicles without the support of major energy companies to help provide the huge infrastructure investments that are needed in terms of refueling stations.

1.2.3 Natural Gas is Still a Fossil Fuel

In a world concerned about climate change and greenhouse gas emissions, some critics of natural gas have taken the view that natural gas is yet another fossil fuel that should *not* play a major role in the world's future energy mix. Critics of natural gas argue that increasing dependence on yet another fossil fuel does not move the world towards a real renewable energy future. These critics point out that natural gas is still a fossil fuel that has some of the same negatives as coal and oil. For example, unlike renewables, natural gas is a fuel resource that we may eventually exhaust. These same critics point out that the recent increase in unconventional shale gas production would not be occurring but for the fact that the US has already exhausted its conventional gas resources. In addition, shale gas drilling comes with its own set of potential environmental risks including potential water contamination and increased greenhouse gas emissions (see Chapter 10).

In terms of emissions, while burning natural gas releases less CO_2 than coal, there are still methane emissions to consider with natural gas production.[16] The main fear of natural gas critics seems to be that the potential dependence on another fossil fuel, even though cleaner burning, could "doom" the world to "another few decades of fossil fuel reliance" at the cost of "making serious inroads in clean energy deployment."[17]

1.2.4 Natural Gas is a "Bridge" Fuel

Some prominent groups have taken the view that, at the very least, natural gas could be a "bridge fuel" to a renewable energy future. This view

---

[15] ExxonMobil is one of the best-known vertically integrated energy companies. See ExxonMobil, 2010 Financial and Operating Review, http://www.exxonmobil.com/Corporate/Files/news_pubs_fo_2010.pdf.

[16] Phil Radford, *"Natural" Gas Fails the Sniff Test*, Greenpeace, June 27, 2011, http://www.greenpeace.org/usa/en/news-and-blogs/campaign-blog/natural-gas-fails-the-sniff-test/blog/35470.

[17] Brian Merchant, *Incoming: A Glut of "Natural Gas is Green" Nonsense*, Treehugger.com, June 9, 2011, http://www.treehugger.com/files/2011/06/incoming-glut-natural-gas-green-nonsense.php.

acknowledges that the abundance of natural gas, particularly US shale gas, creates an opportunity to utilize more natural gas to displace coal or oil, thereby significantly reducing CO_2 emissions.[18] Thus, so long as appropriate low-carbon policies are in place, such as a cap-and-trade system or a carbon tax, natural gas can play an important role as a bridge fuel to a renewable energy future. In the absence of low-carbon policies, however, there is a risk that reliance on natural gas will increase overall energy consumption and displace nuclear or other renewable energy sources for power generation, which ultimately would increase CO_2 emissions.[19]

1.2.5 Natural Gas is Better than Coal . . . But . . .

Other prominent groups have focused their attention on the potential for natural gas to displace coal for power generation. In 2010, researchers at the Massachusetts Institute of Technology (MIT) released the results of a two-year study that analyzed the increased use of natural gas in the US as a short-term substitute for replacing aging coal-fired power plants. The report acknowledged that US energy and climate policy was in a state of flux and cautioned that while natural gas is often touted as a "bridge" to the future, continuing effort is needed to ensure that the bridge has a landing point – such as the expansion of nuclear power or coal power generation using carbon capture and storage (CCS) technology to reduce emissions in the long term. Thus, while the report found that natural gas is less carbon intensive than coal or oil, when one considers the emissions reduction levels required by 2050, the emissions from natural gas start to represent an emissions problem.[20]

More recently, and in the context of opposition to US LNG exports (see Chapter 12), it appears that some environmental groups are gearing up

---

[18] The US based Center for American Progress offers this summary of natural gas as a bridge fuel: "Natural gas is the cleanest fossil fuel – it produces less than half as much carbon pollution as coal. Recent technology advancements make affordable the development of unconventional natural gas resources. This creates an unprecedented opportunity to use gas as a bridge fuel to a 21st-century energy economy that relies on efficiency, renewable sources, and low-carbon fossil fuels such as natural gas." John D. Podesta and Timothy E. Wirth, *Natural Gas, A Bridge Fuel for the 21st Century*, Center for American Progress, Aug. 10, 2009, http://www.americanprogress.org/issues/2009/08/bridge_fuel.html.

[19] Stephen P.A. Brown, Alan J. Krupnick, and Margaret A. Wallis, *Natural Gas: A Bridge to a Low-Carbon Future?* Resources for the Future/NEPI, Dec. 2009, Issue Brief 09-11, http://www.rff.org/RFF/Documents/RFF-IB-09-11.pdf.

[20] *The Future of Natural Gas*, MIT Energy Initiative (MITE), http://web.mit.edu/mitei/research/studies/report-natural-gas.pdf.

to ensure that as coal-fired power plants are retired, they are not replaced with natural gas power plants. For example, the Sierra Club recently announced that it is launching a new "Beyond Gas" campaign that represents a significant expansion of the group's ongoing efforts against other major fossil fuels and is modeled after the decade-old "Beyond Coal" campaign that sought to phase out coal-fired power plants. According to the Sierra Club, it will seek to "prevent new gas plants from being built wherever we can."[21]

1.3 THE GLOBAL ECONOMIC CRISIS AND PROJECTIONS FOR NATURAL GAS LEADING INTO THE COPENHAGEN CLIMATE CHANGE TALKS

In the midst of the debate over the role of natural gas in the future energy supply mix, the global economic crisis hit and between 2008 and 2009, demand for all forms of energy dropped. Demand for natural gas in particular plummeted. At the same time, however, an enormous expansion of gas supply was underway in terms of unconventional or shale gas and LNG. Also in flux was the outcome of climate change negotiations and commitments and their potential impact on world energy markets. All of these issues created unprecedented uncertainty in world energy markets in the late 2000s.

In its *World Energy Outlook 2009*, the International Energy Agency (IEA) noted that the challenges were "urgent and daunting" and that how governments rise to the challenge will have "far-reaching consequences for energy markets."[22] In particular, the IEA noted the upcoming climate change talks to be held in Copenhagen, Denmark, December 7–18, 2009 (COP15) and questioned whether leaders would agree to a successor treaty to the Kyoto Protocol that would put the world on a sustainable energy path.

In terms of demand for natural gas, the IEA noted that, under any scenario, worldwide demand for natural gas was projected to grow in light of constraints under which low-carbon technologies can be deployed. The pace of that demand growth, however, "hinges critically on the strength of

---

21 Amy Harder, *War Over Natural Gas About to Escalate*, National Journal, May 3, 2012, http://www.nationaljournal.com/energy-report/war-over-natural-gas-about-to-escalate-20120503.

22 IEA WEO-2009 at p. 41.

climate policy action."[23] Over the long term, the IEA projected that more stringent policy actions might favor efficiency and low-carbon technologies, thereby reducing natural gas demand.

As the world became mired in economic problems towards the end of 2009, it became increasingly unlikely that world leaders would reach an agreement on a successor treaty to Kyoto at COP15. Ultimately, and just prior to the COP15 conference in Denmark, it was announced that "President Obama and other world leaders have decided to put off the difficult task of reaching a climate change agreement . . . agreeing instead to make it the mission of the Copenhagen conference to reach a less specific 'politically binding' agreement that would punt the most difficult issues into the future."[24]

The result of COP15 was a "political accord" known as the Copenhagen Accord[25] which was negotiated by only a subset of the parties, including the US and China. Since this was not negotiated within the United Nations Framework Convention on Climate Change (UNFCCC)[26] process, it was only "noted" by the COP, which left unclear which governments supported the Accord and the legal and operational significance of the Accord.[27] Needless to say, leading into 2010, global energy markets were in a state of flux with energy and climate change policy in most countries uncertain.

1.4 NATURAL GAS GROWS IN IMPORTANCE IN 2010

By early 2010, the world appeared to be emerging from the worst of the economic crisis and demand for energy resumed its pre-recession upward trajectory.[28] Also starting in 2010 was the growing recognition that regard-

---

23 IEA WEO-2009 at p. 48.

24 Helen Cooper, *Leaders Will Delay Deal on Climate Change*, NY Times, Nov. 14, 2009, http://www.nytimes.com/2009/11/15/world/asia/15prexy.html?_r=1.

25 The Copenhagen Accord is available on the UNFCCC website at http://unfccc.int/resource/docs/2009/cop15/eng/11a01.pdf#page=4.

26 There is a wealth of information and publications about climate change and the climate change conferences and a detailed discussion is beyond the scope of this book. Information on the UNFCCC and the status of climate change discussions can be found at http://unfccc.int/2860.php.

27 Jacob Werksmen, *Associating with the Copenhagen Accord: What Does It Mean?*, World Resources Institute, March 25, 2010, http://www.wri.org/stories/2010/03/associating-copenhagen-accord-what-does-it-mean.

28 As of the date this book goes to print, the economic outlook for the coming

less of the divergent views about natural gas, it would play a greater role in the world's future energy mix for a variety of reasons including demand growth, environmental benefits over other fossil fuels, and energy security. Another reason for the growing importance of natural gas was that in the face of continuing global economic challenges, with most governments facing huge budget deficits, it seemed unlikely that governments, industry and the private sector would make the trillions of dollars' investment needed for renewables. According to the IEA, approximately $18 trillion (in year 2009 dollars) of additional spending is needed on low-carbon energy technologies.[29]

In November 2010, the IEA issued its annual *World Energy Outlook* (IEA WEO-2010), which explicitly highlighted the increased role that natural gas would play in the 21st century. In the WEO-2010, the IEA raised the question "Are we entering the golden age of gas?" and noted that while this may be an exaggeration, natural gas was "certainly set to play a central role in meeting the world's energy needs for at least the next two-and-a-half decades."[30]

In the WEO-2010, the IEA acknowledged at the outset that while the pace of the global economic recovery was key to energy prospects in the near term, it is how governments respond to the "twin challenges of climate change and energy security" that will shape the future of energy in the longer term.[31] The IEA went on to present several policy scenarios that differed according to the level of commitment to these challenges.[32]

The Current Policies Scenario assumes that no policy commitments to meet climate change goals are acted upon. The New Policies Scenario takes account of the broad policy commitments and plans that have been announced by countries around the world, including national pledges to reduce greenhouse gas emissions and phase out fossil energy subsidies, and assumes that governments will actually implement the policies and measures to meet the set goals. The 450 Scenario, which was first presented in the IEA WEO-2008, sets out an energy pathway consistent with the goal of reducing greenhouse gas emissions to around 450 parts per million of

---

years remains uncertain, amid fears of a double-dip recession and burgeoning government deficits. Despite this uncertainty, history has shown that while economic forces may lead to ups and downs in terms of energy demand, over the long term, future energy demand is projected to grow and, along with it, the role of natural gas and LNG in the global energy mix. This book takes this long-term view.

[29] IEA WEO-2010 at pp. 379–416.

[30] IEA WEO-2010 at pp. 179–80.

[31] IEA WEO-2010 at pp. 45, 78–9.

[32] IEA WEO-2010 at p. 79.

CO_2 equivalent (ppm CO_2-eq) in order to limit global temperature increase to 2 °C.[33] For purposes of this discussion, the focus will be on the New Policies Scenario, which was considered to be the most likely scenario.

The IEA WEO-2010 New Policies Scenario encompassed several themes, which are discussed in detail below: world energy demand increases significantly in the coming decades under any scenario; natural gas will play a central role in meeting energy demand; power generation drives demand growth; energy poverty; natural gas for transportation; and climate change emissions targets and the impact on the energy sector.[34]

1.4.1 World Energy Demand Grows under Any Scenario

In the New Policies Scenario, the IEA assumed that world economic growth averages 3.2 percent per year between 2008 and 2035 with non-OECD countries showing the highest growth.[35] World primary energy demand increases by 36 percent between 2008 and 2035, or 1.5 percent per year on average, with non-OECD countries accounting for 93 percent of the projected increase in world primary energy demand, reflecting growth of economic activity, industrial production, population[36] and urbanization.[37]

In particular, the IEA noted that "it is hard to overstate the growing importance of China in global energy markets."[38] In 2009, China overtook the United States to become the world's largest energy user. Between 2000 and 2008, China's energy consumption was more than four times greater than the prior decade and contributed to 36 percent of the growth in global energy use. Even greater growth is projected in the coming decades given that China's per capita energy consumption level remains low compared to the OECD average and that China, with 1.3 billion people, is the

---

33 The IEA devoted much of the discussion in the WEO-2010 to the 450 Scenario, since, in the years prior, policy leaders had expressed willingness to try to reduce emissions sufficiently to limit the global increase in temperature to 2 °C. Since that scenario seems unlikely, a detailed discussion of it is beyond the scope of this book. See IEA, Latest Information, *Prospect of Limiting the Global Increase in Temperature to 2° is Getting Bleaker*, May 30, 2011, http://www.iea.org/index_info.asp?id=1959

34 IEA WEO-2010 at pp. 60–62.

35 IEA WEO-2010 at p. 68.

36 The IEA notes that population growth is an important driver of energy use. World population is projected to grow by 0.9% per year on average, from an estimated 6.8 billion in 2008 to 8.5 billion in 2035. IEA WEO-2010 at p. 64.

37 IEA WEO-2010 at pp. 81–4.

38 IEA WEO-2010 at p. 47.

world's most populous nation. By 2035, China accounts for 22 percent of world energy demand, up from 17 percent today.[39]

As a result of China's importance, global energy projections remain highly sensitive to the key variables that drive energy demand in China, including prospects for economic growth and developments in energy policy. This is a critical factor that will come up again in the IEA's "Golden Age of Gas" Report, which is discussed in detail below in Section 1.5.

India is the second largest contributor to the increase in global energy demand to 2035, accounting for 18 percent of the rise. Outside of Asia, the Middle East experiences the fastest rate of increase at 2 percent per year. In terms of OECD countries, energy demand growth rises slowly to 2035 with the US projected to be the second-largest energy consumer with China the first and India a distant third.[40]

1.4.2 Natural Gas Will Play a Central Role in Meeting Energy Demand to 2035

In terms of gas demand and trends, the IEA WEO-2010 New Policies Scenario highlighted the fact that natural gas is set to play a key role in meeting the world's growing energy needs over the next 25 years.[41] Under each of the three policy scenarios, natural gas is the only fossil fuel for which demand is higher in 2035 than in 2008, although it grows at different rates depending on the scenario (Figure 1.3). In the New Policies Scenario, demand reaches 4.5 trillion cubic metres (tcm) in 2035 – an increase of 1.4 tcm, or 44 percent over 2008 and an average rate of increase of 1.4 percent per year.[42]

Non-OECD countries are the key drivers of demand growth and account for almost 80 percent of the growth in gas demand to 2035, primarily because non-OECD economies and population grow much faster and therefore require more energy use. China's demand grows the fastest at an average rate of almost 6 percent per year and accounts for more than one-fifth of the increase in global demand to 2035.[43] The potential for Chinese gas demand to grow even faster depending on whether coal use is restricted for environmental reasons led the IEA to note that "China could lead us into a golden age for gas."[44]

---

39 IEA WEO-2010 at p. 87.
40 IEA WEO-2010 at pp. 84–8.
41 IEA WEO-2010 at p. 180.
42 IEA WEO-2010 at p. 180.
43 IEA WEO-2010 at pp. 180–81.
44 IEA WEO-2010 at p. 49.

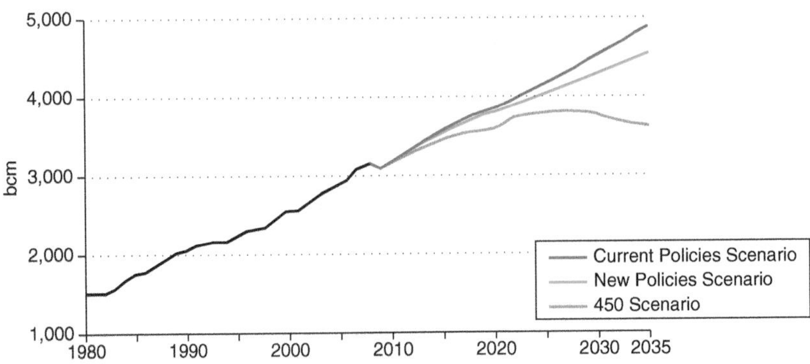

Source: IEA WEO-2010 ©OECD/IEA 2010, Figure 5.1, p. 180

Figure 1.3 World primary natural gas demand by scenario

Although growth in gas demand is the highest in China, somewhat surprisingly demand growth for natural gas in the Middle East increases almost as much as projected in China, primarily driven by the power sector.[45] India is also a key source of demand growth for natural gas.[46] Growth in gas demand in OECD countries is considerably slower than in the non-OECD countries, although the US and Europe remain two of the largest users of natural gas through 2035.[47] (See Figure 1.4.)

1.4.3 Power Generation Drives Demand Growth

In the New Policies Scenario, power generation is the main driver of natural gas demand growth in most regions to 2035 and accounts for almost half of the incremental growth in demand (Figure 1.5). The IEA noted that demand for electricity is expected to grow more strongly than any other final form of energy, growing by 2.2 percent per year between 2008 and 2035 with more than 80 percent of the demand growth occurring in non-OECD countries. In China alone, electricity demand triples between 2008 and 2035 and over the next 15 years China is projected to add generating capacity equivalent to the current total installed capacity of the United States.[48]

45 IEA WEO-2010 at p. 182.
46 IEA WEO-2010 at p. 182.
47 IEA WEO-2010 at p. 183.
48 IEA WEO-2010 at pp. 183–4.

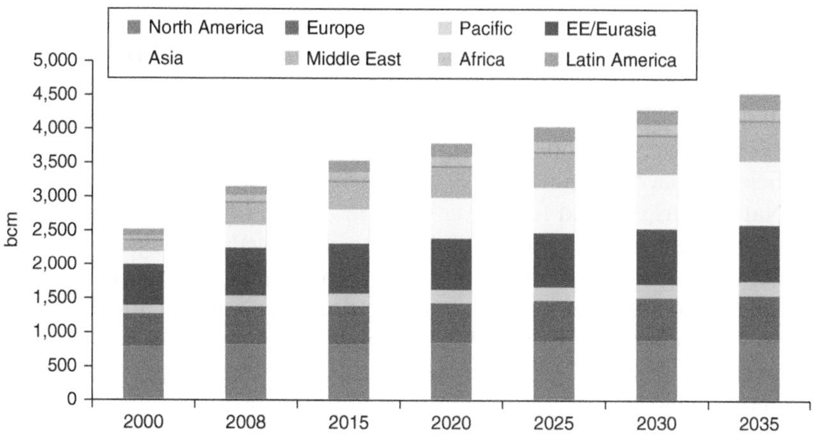

Source: IEA Medium-Term Oil and Gas Markets 2010 ©OECD/IEA 2010

Figure 1.4 Gas demand grows mostly in non-OECD countries, mostly in Asia

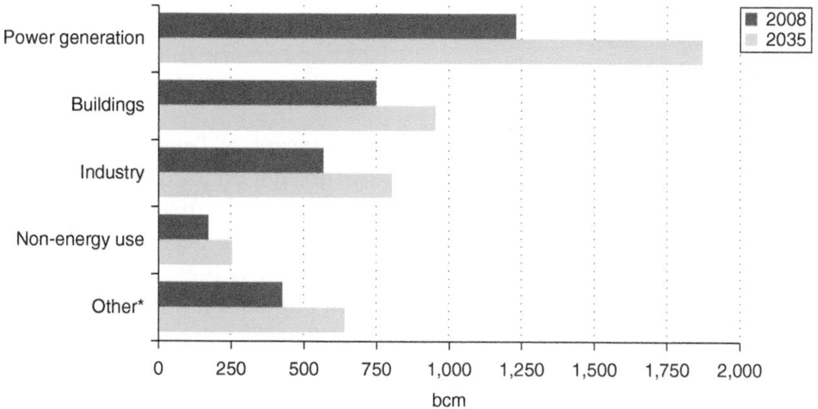

Note: * Includes other energy sector, transport and agriculture.

Source: IEA WEO-2010, ©OECD/IEA 2010, Figure 5.2, p. 184

Figure 1.5 World primary natural gas demand by sector

The world is undergoing a period of profound change in the way electricity is generated as governments shift to low-carbon technologies and fuels to enhance energy security and curb emissions of CO_2. Even assuming slowly rising gas prices, combined-cycle gas turbines (CCGTs) are expected to be the main choice for new power plants in many regions for several reasons. In non-OECD countries, electricity demand is rising rapidly and natural gas fired power plants are easier, less costly and quicker to build than other forms of power generation plants. In OECD countries, natural gas fired power is competitive with coal due to proposed CO_2 prices and policies, which are assumed to be implemented.[49]

Natural gas demand for power generation is lower for those countries with more support for renewables. The use of renewables, including hydro, wind, solar, geothermal, modern biomass and marine energy, triples over the period to 2035 and its share in total primary energy demand increases from 7 percent to 14 percent. Nuclear power increases from 6 percent in 2008 to 8 percent in 2035. The IEA cautioned, however, that the future of renewables hinges critically on strong government support and that the need for such support will increase if natural gas prices are lower than assumed since low natural gas prices will displace more expensive renewables.

The IEA noted that government support for renewables could be justified by the long-term economic, energy-security and environmental benefits renewables can bring but cautioned that attention needs to be given to the cost-effectiveness of support mechanisms.

1.4.4 Energy Poverty

The IEA WEO-2010 also recognized the concept of "energy poverty," which is an emerging issue making its way through policy circles that recognizes that, despite the projected increase in energy use around the world, many households in the developing world still lack access to modern energy services. The numbers are quite striking: the IEA estimates that 1.4 billion people – over 20 percent of the global population – still lack access to electricity and some 2.7 billion people still rely on traditional uses of biomass for cooking. The IEA notes in the New Policies Scenario that energy poverty continues to 2030 and that substantial progress is needed on improving access to energy in the coming decades.[50]

1.4.5 Natural Gas for Transportation

In the New Policies Scenario, natural gas used in the transportation sector accounts for just 4 percent of additional demand during 2008–35.

---

49 IEA WEO-2010 at pp. 183–4.
50 IEA WEO-2010 at pp. 238–71.

Nearly all new gas consumption from natural gas used in vehicles is from compressed natural gas (CNG). Non-OECD Asia, Latin America and North America are responsible for the majority of the increase in demand. The greatest potential may be in North America due to low natural gas prices driven by increased production of shale gas. The scope of demand for natural gas in the transportation sector depends on the future market penetration of natural gas vehicles (NGVs), which today comprise a very small share of the world car fleet (less than 1 percent) and face significant infrastructure hurdles. The greatest potential seems to be with heavy-duty vehicles that are primarily used in fleets and thus face less infrastructure costs.[51]

1.4.6 Climate Change Emissions Targets and the Impact on the Energy Sector

Under the Copenhagen Accord, countries made commitments to reduce their greenhouse gas emissions with the ultimate goal of limiting the global temperature increase to 2 °C. In the WEO-2010, the IEA noted at the outset of its discussion of climate change and the energy sector that the commitments announced under the Copenhagen Accord collectively fall short of what would be required to put the world on a path to achieving the 2 °C goal.

Under the New Policies Scenario, the IEA assumes that countries act upon the commitments in a cautious manner, which has some impact, but that rising demand for fossil fuels continues to drive up energy-related CO_2 emissions through 2035. "Such a trend makes it all but impossible to achieve the 2 °C goal, as the required reductions in emissions after 2020 would be too steep."[52] Nonetheless, the emissions under the New Policies Scenario are a notable improvement from the Current Policies (that is, no action) Scenario, where emissions grow at 1.4 percent per year versus 0.7 percent per year under the New Policies Scenario.

Under the New Policies Scenario emission trends are in line with stabilizing the concentration of greenhouse gases at just over 650 parts per million (ppm) or CO_2-eq, resulting in a likely temperature rise of more than 3.5 °C in the long term.[53] In order to have a reasonable chance of achieving the 2 °C goal, much more vigorous action is needed.

According to climate change experts, in order to achieve this goal, the

---

[51] IEA WEO-2010 at p. 186.
[52] IEA WEO-2010 at pp. 95–7.
[53] IEA WEO-2010 at p. 97.

concentration of greenhouse gases would need to be stabilized at a level no higher than 450 ppm CO_2-eq.[54] The IEA's 450 Scenario describes how the energy sector could evolve to meet this target. Under this scenario, much more ambitious targets than those announced under the Copenhagen Accord are assumed, as is more rapid implementation of the removal of fossil fuel subsidies as agreed by the G-20. These actions bring about a faster transformation of the global energy sector and a correspondingly faster slowdown in global CO_2 emissions. Under the 450 Scenario, oil demand peaks before 2020 at 88 mb/d, coal demand peaks before 2020, and natural gas demand peaks before the end of the 2020s. Renewables and nuclear double their current combined share to 38 percent in 2035. Under the 450 Scenario, additional spending on low-carbon energy technologies (business investment and consumer spending) amounts to $18 trillion (in year 2009 dollars).[55]

1.5 THE GOLDEN AGE OF NATURAL GAS

In early 2011, several significant events transpired which called into question some of the key assumptions in the WEO-2010. As a result of the potential cumulative impact of these events, on June 6, 2011, the IEA released a Special Report titled "Are We Entering a Golden Age of Gas?" (IEA Golden Age Report), which presents a new natural gas focused scenario (GAS Scenario).[56]

The GAS Scenario takes the IEA's WEO-2010 New Policies Scenario as the starting point but incorporates recent assumptions about "policy, prices and other drivers that affect gas demand and supply prospects" over the coming decades.[57] Under the new GAS Scenario, global use of natural gas rises by more than 50 percent from 2010 levels with global

---

[54] IEA WEO-2010 at p. 97.

[55] IEA WEO-2010 at pp. 379–416. In Chapters 13–15 of the WEO-2010, the IEA sets out in detail the climate change goals under the Copenhagen Accord as well as the IEA's 450 Scenario on what is required to achieve those goals and the implications for the energy sector. A detailed discussion of those chapters is beyond the scope of this book, which focuses on the role of natural gas and LNG, but readers interested in climate change and energy are urged to review those chapters for more detail.

[56] IEA World Energy Outlook 2011, Special Report, *Are We Entering a Golden Age of Gas?*, June 6, 2011, http://www.iea.org/weo/docs/weo2011/WEO2011_GoldenAgeofGasReport.pdf (hereinafter IEA Golden Age Report).

[57] IEA Golden Age Report at p. 14.

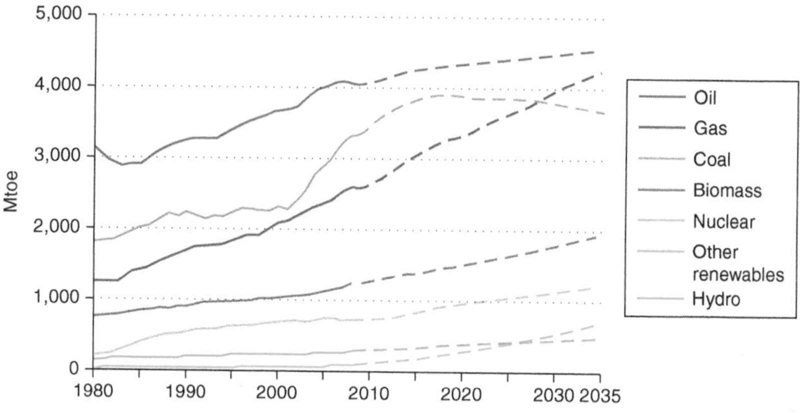

Source: IEA Golden Age Report, ©OECD/IEA 2011, Fig. 1.1, p. 19

Figure 1.6 World primary energy demand by fuel in the GAS Scenario

gas demand increasing nearly 2 percent per year.[58] Natural gas sees the strongest demand growth of all energy sources in the GAS Scenario and overtakes coal before 2030 (see Figure 1.6). By 2035, natural gas comprises 25 percent of the world's fuel mix.[59]

As in the IEA WEO-2010, the GAS Scenario highlights that the largest sector for gas demand continues to be power generation, which along with the industry sector experiences the largest increase compared to the New Policies Scenario.[60] At the outset, the IEA noted that since the IEA WEO-2010 was issued, more recent developments had created considerable opportunities for greater future use of natural gas globally, depending on the interaction between economic and environmental factors and various policy interventions in the market. The report analyzes the key factors that could result in a more prominent role for natural gas in the global energy mix as well as the implications for other fuels, energy security and climate change.

In the Golden Age Report, the IEA indicated that several factors arose in early 2011 that point to a future in which natural gas plays a greater role in the global energy mix. These factors, which will be addressed in detail below, include: (1) increased demand from China as set forth in China's 12th Five-Year Plan; (2) lower growth of nuclear power as a result of the

58 IEA Golden Age Report at p. 19.
59 IEA Golden Age Report at p. 19.
60 IEA Golden Age Report at p. 23.

March 2011 nuclear crisis at Japan's Fukushima Daiichi power plant; (3) more planned use of natural gas in transportation; and (4) continued increase of availability of gas, mainly through increased shale gas production, which lowers average gas prices.

The Report strikes a cautious note about the role of natural gas to meet climate change targets and notes that although natural gas is the cleanest-burning fossil fuel, an expansion of natural gas is not enough on its own to put the world on the agreed path of limited carbon emissions consistent with a temperature rise of no more than 2 °C.

1.5.1 Demand Drivers in the Golden Age of Gas

(1) Increased demand from China – China's 12th Five-Year Plan

One of the key policy drivers noted in the GAS Scenario was China's recently announced (March 2011) 12th Five-Year Plan (FYP) for 2011–15, which outlines a path for a more sustainable energy future, focused on energy efficiency and the use of cleaner energy sources to mitigate environmental impacts.[61] Given its enormous demand for energy, China is the most important country in shaping future energy markets and thus energy policy in China matters and can affect the trajectory of global gas demand.[62]

China's 12th FYP sets out targets for China's primary energy mix and has a strong focus on natural gas, which is targeted to comprise an 8.3 percent share in the primary energy mix in 2015 (260 bcm annually) – up from 85 bcm consumed in 2008 (3.8 percent of energy use). This is a significant upward revision from the IEA WEO-2010 New Policies Scenario, in which China's demand was projected to reach 170 bcm in 2015.[63] China is encouraging natural gas in all sectors but the near-term priority is in power generation.

Other key growth regions noted in the IEA's Golden Age Report include the Middle East North Africa (MENA) Region, which sees an increase in gas demand from 300 bcm to 630 bcm by 2035. Demand for natural gas in India and Latin America also sees significant growth.

As shown in Figure 1.7, non-OECD countries account for 80 percent of demand growth to 2035, with China alone making up nearly 30 percent of global growth in demand for natural gas. By 2034, China will use as much natural gas as the EU.[64]

---

61 IEA Golden Age Report at p. 15.
62 IEA Golden Age Report at p. 14.
63 IEA Golden Age Report at p. 14.
64 N. Tanaka & J. Corben, IEA Presentation to the Center for Strategic and

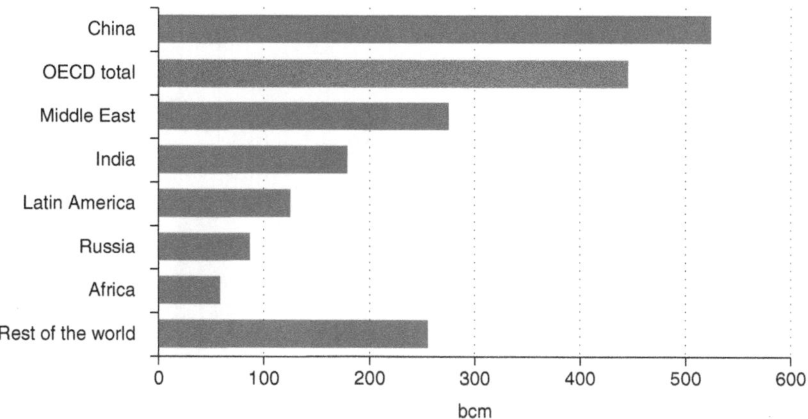

Source: IEA CSIS, ©OECD/IEA 2011

Figure 1.7 Increase in natural gas consumption in the GAS Scenario, 2010–35

(2) Lower growth of nuclear power – Japan's Fukushima Daiichi crisis
Another key event relevant to the IEA's GAS Scenario was the March 2011 disaster at the Fukushima Daiichi nuclear power plant in Japan.[65] As a result of that incident, many countries around the world are re-thinking, and in some cases, suspending, their nuclear programs.[66] Thus, the IEA's GAS Scenario assumes that there will be lower global nuclear power generation capacity than projected in the WEO-2010 New Policies Scenario (see Figure 1.8).[67]

International Studies (CSIS), June 8, 2011, available at http://csis.org/files/attachments/110608_EnergyIEAGas.pdf (IEA CSIS, ©OECD/IEA 2011).

65 On March 11, 2011, Japanese authorities informed the International Atomic Energy Agency (IAEA) that an earthquake and tsunami had struck Japan, resulting in damage to Japan's Fukushima Daiichi nuclear power plant. Flooding caused by the tsunami disabled diesel generators intended to provide back-up electricity to the plant's cooling system and Japanese officials declared a nuclear emergency situation. IAEA, *Fukushima Nuclear Accident Update Log*, March 11, 2011, http://www.iaea.org/newscenter/news/2011/fukushima110311.html.

66 For example, on May 30, 2011, Germany announced that it would phase out all of its nuclear power plants by 2022. The announcement followed mass anti-nuclear protests in Germany in response to Japan's nuclear crisis. BBC News Europe, *Germany: Nuclear Power Plants to Close by 2022*, May 30, 2011, http://www.bbc.co.uk/news/world-europe-13592208.

67 IEA GAS Scenario at p. 20.

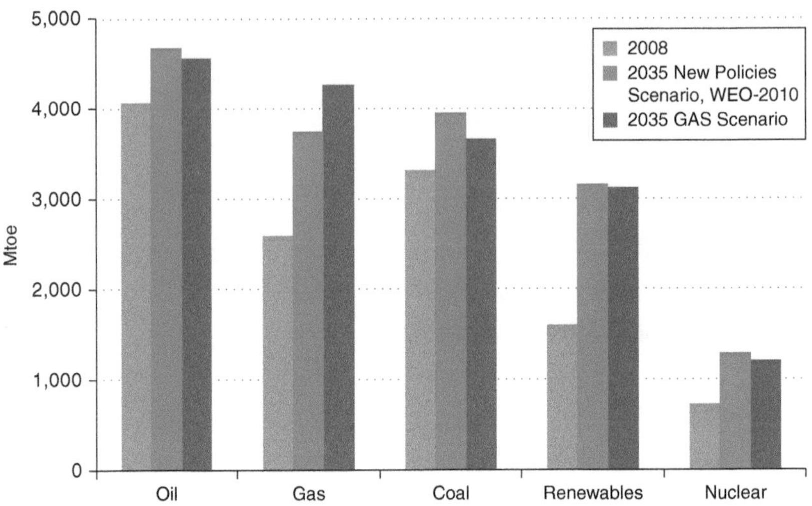

Source: IEA Golden Age Report, ©OECD/IEA 2011, Fig. 1.2, p. 20

Figure 1.8 World primary energy demand by fuel and scenario

This lost nuclear power generation will most likely be replaced by gas-fired power generation, leading to an increase in natural gas demand.[68] As will be discussed in more detail in Chapter 3, Japan's nuclear crisis has reverberated through the LNG markets, as Japan has had to import record amounts of LNG to make up for the nuclear power lost in the wake of the crisis. Japan's imports of LNG for April 2011 were 23 percent higher than for April 2010 and many analysts assume this elevated demand will continue through 2011–12. Analysts also assume that Japan's increased use will absorb any excess supply of LNG and may possibly even lead to a global LNG shortage, which will drive up LNG prices in other markets, most notably Europe.[69]

---

[68] The IEA has cautioned that Germany's moratorium on nuclear power generation adds around 25 million metric tons a year to the country's carbon dioxide emissions, which will have to be offset elsewhere by replacing coal-fired power with cleaner gas-burning plants. James Herron, *IEA Warns on Impact of German Nuclear Halt*, Wall. St. J., May 27, 2011, http://online.wsj.com/article/SB10001424052702304520804576348943486991956.html.

[69] James Herron, *Japan LNG Imports Surge, Supporting Global Prices*, MarketWatch, May 17, 2011, http://www.marketwatch.com/story/japan-lng-imports-surge-supporting-global-prices-2011-05-17. See also *Japan LNG Imports*

Table 1.1 Natural gas import price assumptions by scenario (in year 2009 dollars per MBtu)

	Gas Scenario						New Policies Scenario, WEO-2010				
	2009	2015	2020	2025	2030	2035	2015	2020	2025	2030	2035
United States	4.1	5.6	6.1	6.4	7.0	8.0	7.0	8.1	9.1	9.9	10.4
Europe	7.4	9.0	9.5	9.7	10.1	10.9	10.6	11.6	12.3	12.9	13.3
Japan	9.4	11.5	11.7	11.9	12.3	12.9	12.2	13.4	14.2	14.9	15.3

Source: IEA Golden Age Report, p. 17

(3) Natural gas in transportation

While no new events or policies are introduced in the GAS Scenario, the IEA nonetheless assumes that governments in some countries will encourage the greater use of natural gas vehicles than in the New Policies Scenario. The New Policies Scenario projected around 30 million NGVs by 2035, and the GAS Scenario projects around 70 million.[70] For a more detailed discussion about the possible role of LNG in transportation, see Chapter 12.

(4) Price and supply of natural gas

In the GAS Scenario, the IEA noted that price is a key determinant of the level of future global gas demand.[71] The price assumptions for natural gas in the GAS Scenario are markedly different from those in the New Policies Scenario (see Table 1.1). In the GAS Scenario, the rate of increase slows around the middle of the Outlook period (2020–25) before accelerating again as it approaches 2035. The price path set out by the IEA reflects expectations of demand and supply but primarily represents a more optimistic assumption relating to increases in future gas supply, largely driven by availability of unconventional shale gas at relatively low cost.

to Increase After Quake Hits Energy Infrastructure, LNG World News, March 12, 2011, http://www.lngworldnews.com/japan-lng-imports-to-increase-after-quake-hits-energy-infrastructure/; *Japanese Utilities Prepare to Step Up LNG Imports to Meet Summer Electricity Demand,* IHS Global Insight, June 10, 2011, http://www.ihs.com/products/global-insight/industry-economic-report.aspx?ID=1065929775.

70 GAS Scenario at p. 16.

71 IEA Golden Age Report at p. 17.

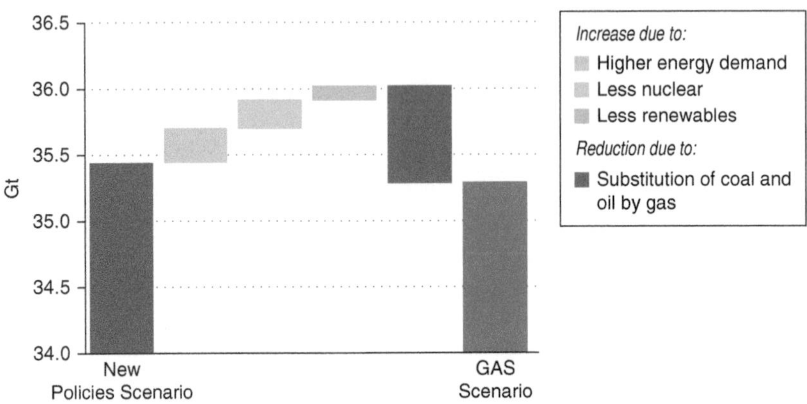

Source: IEA Golden Age Report, ©OECD/IEA 2011, Fig. 1.16, p. 38

Figure 1.9 CO_2 emissions in the GAS Scenario relative to the New Policies Scenario, 2035

1.5.2 Climate Change and the Role of Natural Gas in the GAS Scenario

At the UN climate change talks in Cancun 2010, global leaders agreed to a target of limiting temperature increase to 2 °C. For this goal to be achieved, the long-term concentration of greenhouse gases in the atmosphere must be limited to around 450 parts per million of CO_2-equivalent which is only a 5 percent increase compared to an estimated 430 parts per million in 2000.[72]

In the GAS Scenario, the IEA notes that natural gas is the cleanest-burning fossil fuel and thus has emission and environmental benefits when compared to other fossil fuels, especially coal.[73] In the GAS Scenario, energy-related CO_2 emissions follow a similar path to that in the WEO-2010 and reach 35.3 gigatonnes (Gt) in 2035, which is just 160 million tonnes (Mt) lower than emissions in the New Policies Scenario in that year (see Figure 1.9).[74]

In the Golden Age Report, the IEA notes the competing interactions between natural gas emissions benefits, prices and renewables. While low natural gas prices encourage displacement of more carbon

[72] IEA, Latest Information, *Prospect of Limiting the Global Increase in Temperature to 2° is Getting Bleaker*, May 30, 2011, http://www.iea.org/index_info.asp?id=1959.

[73] IEA Golden Age Report at p. 9.

[74] IEA Golden Age Report at p. 37.

intensive fuels such as coal and oil, in the absence of a global cap on CO_2 emissions low natural gas prices may also displace more expensive low-carbon fuels such as nuclear and renewables.[75] As a result, the IEA specifically notes that an increased share of natural gas in the global energy mix is not enough on its own to put the world on a carbon emissions path consistent with an average global temperature rise of no more than 2 °C. To meet this target, there needs to be a greater shift to low-carbon energy sources, increased efficiency, and new technologies such as CCS.

The GAS Scenario assumes that governments will continue to provide regulatory and financial support for renewables (the WEO-2010 estimated $57 billion of support for renewables and biofuels) but notes that "lower gas prices may put pressure on some governments to review their policies and level of support."[76] Thus, it remains to be seen whether there will be any net benefit from an increase in natural gas use over other more carbon intensive fuels such as coal and oil.

1.5.3 Are We Entering a Golden Age of Natural Gas?

The above sections highlight the competing interactions at work as natural gas struggles to find its role in the future energy supply mix. On the one hand, natural gas is a clean-burning, flexible fuel that can be used extensively in power generation and other sectors to help reduce emissions by displacing other fuels, such as coal and oil. Natural gas resources are abundant and the prospects of global shale gas development imply that the world will be well supplied with natural gas in the 21st century.

On the other hand, the emissions benefits of natural gas, on their own, will not be enough to meet global climate change goals, especially if low natural gas prices lead to displacement of other cleaner fuels such as nuclear and renewables. After weighing these factors and recognizing that there are many uncertainties that may tip the scales, the IEA noted in the GAS Scenario that with natural gas demand expected to "rise by more than 50% and account for over 25% of the world demand in 2035," the Golden Age of Natural Gas is upon us.[77]

---

[75] IEA Golden Age Report at p. 37.

[76] IEA Golden Age Report at p. 18.

[77] IEA Golden Age Report at p. 9. The IEA noted that the rise in natural gas demand of more than 50%, accounting for 25% of world energy demand by 2035, was "surely a prospect to designate the Golden Age of Gas." Ibid.

1.6 THE ROLE OF LNG IN THE GOLDEN AGE OF GAS: LNG IS THE "GLUE" LINKING GLOBAL GAS MARKETS

The Golden Age of Gas would not be possible were it not for liquefied natural gas – LNG. Most natural gas is consumed in the same region in which it is produced due to the costs and impracticalities of transporting natural gas via pipeline over long distances. LNG is natural gas that has been cooled to approximately −161 °C, at which point it condenses to a liquid that occupies approximately 1/600th of the volume of natural gas, thereby allowing it to be shipped via LNG tanker or stored.[78] Of primary significance is the fact that LNG provides a sea-borne solution to the impracticality of serving distant natural gas markets via pipeline or for exploiting otherwise "stranded" gas reserves.[79] Since the majority of the world's natural gas reserves are located far away from key demand markets, LNG offers an important solution for the global gas markets in terms of moving natural gas to markets where it is most needed.

Between 2002 and 2007, global LNG trade expanded by around 50 percent, followed by almost no growth in 2008–09 due to upstream issues in producing countries and the fall in demand due to the global economic recession.[80] Trade in LNG resumed its upward trajectory in 2010 as the global economy showed signs it was coming out of the recession. According to recent IEA projections, international trade in natural gas is set to grow significantly in the coming decades, with more than half of that growth in the form of LNG.[81] The significant increase in LNG trade, particularly between the historically distinct regions, has led many to question in recent years whether the gas markets are "globalizing."[82] While this issue will be discussed in more detail in Chapter 6, the general consensus that seems to have emerged is that while the gas markets are "globalizing," they are not yet globalized since approximately two-thirds

---

[78] The LNG value chain is a complicated series of interactions that will be discussed in detail in Chapter 3.

[79] Foss, Michelle Michot (2007), *Introduction to LNG*, Centre for Energy Economics, University of Texas at Austin, pp. 7–8, available at, www.beg.utexas.edu/energyecon/lng. Natural gas resources are considered "stranded" when they are located far from any market. Transportation of LNG by ship is one method to bring this stranded gas to market.

[80] IEA Medium-Term Oil and Gas Markets 2010 (MTOGM 2010) at pp. 158, 168.

[81] IEA WEO-2010 at p. 192.

[82] IEA MTOGM 2010 at pp. 158, 168.

of global gas is still consumed in the country where it is produced and also because there is not yet a single pricing structure for LNG.[83] Nonetheless, there is widespread recognition that LNG is the "glue" linking global gas markets.[84]

83 IEA MTOGM 2010 at p. 158–60.
84 IEA MTOGM 2010 at p. 158.

2. The LNG value chain

2.1 OVERVIEW OF THE LNG VALUE CHAIN

The intense interest and growth in LNG trade over the past decade warrants a brief discussion of the entire LNG value or supply chain since the LNG "boom" has affected every aspect of the business. In the past decade, the pressure on the LNG industry to successfully navigate the numerous operational, commercial and environmental issues related to the LNG value chain has been incredibly intense. On the construction end, there was intense pressure on the industry to construct new LNG infrastructure to meet rising demand.

The demand for new liquefaction and regasification facilities led to constraints in engineering services and bottlenecks of delivery of certain equipment such as cryogenic storage. The intense demand affected the cost and schedule of construction and operations and many in the industry complained of skyrocketing costs and equipment delays. Nonetheless, during this time period, liquefaction capacity witnessed significant growth, as did the number of receiving terminals in construction, expansion projects and proposals.

As demand for LNG grew, so too did the number of players seeking to enter the fast-growing LNG market. The growth led to a more competitive global LNG market, which in turn drove companies to seek cost savings and efficiencies along the entire value chain.

The LNG value chain comprises a complex set of activities, all of which are capital intensive and require specialized knowledge in order to execute successfully.[1] The LNG value chain begins with the drilling and production of natural gas from subsurface gas reservoirs – usually offshore. The natural gas is then piped from the offshore platform to an onshore treat-

[1] According to ExxonMobil, to implement the LNG value chain successfully, there must be "unwavering focus on: strong partnerships; long-term vision; technological innovation; extraordinary fiscal investments, safety and environmental leadership; significant experience in developing and operating each link of the value chain; technical discipline; project execution and operational excellence; and detailed knowledge of the market." ExxonMobil, *LNG – Fueling the Future*, http://www.exxonmobil.com/corporate/files/corporate/LNG_Brochure.pdf.

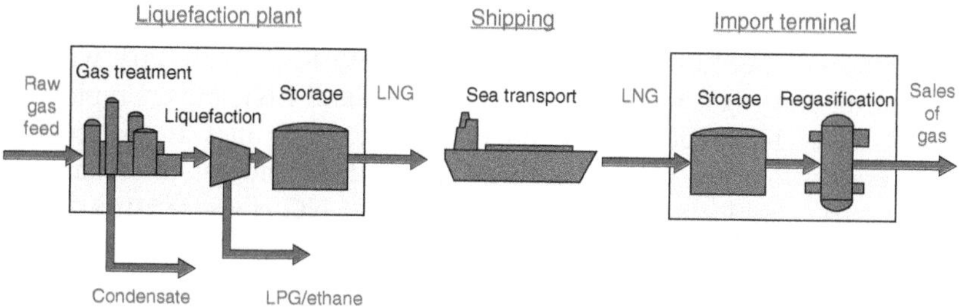

Source: GIIGNL, Information Paper 2 – "The LNG Process Chain"

Figure 2.1 LNG value chain

ment plant, which removes impurities and liquids. The next step is LNG liquefaction, where the natural gas is condensed into a liquid at atmospheric pressure by cooling it to −162 °C (−260 °F). After liquefaction, the LNG is transported on special LNG carriers (ships) to the regasification terminal, where it is warmed back to its gaseous state and fed into natural gas pipelines for ultimate distribution to consumers. (See Figure 2.1.) The three main stages of the LNG value chain – liquefaction, shipping and regasification – are discussed in more detail below.

2.2 LIQUEFACTION

Extracting the natural gas from the ground is the first step in the LNG value chain. The gas supply that comes from the production field is called "feed" gas and the feed gas is first sent to the onshore processing plant[2] for treatment prior to liquefaction. It is important to note that raw natural gas has to be purified before it is suitable for use by consumers. While the natural gas used by consumers is almost entirely methane, natural gas is associated with a variety of other compounds and gases such as ethane, propane, butane,[3] carbon dioxide, and sulfur, as well as oil, water, and

---

[2] The processing plant is considered part of the overall liquefaction process.

[3] Ethane, propane, butane, and others are referred to as natural gas liquids (NGLs) and these NGLs are typically collected as valuable products in their own right. Bob Shively, John Ferrare, and Belinda Petty, *Understanding Today's Global LNG Business*, Energy Dynamics Publications (www.enerdynamics.com) at pp. 25–6. "Natural gas is referred to as 'wet' when hydrocarbons other than methane are present, 'dry' when it is almost pure methane, and 'sour' when it contains significant amounts of hydrogen sulphide." IEA, *Natural Gas*, www.iea.org.

other substances. All of these must be removed prior to the liquefaction process.[4] Because the composition of natural gas can vary widely depending on the gas field, pipeline companies and LNG buyers typically specify allowable ranges of components and heating values. In particular, the specifications for higher heating value (HHV) and gas interchangeability vary significantly worldwide, as do the LNG characteristics from worldwide sources.[5]

Historically, LNG product specifications were not major plant design issues since plant designs were based on long-term contracts with specified buyers so there was little need for flexibility in the plant designs. This situation has changed in recent years as LNG trade has become more global, with owners of liquefaction plant often targeting more than one market. Moreover, the growing spot market for LNG provides more opportunities for buyers and sellers who have the ability to be flexible on product specifications. As a result, various technical solutions have been developed for conditioning LNG on liquefaction and receiving ends. For example, LPG extraction is a common method for adjusting LNG HHV downward on both the receiving and liquefaction ends. Nitrogen injection is also used in existing terminals to reduce HHV.[6]

2.2.1 Liquefaction Technology/Processes

Once the impurities and liquids are removed, the natural gas is ready to be "liquefied" at the liquefaction plant – the "heart" of the LNG project and the "largest single investment" in the LNG value/process chain.[7]

---

4 Groupe International Des Importateurs De Gaz Naturel Liquéfié (the International Group of Liquefied Natural Gas Importers) (GIIGNL), Information Paper No. 2 – *The LNG Process Chain*, available at www.giignl.org.

5 For a table of LNG characteristics worldwide, see GIIGNL, *The LNG Industry 2011*, at p. 12, www.giignL.org.

6 David Coyle, Felix F. de la Vega, Charles Durr, *Natural Gas Specification Challenges in the LNG Industry* (Jan. 2007), KBR, Liquefied Natural Gas (LNG) Publications, http://www.kbr.com/Newsroom/Publications/Technical-Papers/Natural-Gas-Specification-Challenges-in-the-LNG-Industry.pdf. KBR has led the study, design and construction of many of the world's LNG facilities since 1976 and has also achieved many LNG firsts along the way including the first fully modular LNG plant design in Western Australia, the first liquefaction train, and the first large-scale gas turbine plant. www.kbr.com.

7 Michael D. Tusiani & Gordon Shearer, LNG: A Nontechnical Guide 71 (PennWell 2007). In the later part of the 20th century, the estimated cost to build a large liquefaction plant of around 8 Mtpa was approximately $1.5–2.0 billion. Costs began to escalate in the mid- to late 2000s due to the rising cost of materials and limited number of EPC contractors. Ibid. at 80. As discussed in detail in Chapters 11

At the liquefaction plant, the natural gas is condensed into a liquid at atmospheric pressure by cooling it to −162 °C (−260 °F). In its liquefied form, LNG takes up about 1/600th of the space of the gaseous form, which makes it more feasible to transport and store natural gas.[8]

Liquefaction plants are typically set up as a number of parallel processing units, called trains. Each train is a complete stand-alone processing unit but typically there are multiple trains built side by side. The capacity of each train is determined by the size of the equipment used to drive the compressors needed for the liquefaction process. Over the past decade, the size of train capacities has significantly increased from 1 MMt/y (1.4 Bcm/y) capacity in the early plants (Libya, Algeria, Alaska and Brunei) to almost 8 MMt/y (more than 11 Bcm/y) in the new projects in Qatar.[9]

The main equipment in the liquefaction process consists of heat exchangers and compressors driven by steam or gas turbines. While there are a number of proprietary processes, the basic cooling and liquefaction principles of each process are the same, with the heat exchangers being one of the most significant processes. In this process, a cold liquid refrigerant is passed through cooling coils. The natural gas is then flowed over the cooled coils, resulting in the cooling of the natural gas. As the natural gas temperature gradually cools to −162 °C (−260 °F), it becomes a liquid.[10]

Another important technology in the liquefaction process is the drive

---

and 12, Cheniere's total expected costs (before financing) to build the first phase of its proposed LNG export facilities, which includes two liquefaction trains capable of producing 9.0 Mtpa, is $4.5–5.0 billion. The total contract price of the engineering, procurement, and construction (EPC) contract with Bechtel is $3.9 billion. Cheniere Energy Partners, Press Release, *Cheniere Partners Enters into Lump Sum Turnkey Contract with Bechtel* (Nov. 14, 2011), available at http://phx.corporate-ir.net/phoenix.zhtml?c=207560&p=irol-newsArticle&ID=1629831&highlight=.

[8] Liquefaction also permits the storage of natural gas for use during high demand periods in areas where underground storage facilities are not feasible and plays an important role in markets where "peakshaving" occurs. "Peakshaving" is used by utility companies that typically liquefy natural gas when it is abundant and available at off-peak prices and later, during peak demand periods, the stored LNG is converted back to its gaseous state for use. Michelle Michot Foss, *Introduction to LNG*, Centre for Energy Economics, University of Texas at Austin (2007) at pp. 7–8, available at www.beg.utexas.edu/energyecon/lng.

[9] Michael D. Tusiani and Gordon Shearer, LNG: A Nontechnical Guide 71 (PennWell 2007).

[10] Bob Shively, John Ferrare, and Belinda Petty, Understanding Today's Global LNG Business, Energy Dynamics Publications (www.enerdynamics.com) at p. 27. There is some slight variance in terms of the temperature when the natural gas becomes a liquid and some references use −163 °C. See Tusiani and Shearer at 71. Some organizations, such as the IEA, use −161 °C. This book generally uses −161 °C or −162 °C depending on what the original sources uses.

train for the compressor that pressurizes the refrigerant. Most of the early liquefaction plants were built using steam turbines but since the late 1980s most LNG facilities have been constructed using gas turbines.[11]

Compressor sizing, layout, and liquefaction process are dependent on the specific design for the facility. The two main liquefaction processes used worldwide are the APCI MCR (Air Products and Chemicals, Inc. Multi-Component Refrigerant) process and the ConocoPhillips Optimized Cascade process. The processes differ by the type of refrigerant used and the design of the cooling stages. Since each process is proprietary, the decision to use one or the other ties the LNG plant owner to a specific technology license for the life of the plant.[12]

Over 80 percent of the liquefaction plants worldwide use the APCI MCR process, which uses pre-cooling with propane followed by final cooling with the proprietary refrigerant.[13] For Qatar's massive LNG projects, ExxonMobil and Qatar Petroleum first used a variation called the APCI AP-X process at the two-train 15.6 MMtpa Qatargas II project. The same design was repeated for all the 7.8 MMtpa mega trains used in Qatar.[14]

The Snohvit LNG plant in Norway uses a new process developed by Linde/Statoil called the Mixed Fluid Cascade (MFC) process, comprising three refrigeration cycles in a series. The novel project design also uses electrically driven compressors and seawater for cooling. Another feature is that carbon dioxide present in the feed gas is removed and re-injected underground.[15]

The Sakhalin project in Russia is using the Shell Dual Mixed Refrigerant (DMR) process. This novel process uses two mixed refrigerant cycles in series and the process is air-cooled for process and environmental reasons. The Shell DMR process can be used in a range of temperature environments and was particularly suited to Sakhalin's sub-arctic environment.[16]

After the liquefaction process, the LNG is then stored at subzero temperatures in insulated tanks until it is loaded onto LNG carriers ready to be shipped to the designated destination.

---

11 Shively et al., Understanding Today's Global LNG Business at p. 27.

12 Ibid.

13 Ibid.

14 International Gas Union (IGU) World LNG Report 2010, at p. 18, available at http://www.igu.org.

15 Ibid.

16 Ibid. at p. 19.

2.2.2 Global LNG Liquefaction Plants and Capacity

During the past decade, the pace of growth in the global LNG industry has been remarkable in terms of the liquefaction capacity added worldwide. LNG liquefaction capacity during this time frame has almost doubled with new liquefaction capacity added at an average annual rate of 10 percent during 2006–2010 compared to an average of 5 percent during 1990–2000.[17]

As of the end of 2011, there were 25 liquefaction facilities in operation in 18 countries around the world.[18] In terms of LNG capacity, at the end of 2011, there were 95 liquefaction trains in operation with total global liquefaction capacity of 278.7 MMtpa, compared to 171.4 MMtpa at the end of 2005.[19] Appendix A provides a listing of all LNG liquefaction terminals worldwide, including those under construction, approved and in the permitting stage.

In terms of liquefaction capacity by country, since 2005, five countries have commissioned new LNG plants – Equatorial Guinea, Norway, Peru, Russia and Yemen. Another seven countries expanded existing liquefaction capacity: Australia, Egypt, Indonesia, Malaysia, Nigeria, Oman and Qatar.

Historically, Indonesia was the largest LNG producer but in recent years Qatar has been the driving force behind liquefaction capacity growth. Over the next decade, Australia's liquefaction capacity is set to grow significantly and the majority of the remaining LNG projects under construction are located in Australia.[20]

2.3 LNG SHIPPING

Once the natural gas is liquefied, it is ready to be transported via specialized LNG ships/carriers to the regasification facility. LNG carriers are double-hulled ships specially designed to contain the LNG cargo at or near atmospheric pressure at a cryogenic temperature of approximately −162 °C (−259 °F).[21]

---

17 Ibid. at p. 14.

18 IGU World LNG Report 2011 at p. 20.

19 Ibid.

20 Ibid. at p. 22.

21 LNG can also be transported via special double-skinned tank trucks where the liquefaction plant is close to the regasification facility. LNG is regularly transported by tank truck in several countries including Australia, Belgium, Brazil, China, Germany, Japan, Korea, Norway, Portugal, Turkey, the UK and

Table 2.1 Liquefaction capacity by country, 2011

Country	MMtpa
Qatar	77.0
Indonesia	34.1
Malaysia	25.0
Nigeria	21.9
Australia	19.9
Algeria	18.4
Trinidad	15.5
Egypt	12.2
Oman	10.8
Russia	9.6
Brunei	7.2
Yemen	6.7
UAE	5.8
Norway	4.5
Peru	4.5
Equatorial Guinea	3.7
US	1.3
Libya	0.7
Total capacity	278.7

Source: IGU World LNG Report 2011

LNG shipping and carriers play a critical role in the value chain since shipping is often the only option to connect producers of LNG with buyers or consumers of LNG. For example, Qatar, the biggest LNG supplier, is about 5,000 miles away from Japan, the largest consumer of LNG. This distance makes shipping Qatari LNG to Japan the only viable option. While there is often discussion of pipeline gas competing with LNG ("gas on gas competition"), in reality, both sources usually serve different purposes with LNG often being the only viable option. For example, one expert has opined that all of the planned pipeline projects into Europe combined would not be enough to meet anticipated demand, making the importation of LNG into Europe a necessity.[22] Moreover, constructing

the US. See GIIGNL, Information Paper No. 2 – *The LNG Process Chain*, at pp. 3–4.

[22] Moming Zhou and Alistar Holloway, *LNG-Tanker Rates Doubling as Ship Glut Erodes*, Bloomberg, Feb. 16, 2011, http://www.bloomberg.com/news/print/2011-02-16/lng-tanker-rates-seen-doubling-as-ship-glut-disappears-freight-markets.html.

Source: GIIGNL, Information Paper No. 3 – "LNG Ships"

Figure 2.2 LNG carrier types: membrane design (top) and Moss sphere design (bottom)

pipelines between sovereign countries is often fraught with geopolitical tensions, making LNG often the better solution.

LNG carriers blend conventional ship design with specialized materials and advanced systems for handling cryogenic cargos. There are two types of LNG carriers – the membrane design and the Moss sphere design, which is easily identified by its spherical tanks (Figure 2.2).[23] Most recently built tankers are membrane design since, unlike the earlier-designed domed Moss LNG tankers, the newer low-elevation membrane ships can

[23] GIIGNL, Information Paper No. 3 – *LNG Ships*, available at www.giignl.org.

pass under most bridges, making more ports accessible.[24] In addition, the newer membrane ships are being built to meet the highest international standards with sophisticated onboard monitoring and control features for safely receiving, transporting and discharging the LNG.

In compliance with guidelines from the International Gas Codes (IGC), the containment tanks on LNG ships have layers of insulation that isolate the LNG cargo from the hull by ensuring that there is a minimum distance from the sides and bottom of the hull. The insulation system serves to isolate the cargo in the event of grounding or collision and also limits the amount of LNG that boils off or evaporates during the voyage. On many LNG vessels, the boil-off gas is used to supplement fuel during the voyage.[25]

In addition to various design regulations, there are stringent international regulations governing the construction and operation of LNG carriers at sea and at port. The International Maritime Organization (IMO) is the United Nations agency responsible for adopting and updating international treaties (conventions) for shipping safety and security. The IMO has adopted approximately 40 conventions and protocols including the International Code for the Construction and Equipment of Ships Carrying Liquefied Gases in Bulk (IMO Gas Code) and the International Maritime Dangerous Goods (IMDG) Code.[26]

In recent years, the IMO has been working on new guidelines to reduce greenhouse gas emissions from ships with the aim of building a binding international regime for adoption.[27] This is discussed in more detail in Chapter 12.

---

[24] Frank Baker, *Crest of the Wave* (BP Publication), http://www.bp.com/liveassets/bp_internet/globalbp/STAGING/global_assets/downloads/B/BPM_05two_wave.pdf. For example, three of BP's current LNG fleet – the *British Trader*, *British Innovator* and *British Merchant* – are all dual membrane type LNG tankers.

[25] GIIGNL, Information Paper No. 2 – *The LNG Process Chain*, at p. 3.

[26] Ibid.

[27] United Nations Conference on Trade and Development (UNCTAD), Review of Maritime Transport 2008, http://www.unctad.org/Templates/Page.asp?intItemID=4658&lang=1. Published since 1968, the Review of Maritime Transport is one of UNCTAD's flagship publications reporting on the worldwide evolution of shipping, ports, and transportation related to the trade of liquid bulk, dry bulk and container ships.

2.3.1 Ship Registration and Flag States

Shipping is one of the most significant industries in the world and is vital to the global economy since more than 80 percent of international trade in goods is carried by sea.[28] The ownership and subsequent management chain of any ship often comprise many countries and nationalities, with the economic and physical life of a ship spent moving between many different jurisdictions. As such, another important aspect of LNG shipping pertains to ship registration.

The registration of ships is a time-honored practice that historically was a means of controlling ships entitled to carry cargoes within the maritime empires of Europe. "Today, registration confers nationality on a ship and brings it within the jurisdiction of the law of the flag State."[29] The flag state establishes rules and requirements for ships that fly its flag and also has the primary responsibility for ensuring that registered flag state vessels meet all national and international laws and regulations.

Since the 1980s, "open registries" have represented more than 50 percent of the world shipping markets. Established after World War II to eliminate barriers to free trade, "open registry" offers registration to shipowners from all nationalities and provides political neutrality. One of the largest open registries in the world is the Marshall Islands Registry, which is administered by International Registries, Inc. (IRI), through a legislatively endorsed joint venture with the Government of the Republic of the Marshall Islands.[30] The Republic of the Marshall Islands maintains a comprehensive legislative and regulatory framework that ensures the Registry's full compliance with international conventions and the regulations of the IMO.[31]

The Marshall Islands ship registry is the third-largest open registry in the world, surpassing the 70 million gross ton mark in the first half of 2011.[32] Vessel types include oil tankers, bulk carriers, mobile offshore

---

[28] UNCTAD, Review of Maritime Transport 2010, http://www.unctad.org/en/docs/.pdf.

[29] International Registries, Inc. (IRI), *Corporate Brochure*, available at www.register-iri.com.

[30] The Republic of the Marshall Islands is located midway between Hawaii and the Philippines and is the easternmost island group in Micronesia. The Republic of the Marshall Islands gained its independence in 1986 after ending a UN–US Trusteeship Agreement by signing a Compact of Free Association (Compact) with the United States. Under the Compact, the Republic of the Marshall Islands is fully sovereign in domestic and foreign affairs, but gives responsibility for its defense to the US.

[31] IRI, *Corporate Brochure*, available at www.register-iri.com.

[32] According to UNCTAD, the largest flag of registration country continues

drilling units (MODUs), container ships, passenger vessels and yachts.[33] According to UNCTAD, in 2008 the Marshall Islands recorded double-digit growth (+14.9 percent) in terms of the dead weight tonnage (dwt) of ships registered under their flag.[34] Some of this growth was no doubt attributable to the registration of the new LNG vessels built to transport LNG from the massive Qatar LNG projects. For example, in April 2008, the Marshall Islands Registry announced that eight Japanese-owned LNG vessels for Qatar would register with the Marshall Islands with the AL AAMRIYA, a 135,100 gross ton gas carrier being built by the Daewoo Shipbuilding and Marine Engineering Co., registering that month.[35] In total, 40 Qatar LNG vessels will join the Marshall Islands Registry over the four years between 2008 and 2012.[36]

2.3.2 Recent Developments and Expansion of the Global LNG Fleet

The year 2009 was a major milestone for the LNG shipping industry with several significant events occurring. The first was the celebration of the

---

to be Panama with 274 million dwt (dead weight tonnage), followed by Liberia (124 dwt) and the Marshall Islands (68 dwt). UNCTAD, Review of Maritime Transport 2009, Chapter 2, available at http://www.unctad.org/en/docs/rmt2009ch2_en.pdf.

[33] IRI, *Corporate Brochure*, www.register-iri.com.

[34] UNCTAD, Review of Maritime Transport 2009, Chapter 2, http://www.unctad.org/en/docs/rmt2009ch2_en.pdf.

[35] According to UNCTAD, shipyards in Korea, and to a lesser extent, China and Japan, build most LNG carriers. UNCTAD, Review of Maritime Transport 2010, http://www.unctad.org/en/docs/rmt2010ch4_en.pdf. The world's three largest shipyards, Hyundai Heavy Industries, Co., Daewoo Shipbuilding & Marine Engineering Co. and Samsung Heavy Industries, are located in South Korea. Kyunghee Park, *Korea Shipyards' LNG Skill Beats China Bulk Focus Freight*, Businessweek, Sept. 30, 2011, http://www.businessweek.com/news/2011-09-30/korea-shipyards-lng-skill-beats-china-bulk-focus-freight.html. In 2008, it was announced that South Korea's three major shipbuilders, Hyundai, Samsung and Daewoo, would construct the massive $16B order to build 54 new ships for Qatar. Most of the new ships were the new, larger LNG carriers – 31 Q-Flex category – 14 are Q-Max and 9 conventional carriers. John Pratap, *South Korea Building 54 Ships for Qatar*, Gulf Times, Feb. 24, 2008, http://www.gulftimes.com/site/topics/article.asp?cu_no=2&item_no=203504&version=1&template_id=57&parent_id=56.

[36] The Marshall Islands Registry, Press Release, *Japanese-Owned Qatar LNG Vessels to Register with The Marshall Islands*, April 8, 2008, available at http://www.register-iri.com/forms/upload/JAPANESE-OWNED%20QATAR%20LNG%20VESSELS%20TO%20REGISTER%20WITH%20THE%20MARSHALL%20ISLANDS%208%20April%202008.pdf#search=lng.

50th anniversary of the world's first LNG shipment, which took place on January 28, 1959, when the *Methane Pioneer* carried LNG from Lake Charles in the United States to Canvey Island in the United Kingdom, thereby demonstrating that LNG cargoes could be transported over long distances by ship. Also in 2009 was the celebration of the 40th anniversary of the Moss design and the 150th anniversary of the Suez Canal.[37]

The LNG shipping industry has also experienced major milestones in terms of the expansion of the global LNG fleet. In the past decade, the global LNG fleet expanded rapidly from 195 ships at the end of 2005 to 360 LNG ships at the end of 2010 with a combined capacity of 53 MMcm. A record number of LNG vessels (47) were delivered in 2008, most of which went to Qatar. Appendix B provides a current listing of the worldwide LNG carrier fleet compiled by EA Gibson.[38]

In more recent years, the global expansion of the LNG fleet exceeded the global LNG trade, a situation exacerbated by the global recession in 2008–09. This resulted in significant overcapacity of LNG shipping capacity. As demand recovered in 2010, the overcapacity narrowed and it is expected to narrow in the future. An increase in inter-regional trade is also expected to lead to higher utilization of the additional shipping capacity. As this book goes to print, it appears in fact that the LNG shipping glut that persisted between 2008 and 2011 is disappearing due to the rise in global LNG demand, especially from China and India and also emerging from Latin America and the Middle East.[39]

Some experts predict that by 2020 another 100 LNG ships will be needed and that by 2035 the global LNG fleet has to double in order to meet expected demand.[40] It remains to be seen whether the industry will

---

[37] GIIGNL, The LNG Industry in 2009.

[38] EA Gibson is recognized as one of the world's leading shipbrokers and also compiles ship and charter information databases, as well as in-depth market analysis, research and consultancy. EA Gibson publishes an annual Gas Carrier Register, which contains details of all ocean-going gas carriers, as well as new-building vessels on order with a liquefied gas carrying capacity of 1,000 cubic meters and above. The publication contains the following sections: vessels by owner, name and size, pressure LPG, pressurized/refrigerated LPG, fully refrigerated LPG, ethylene, LNG, Ice Class, former vessel names, and orderbook by LPG and LNG. More information on EA Gibson's 2012 Annual Tanker & Gas Registers is available at http://www.gibsons.co.uk.

[39] Moming Zhou and Alistar Holloway, *LNG-Tanker Rates Doubling as Ship Glut Erodes*, Bloomberg, Feb. 16, 2011, http://www.bloomberg.com/news/print/2011-02-16/lng-tanker-rates-seen-doubling-as-ship-glut-disappears-freight-markets.html.

[40] Ibid.

meet this challenge and one reason is the significant expense in building LNG carriers. The current cost of an LNG ship is approximately $210 million, compared with $99 million for a supertanker.[41]

As this book goes to print, orders for new LNG carriers in 2011 "will expand LNG shipping capacity to 380 billion cubic meters (13.4 trillion cubic feet) by 2015 from 300 billion cubic meters today." According to data from Clarkson Research Services Ltd., 24 LNG tankers have been ordered at shipbuilders in South Korea and China in 2011 and there are 44 ships contracted and 362 LNG carriers in service.[42]

In addition to the increase in the sheer number of LNG carriers, the actual size of LNG carriers has also increased significantly in the past decade. For example, the new LNG mega ships that were designed for Qatar – the Q-Flex and the Q-Max – are significantly larger than conventional LNG carriers (Figure 2.3). A majority of the world's LNG terminals can accommodate vessels with LNG carrying capacity over 155,000 cm and a growing number of terminals are upgrading facilities to accommodate larger ships.

Despite the recent trend in building larger-capacity LNG carriers, the size of LNG carriers in the global LNG fleet varies significantly. In 2010, the global LNG fleet averaged 146,686 cm of capacity per carrier. The smallest vessels (typically 18,000–40,000 cm) are used primarily to transport LNG from Southeast Asia to smaller terminals in Japan. There are even smaller carriers (7,500 cm and below) that are used in domestic and coastal areas, thereby facilitating delivery of LNG to remote areas.

2.3.3 The Future of LNG Shipping and New Technology

In June 2011, Sovcomflot (SCF Group) announced that it had ordered two ultramodern ice-class LNG carriers, to be used by Gazprom Global LNG Limited (GGLNG) for LNG transportation. GGLNG, a subsidiary of Gazprom, confirmed that the SCF Group will operate the two new LNG carriers on a long-term time charter, for a minimum of 15 years, and that construction will take place at the STX Offshore & Shipbuilding shipyard, South Korea, with assistance from the Russian Shipbuilding Corporation. Delivery of the first vessel is expected during the fourth

---

41 Ibid.

42 Michelle Wiese Bockmann, *LNG Tanker-Fleet Surge Risks Worker Shortage, Recruiter Says*, Bloomberg Businessweek, June 20, 2011, http://www.businessweek.com/news/2011-06-20/lng-tanker-fleet-surge-risks-worker-shortage-recruiter-says.html. Clarkson Research Services Ltd., a unit of Clarkson Plc, is the world's largest shipbroker. Ibid.

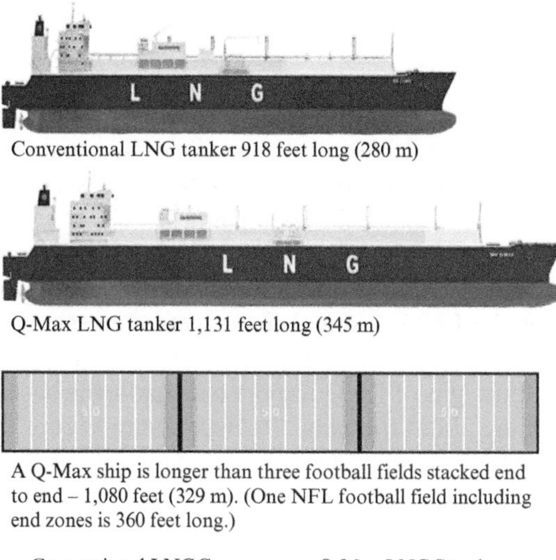

Conventional LNGC		Q-Max LNGC tanker	
Cargo capacity	145,000 m^3	Cargo capacity	250,000 m^3
Breadth	44 m	Breadth	55 m
Depth	26 m	Depth	27 m
Draft	11.5 m	Draft	12 m

Source: Graphic courtesy of ExxonMobil Corporation

Figure 2.3 Conventional and Q-Max LNG tanker size comparison

quarter of 2013 while the second vessel will follow during the second quarter of 2014.[43]

The proposed new ships have the potential to revolutionize the delivery of LNG from extremely harsh environments and are expected to offer year-round gas exports from Russia's first LNG project – Sakhalin-2.[44]

2.4 LNG REGASIFICATION

The third major step in the LNG value chain is regasification, in which the LNG is returned to its original gaseous form at the LNG import or regasification terminal. Once returned to its gaseous state, the natural

---

[43] Ship Management International, *Sovcomflot Orders Two Ultra-Modern LNG Carriers*, June 6, 2011, available at http://www.shipmanagementinternational.com/2011/06/sovcomflot-orders-two-ultra-modern-lng-carriers/

[44] Ibid.

gas is then injected into the domestic pipeline grid for delivery to end-users.[45]

Regasification usually occurs at an onshore import terminal that includes docking facilities for the LNG carrier, one or more cryogenic storage tanks to hold the LNG until regasification capacity is available, and a regasification plant.[46] In some cases, LNG is regasified offshore and then transferred onshore via undersea pipelines.

Typically, after an initial feasibility study, an engineering firm is hired to perform detailed FEED (front-end engineering and design) work to determine a design for the LNG terminal that meets the desired requirements and provides sufficient detail for regulatory review and approval.[47] While there are different designs of LNG import terminals, the basic process and major equipment are about the same. As depicted in Figure 2.4, the major equipment components of an LNG import and regasification terminal are: unloading arms; cryogenic pipelines; storage tanks; low pressure pumps, boil-off gas compressors and recondensers; high pressure pumps; and vaporizers.

2.4.1 The Regasification Process

The majority of LNG shipped today is unloaded at onshore LNG import terminals. In order to unload the LNG cargo, the carrier is parked and moored at a berth alongside the LNG import terminal, which means the water in the berth must be deep enough to accommodate the size of the LNG carrier and usually requires a depth of 40 feet or more. For this reason, the location of the import terminal is a critical factor so that the maximum types of ship sizes can access the terminal.

Once the LNG ship is moored at the import terminal, unloading operations use articulated arms to transfer the LNG cargo safely from the ship to the terminal. The arms serve as the connection between the ship's piping connection (manifold system) and the terminal. The unloading arms are gradually chilled to −162 °C (−259 °F) prior to commencing unloading operations and the unloading arms are specifically designed to withstand the expansion and contraction that results from changes in temperature. Since there is a risk during the unloading process that the ship may move and cause the potential extension and rupture of the unloading arms, a

---

[45] GIIGNL, Information Paper No. 2 – *The LNG Process Chain*, at p. 4. Since natural gas is odorless, odorization of the regasified natural gas is required in many regions and countries before the natural gas can be distributed to consumers.

[46] Shively et al. at p. 49.

[47] Tusiani and Shearer at p. 167.

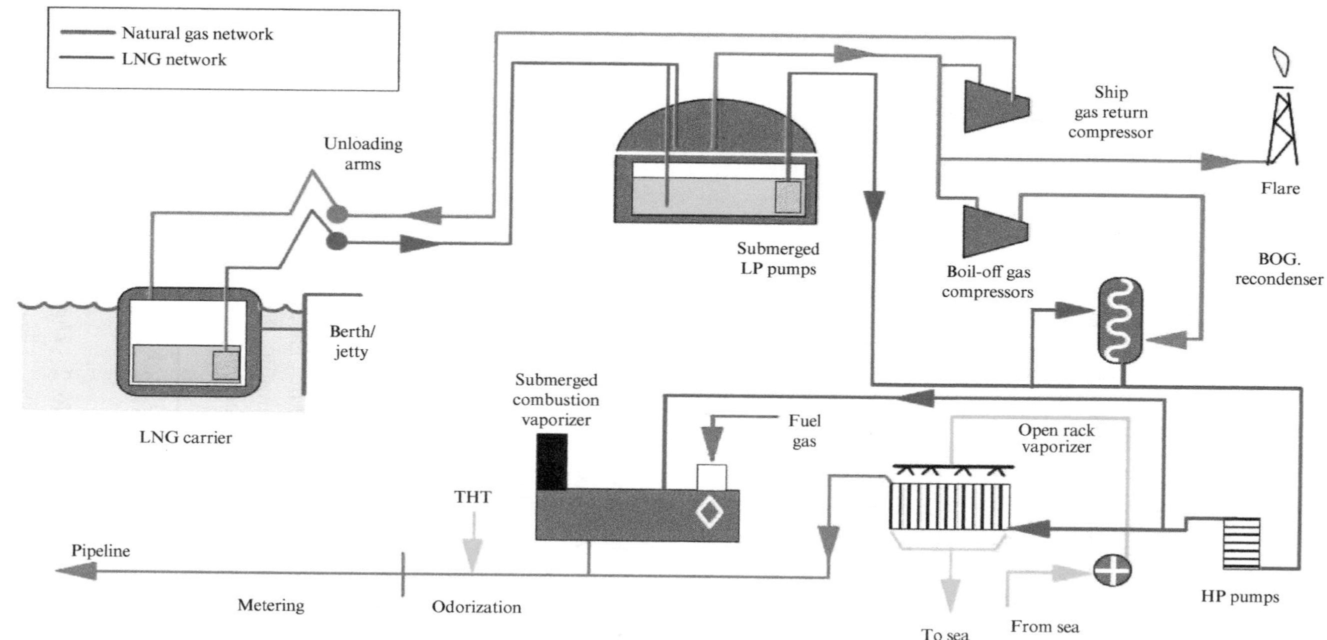

Source: GIIGNL, Information Paper 2 – "The LNG Process Chain"

Figure 2.4 Example LNG import/regasification terminal process

power emergency release coupler (PERC) is fitted into most arm installations. In addition, there are other safety and emergency procedures and equipment that allow for the rapid disconnection of the unloading arms and the emergency shutdown of cargo transfer.[48]

Once the LNG cargo is unloaded, the LNG is transferred via cryogenic pipelines to insulated storage tanks (cryogenic storage tanks) specifically built to hold LNG. Import terminals must have specialized, full containment cryogenic storage tanks that are capable of holding a minimum of one shipload of LNG. Typical storage tanks range in size from 55,000 to 180,000 m^3 but in recent years economies of scale have resulted in larger tanks being built, with tank sizes of 200,000 m^3 now becoming more common.[49]

The storage tanks represent the largest single piece of equipment needed for an onshore import terminal and also represent a significant portion of the capital costs of the terminal. For example, a 160,000-m^3 full containment storage tank is estimated to cost $100 million or more. If a facility is being constructed to handle 145,000 m^3 ships, then three storage tanks may be required, for a total cost of up to $500 million.[50]

2.4.2 Regasification and Vaporizers

The LNG that is stored in the tanks is ultimately sent to vaporizers that warm and regasify the LNG back into its original gaseous state. The main types of vaporizers used in the industry are open rack vaporizers, submerged combustion vaporizers, and ambient air vaporizers.[51]

Open rack vaporizers derive the heat necessary to vaporize the LNG from seawater. The water is first filtered to remove impurities then run over panels of tubes containing LNG and then gathered in a trough underneath before being discharged back into the sea. The LNG is heated and vaporized as it passes through the tubes that are specifically designed to optimize heat exchange.[52]

Submerged combustion vaporizers burn natural gas produced by the import terminal and then pass the hot gases into a water bath containing the tubular heat exchanger that contains the flowing LNG.[53]

Ambient air vaporizers use ambient air, either a natural draft mode or

---

48 GIIGNL, Information Paper No. 2 – *The LNG Process Chain*, at pp. 4–5.
49 Shively et al. at p. 50.
50 Ibid. at p. 55.
51 GIIGNL, Information Paper No. 2 – *The LNG Process Chain*, at p. 6.
52 Ibid.
53 Ibid.

a forced draft mode, to vaporize the LNG. The advantage of these vaporizers is that they have the least environmental impact of any design since there is not air or water discharge. The disadvantage is that they require large sites and are more suitable for service in warm climates.[54]

2.4.3 Global Regasification Terminals and Capacity

Over the past decade, the number of markets turning to LNG to meet natural gas demand has grown dramatically, as evidenced by the more than doubling of the number of countries with LNG receiving capacity between 2000 and 2010. In just five years (2006–10), 30 terminals started operation, bringing the total number of LNG import terminals to 83 as of the end of 2010. Ten of these terminals are floating terminals, nine of which use floating regasification vessels and one of which is a gravity-based structure.[55] Appendix C provides a listing of the worldwide existing and planned LNG regasification terminals.

Since 2005, South America and the Middle East have emerged as new LNG importers with the following five countries, along with Canada, entering the list of LNG importers: Argentina, Brazil, Chile, Kuwait, and the UAE. The world's other 17 LNG importing countries are Belgium, China, the Dominican Republic, France, Greece, India, Italy, Japan, Mexico, Portugal, Puerto Rico, South Korea, Spain, Taiwan, Turkey, the UK and the US.[56]

According to the IGU, the global annual send-out capacity of the world's regasification terminals has increased 70 percent since 2005, for a total capacity of 572 MMtpa at the end of 2010. An additional 110 MMtpa is under construction and is due online in 2015. Indonesia, Malaysia, the Netherlands, Poland, Singapore, Sweden and Thailand currently have their first regasification facilities under construction and these countries are expected to join the ranks of LNG importers by 2015, bringing the total number to 30.[57]

In addition to the 30 countries with existing or under-construction LNG regasification terminals, at least 11 countries in Europe, 7 in South American and the Caribbean, 5 in South and Southeast Asia, 4 in MENA and 2 in Africa are studying or planning LNG imports to meet domestic gas needs going forward.[58]

---

[54] Tusiani and Shearer at p. 179.
[55] IGU World LNG Report 2010.
[56] Ibid.
[57] Ibid.
[58] Ibid.

2.4.4 Receiving Terminals by Country and Region

Due to a lack of indigenous natural resources, Japan has the most receiving terminals in the world with 28 terminals in operation at the end of 2010. Japan is followed by the US with ten terminals at the end of 2010 (an eleventh terminal, Golden Pass, came online in 2011). In terms of regasification capacity, Japan has more than one-third of the total global regasification capacity (31.5 percent), followed by the US (19 percent), Korea (14.6 percent), Spain (6.8 percent), the UK (5.8 percent), with the rest dispersed among numerous other countries.

As of the end of 2010, East Asia, which includes the traditional major LNG importers of Japan, Korea and Taiwan, as well as fast-growing China, held the majority of the world's regasification capacity, at 51 percent or 280 MMtpa. While East Asia historically held the majority of regasification capacity throughout the 1990s and early 2000s, significant regasification capacity was added in the US in the early to mid-2000s, which altered the historic breakdown. Europe also saw regasification capacity increase since the mid-2000s and the next projected increases are expected to come from the emergence of LNG importing countries in South Asia, South America and the Middle East.[59]

According to the IGU, at the end of 2011 there were 89 regasification terminals around the world with a total regasification capacity of 608 MTPA.[60]

2.4.5 LNG Storage

In terms of LNG storage capacity (the combined capacity of the LNG storage tanks), the top five countries claim 78 percent of the global LNG storage capacity: Japan and Korea account for 54 percent, followed by the US (11.6 percent), Spain (6.8 percent) and the UK (5.5 percent).

2.5 LNG PROJECT STRUCTURES[61]

A primary decision to make when developing an LNG project is how to structure the project since this decision will in turn affect the risk profile

---

59 Ibid.

60 IGU World LNG Report 2011 at p. 35.

61 This section is adapted from the following article: Thomas E. Holmberg, *Comparison of Project Structures in an LNG Liquefaction Plant*, Oil & Gas Financial Journal, March 2012, www.ogfj.com.

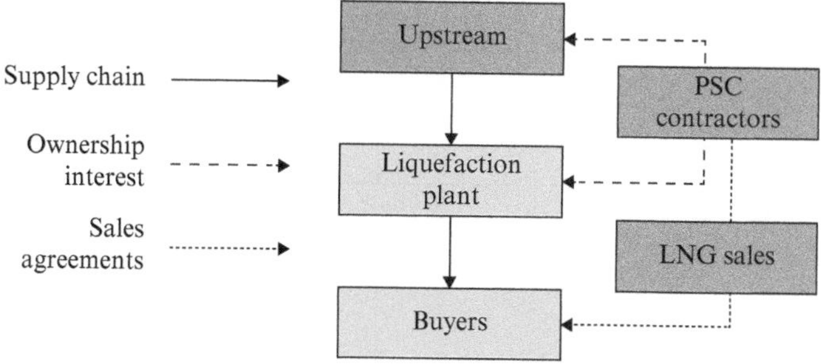

Source: Holmberg, *Oil & Gas Financial Journal*, March 2012

Figure 2.5 Integrated project structure

of the project, financing, and the contractual relationships between the project sponsors. There are various project structures for LNG liquefaction projects and the precise structure will depend on a number of factors such as the identity of the participants throughout the value chain, the tax and legal regimes, risk, and financing. While there is not a "standard" structure, there are several basic structures that exist: an integrated project model, a project company or merchant model, and a tolling model.

In an integrated model (Figure 2.5), the participants share a unity of interest in the LNG value chain from the production of natural gas through the liquefaction of the LNG. One or more of the investors owns the rights to the natural gas reserves and the production of the reserves is pursuant to a production sharing contract (PSC) or similar agreement. The parties to the PSC build the infrastructure to monetize the gas reserves and each PSC contractor holds an interest in the LNG plant in proportion to its ownership of the rest of the project. If the PSC contracts have the capital to construct the project, then an integrated structure may offer many benefits such as the alignment of interests between the parties and the ability to share costs across the entire LNG value chain, which may also have various tax and accounting benefits.

Under a company or merchant model (Figure 2.6), the company that owns the LNG facility does not have an interest in the upstream assets and instead purchases natural gas as feedstock for the plant, processes it into LNG, and then sells the LNG to one or more buyers on a CIF, FOB, or DAP.[62] There are many variations of the project company

[62] In a "cost, insurance and freight" or CIF sale, the seller delivers the goods to the buyer on board the vessel and risk of loss passes from the seller to the buyer

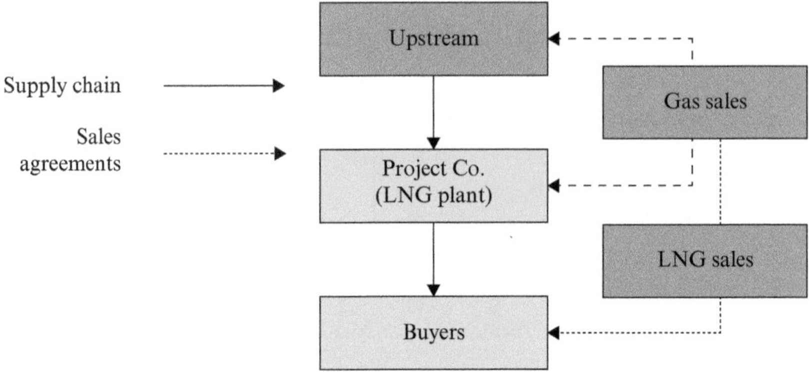

Source: Holmberg, *Oil & Gas Financial Journal*, March 2012

Figure 2.6 Project company (merchant) structure

model, including one where a governmental entity may own the LNG facility (directly or through the state-owned national oil company) while an operating company owned by the upstream PSC contractors operates the facility.

Under a tolling model (Figure 2.7), the company that owns the LNG liquefaction facility does not take title to the gas and receives a fee to process the natural gas into LNG for the buyer. A principal benefit of this model is that it minimizes the market risk to the facility owner by providing for predictable payments of the tolling fee. A second fee (for example, a commodity charge) may also be assessed per unit of natural gas processed.

---

on board the vessel. The seller is responsible for the costs and freight to bring the vessel to the port of destination and is also responsible for insurance to cover buyer's risk of loss to the goods. In a "free on board" or FOB sale, the buyer is responsible for providing LNG shipping, and title and risk of loss pass from the seller to the buyer as the LNG is loaded onto the LNG ship from the liquefaction facility. In a "delivered at place" or DAP sale, the seller is responsible, at its own expense, for providing shipping for the LNG, the title to and risk of loss of which transfer from the seller to the buyer as the LNG is unloaded from the LNG ship to the receiving terminal. Thomas E. Holmberg, *Comparison of Project Structures in an LNG Liquefaction Plant*, Oil & Gas Financial Journal, March 2012, www.ogfj.com.

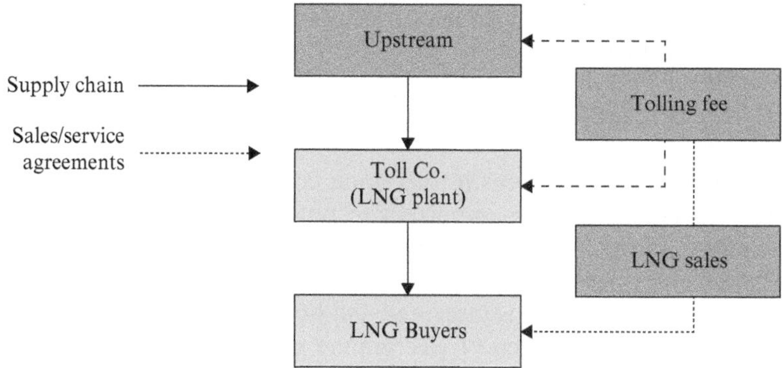

Source: Holmberg, *Oil & Gas Financial Journal*, March 2012

Figure 2.7 Tolling structure

2.6 THE COMPLICATED WORLD OF LNG MEASUREMENTS[63]

The units of measurement in the LNG industry are complex and vary according to whether used to measure natural gas in its gaseous form or its liquid form. To add to the complexity, since LNG is a global business, both English and metric units are used depending on the location.

LNG is usually measured in metric tons when it is in a liquefied form and in cubic feet or cubic meters when converted back to a gaseous state. Annual capacities are often stated in terms of million tonnes per annum or year with a variety of abbreviations used – mtpa, MMtpa, MMt/y, or million t/yr.

To make it even more confusing, tanker capacities may be stated in cubic meters of liquid and recall that when liquefied the volume is reduced by a factor of 610. A typical tanker holds between 120,000 and 145,000 m^3, which is equivalent to 2.7 to 3.2 Bcf (76.5 million to 90.6 million m^3) of gas or 55,400 to 65,700 tonnes of LNG. In terms of comparing what this means from a practical standpoint, one average size LNG tanker holds enough natural gas to serve approximately 5 percent of the United States',

63 Shively et al., UNDERSTANDING TODAY'S GLOBAL LNG BUSINESS. This section is adapted from an excellent discussion of natural gas and LNG measurements found in the book at pp. 5–6. The book also discusses energy content, which has been omitted from this discussion.

30 percent of the United Kingdom's or 40 percent of Japan's average daily consumption of natural gas.

2.6.1 The IEA and What's in a bcm?[64]

Since many of the references in this book come from the International Energy Agency (IEA), it is important to note that the IEA uses billion cubic metres (bcm). According to the IEA, a bcm of natural gas is a commonly used measure of gas production and trade, but what a "bcm" represents depends on how it is measured and how much energy it contains. The IEA standard is to report gas volume as actual physical volumes, measured at 15 °C and at atmospheric pressure. This means that a bcm of Russian gas, in terms of energy content, can have a different value from a bcm of gas from another country. Since this is a difficult process, the IEA's approach is to keep the underlying balances for each country on an energy basis (rather than a volume basis) and to maintain a database of the different energy content of gas imports, exports, production and consumption for each country.

---

64 IEA WEO-2011 at p. 304.

3. The evolution of LNG markets and primary demand regions

3.1 A BRIEF HISTORY OF LNG

The rapid growth of and intense interest in LNG over the past decade might lead some to believe that LNG is a newly discovered fuel. In fact, LNG is not a new fuel at all; it is merely a means of delivering an old fuel – natural gas.[1] For decades, natural gas was viewed as an unwanted byproduct of drilling for oil.[2] Whereas oil was relatively easy to store, load, and transport, transporting natural gas presented a significant technical challenge that would take decades to solve. The challenge of course is that natural gas has to be chilled through a process known as liquefaction before it can be turned into a liquid that can be transported.

Natural gas liquefaction dates back to the 19th century when British chemist and physicist Michael Faraday experimented with liquefying different types of gases, including methane/natural gas. German engineer Karl von Linde facilitated advances in gas liquefaction on an industrial scale when he invented a heat exchanger and later the first practical compressor refrigerator machine. However, it was not until 1914 that Godfrey Cabot registered the first US patent for transporting liquefied natural gas on a barge, although there is no material proof that this was actually done.[3]

---

[1] Warren R. True, *What LNG Was, Isn't, and Is*, Oil & Gas Journal, Sept. 19, 2011, Vol. 109, Issue 38 ("LNG is only the means of storing and moving the fuel, which is natural gas"), http://www.ogj.com/articles/print/volume-109/issue-38/regular-features/journally-speaking/what-lng-was-isn-t-and-is.html.

[2] Many years ago, when oil companies drilled for oil and found natural gas, their efforts were deemed a failure and the natural gas often had to be reinjected, flared, or left for another day. Fred Bosselman et al., ENERGY, ECONOMICS AND THE ENVIRONMENT 526 (2nd edn, 2006). In fact, oil wildcatters used to joke, "The bad news is we didn't hit oil, the good news is we didn't find gas." *Vapour Trails: Oil Companies Dash for Gas*, The Economist, July 1, 2010, http://www.economist.com/node/16488892.

[3] Nevertheless, the company that he founded, Godfrey L. Cabot, Inc., later became the Cabot Corporation, whose subsidiary, Cabot LNG, ultimately built

It would take over two decades before the first commercial LNG peakshaving plant was built in West Virginia in 1939. In 1941, a second peakshaving facility was built in Cleveland, Ohio, by the East Ohio Gas Company. This facility was subsequently expanded to add a larger storage tank but, owing to metal shortages during World War II, the storage tank was constructed of steel instead of stainless steel alloys, which could withstand very cold temperatures without fracturing. In 1944, the tank ruptured causing natural gas to leak into the sewer system, and then into people's homes where it ignited and ultimately killed 128 people. This incident effectively put LNG development on hold for another decade.[4]

In the 1950s and 1960s, the Union Stock Yards of Chicago was facing escalating electricity rates and began to explore the possibility of liquefying natural gas in Louisiana and barging it up the Mississippi River to Chicago. At the same time, the British Gas Council was also looking into ways to transport natural gas. The Union Stock Yards and the British Gas Council, along with Continental Oil Company, eventually joined forces to convert an old World War II dry bulk carrier into the world's first LNG carrier – the *Methane Pioneer*.[5]

In January 1959, the *Methane Pioneer* set sail from Lake Charles, Louisiana, with a cargo of LNG destined for Canvey Island, UK. This first ever US–UK shipment of LNG demonstrated that large quantities of LNG could be transported safely across the ocean and opened up the possibility of transporting large volumes of natural gas from otherwise stranded fields to distant destinations based on consumer demand.[6] The *Methane Pioneer* subsequently carried seven additional LNG cargoes to Canvey Island.[7]

In the early 1960s, a major discovery of natural gas in Algeria led Great Britain to sign the first LNG contract with Algeria in 1961, thus establishing the "first commercial-scale LNG 'chain' with a liquefaction plant

---

the first US LNG import terminal (Everett) in Boston in 1971. Stream Repsol, *The History of LNG*, http://www.streamrgn.com/servlet/ContentServer?gnpage=3-140-0¢ralassetname=3-140-Articulo-HistoriaGNL¢ralassettype=Articulo.

4 Michael D. Tusiani and Gordon Shearer, LNG: A NONTECHNICAL GUIDE (Pennwell, 2007) at p. 14.

5 Tusiani and Shearer, LNG: A NONTECHNICAL GUIDE at p. 14.

6 Michelle Michot Foss, *Introduction to LNG*, Centre for Energy Economics, University of Texas at Austin (2007), pp. 7–8, available at www.beg.utexas.edu/energyecon/lng. Natural gas reserves are considered "stranded" where there is no significant market where the reserves are located or where pipeline options are limited. Ibid. at p. 12.

7 Ibid.

at Arzew, Algeria" and a receiving terminal in the UK.[8] In 1962, the French signed a similar deal to buy LNG from Algeria.[9]

After the concept was shown to work in the United Kingdom, additional LNG liquefaction plants and import terminals were built in both the Atlantic and Pacific regions.[10] During the 1960s and 1970s, liquefaction facilities were built in countries that had natural gas to export, including Abu Dhabi, UAE, Brunei, Libya, Indonesia, and Algeria. In addition, the first, and to date, the only liquefaction terminal in the United States was built in Kenai, Alaska.

Alaska's Kenai LNG plant began LNG deliveries to Japan's Tokyo Gas and Tokyo Electric Power Company (Tepco) in 1969. In 1972, Brunei became Asia's first producer, bringing on stream an LNG plant at Lumut that now has a capacity of 6.5 mtpa and supplies Korea as well as Japan. Libya's plant at Marsa el Brega began deliveries to Spain in 1970 and also began supplying LNG to Italy. In the 1970s, import terminals were also constructed in a number of countries including Japan, Korea, Taiwan, Belgium, Italy, France, Spain, and the United States.

In 1972, US imports from Algeria were approved with Boston's Distrigas committing to buy 50 million standard cubic feet per day (mmscfd) from the Skikda Algeria plant over a 20-year period. In 1979, the first LNG contract expired when the 15-year contract between Algeria and the UK came to an end. Deliveries from Algeria to the UK continued into the 1980s but were eventually terminated.

In 1979, the US market was shaken by disputes over pricing between the US buyers and Algeria's Sonatrach, which eventually resulted in the termination of the LNG contracts. As a result of the termination of various contracts, six LNG carriers were retired (three of which were subsequently scrapped) and two of the US's four LNG terminals were mothballed. During the 1980s, most of the world experienced an oversupply of natural gas and only two export projects (in Australia and Malaysia) were added during that decade.[11] However, demand for LNG in Asia continued to rise and Malaysia entered the LNG market in 1983, followed by Australia in 1989.

By the early 1990s, demand for natural gas had picked up, especially in the Western hemisphere, which led to a rebirth of LNG projects targeting those markets. A number of LNG projects went online in the late 1990s

---

[8] Tusiani and Shearer at pp. 14–15.

[9] Sempra LNG, *History of LNG*, http://www.sempralng.com/pages/About/History.htm.

[10] Michelle Michot Foss, *Introduction to LNG*.

[11] Tusiani and Shearer at p. 14.

to serve new demand. Qatar became the second Middle Eastern LNG producer with the delivery of its first cargo of LNG from the Qatargas LNG plant in January 1997. Several other plants also came online in the late 1990s including: Trinidad, April 1999; Ras Laffan, Qatar, May 1999; and Nigeria in October 1999. In April 2000, Oman commenced production with a plant of design capacity of 6.6 mtpa delivering its first cargo to Korea.

As the LNG markets evolved over the decades, they tended to develop in regional isolation from one another, primarily due to the high cost of natural gas transportation. Historically, two distinct LNG trade regions developed – the Asia-Pacific region and the Atlantic Basin region, which included North America, South America and most of Europe.[12] Until Qatar began to export LNG to both regions in the mid-1990s, the two regions were largely separate, with unique suppliers, pricing arrangements, project structures, and terms. As will be discussed in detail in Chapter 6, the increase in inter-regional trade in more recent years, as well as the development of a more active spot market, has tended to blur the distinction between the two main regions.

For purposes of this book, the regional markets will be analyzed as the Asia-Pacific region (Section 3.2), the European region (Section 3.3), and the North American/Atlantic Basin region (Section 3.4), which includes North America, South America and Latin America.

3.2 THE ASIA-PACIFIC REGION

The Asia-Pacific region has historically been the largest market for LNG. Japan is the world's largest LNG importer, followed by South Korea and Taiwan. These three countries have few indigenous resources and rely almost exclusively on LNG for their natural gas supply. China and India have recently emerged as LNG importers and could become significant buyers of LNG over time. Smaller, niche markets are also developing as other countries such as Thailand, Singapore, Chile and Pakistan look to become LNG importers (see Chapter 5).

---

[12] The Atlantic Basin LNG region historically was defined to include exporting countries such as Algeria, Libya, Egypt, Nigeria, Oman, Qatar and Trinidad and importing countries such as the US, Mexico, Spain, France, Italy, Turkey, Greece, Portugal, the UK and Belgium. Ken Snodgrass, Shell Gas & Power International, Presentation to the 23rd World Gas Conference, Amsterdam, 2006: *Atlantic Basin LNG: Diverse Drivers that Require Diversified Marketing*, http://www.igu.org/html/wgc2006/pdf/paper/add10832.pdf.

3.2.1 Japan – The World's Largest LNG Importer

Today's global LNG industry was really founded in the 1970s based on a now 40-plus-year relationship between Japan and Kenai, Alaska.[13] The development of LNG technology coincided with the discovery of major natural gas reserves in the Cook Inlet Area of Alaska in the late 1950s and early 1960s. The lack of local demand for the gas led Phillips Petroleum (now ConocoPhillips) and Marathon to consider international LNG projects as a means of securing a home for the newly discovered natural gas. At the same time, Japan was beginning to consider LNG as a means to secure significant energy supplies to compensate for its almost total lack of indigenous energy resources.[14] Phillips and Marathon drew up plans for an LNG project that included a liquefaction plant and deepwater docking and loading facilities in Kenai, Alaska, and a receiving and regasification plant in Japan. Two new LNG tankers were also commissioned to ship the LNG from the new Kenai LNG plant to Japan.[15]

In 1967, Phillips and Marathon signed an agreement with Tokyo Electric and Tokyo Gas to supply LNG to Japan and received authorization from the US regulatory agency to export LNG to Japan.[16] The first LNG tanker docked in Japan two years later, in November 1969, and marked the first LNG export from the Western Hemisphere and the first LNG imported into Asia.

---

[13] Amy Burnett, *The Kenai LNG Plant Celebrates 40 Years*, Spirit Magazine, http://www.conocophillips.com/EN/about/worldwide_ops/country/north_america/Documents/Kenai_spirit.pdf.

[14] Ibid.

[15] Ibid.

[16] Authorization to export LNG to Japan was originally granted to Phillips Petroleum Company (Phillips) and Marathon by the Federal Power Commission in 1967. See Phillips Alaska Natural Gas Corporation and Marathon Oil Company, 37 FPC 777 (April 19, 1967). Phillips and Marathon were specifically authorized to export LNG from the State of Alaska to supply Tokyo Electric Power Company Inc. (Tokyo Electric) and Tokyo Gas Company Limited (Tokyo Gas) for a 15-year period terminating on May 31, 1984. The order also authorized Phillips and Marathon to construct proposed liquefaction and marine terminal facilities in the Cook Inlet Basin near Kenai, Alaska, necessary to support the export of LNG to Japan. The long-term export authorization was subsequently amended and extended numerous times. See DOE/ERA Opinion and Order No. 49 (1 ERA ¶ 70, 116, December 14, 1982) (extended export authority); DOE/ERA Opinion and Order No.49-A (1 ERA ¶ 70,127, April 3, 1986) (transferred authorization from Phillips Petroleum Company to Phillips 66 Natural Gas Company); DOE/ERA Opinion and Order No. 206 (1 ERA ¶ 70,128, November 16, 1987) (amended pricing formula).

At the time, the Kenai LNG plant was only the second LNG plant in the world and the project was the largest undertaking in Phillips' history, completed in 1969 at a cost of $200 million.[17] Kenai LNG also resulted in the development of significant LNG technology – namely the ConocoPhillips Optimized Cascade® process. When Bechtel Corporation built the Kenai LNG liquefaction plant in 1969, ConocoPhillips decided to retain the rights to the new and advanced Optimized Cascade® process for use in future LNG projects, which has served the company well in the years since.[18] Over the next four decades, Kenai LNG would ship more than 1,300 LNG cargoes to Japan,[19] successfully filling two 20-year contracts.[20]

Today, Japan is the largest importer of LNG in the world. In 2002, Japan imported 2.6 Tcf (54.6 million tonnes) of LNG[21] and by 2010, Japan's LNG imports had grown to 70 million tonnes.[22] With very limited domestic gas production and no pipeline imports, Japan's LNG imports account for approximately 35 percent of the global trade in LNG.[23] Japan imports LNG primarily under long-term contracts from

---

17 Amy Burnett, *The Kenai LNG Plant Celebrates 40 Years*, Spirit Magazine.

18 Ibid. Since the Kenai LNG project, ConocoPhillips has licensed the technology for the Optimized Cascade® process in ten LNG plants worldwide, representing almost half of the grassroots LNG projects.

19 Ibid.

20 The most recent application for Kenai LNG exports was granted in June 2008 by DOE/FE Order No. 2500. In that Order, the Department of Energy's Office of Fossil Energy (DOE/FE) authorized the export of LNG from Kenai LNG in Alaska to Japan and/or one or more countries on either side of the Pacific Rim after considering the domestic needs for the gas as well as the policy of promoting free trade. DOE authorized the export of "up to 99 Trillion British thermal units (TBtus) of liquefied natural gas (LNG) (the equivalent of 98.1 Billion cubic feet (Bcf) of natural gas), to Japan and/or one or more countries on either side of the Pacific Rim . . . pursuant to transactions that have terms of no longer than two years. Th[e] authorization [was] effective for a two-year term beginning on April 1, 2009, and extending through March 31, 2011".

21 Energy Information Administration (EIA), *The Global Liquefied Natural Gas Market: Status and Outlook*, December 2003, http://www.eia.gov/oiaf/analysispaper/global/pdf/eia_0637.pdf.

22 Kwok W. Wan, *IEA Forecasts Japan LNG Worst-case*, Petroleum Economist, Sept. 28, 2011, http://www.petroleum-economist.com/Article/2908167/IEA-forecasts-Japan-LNG-worst-case.html.

23 Poten & Partners, *2015–2035 LNG Market Assessment Outlook for the Kitimat LNG Terminal*, Prepared for KM LNG Operating General Partnership, Oct. 2010, https://www.neb-one.gc.ca/ll-eng/livelink.exe/fetch/2000/90466/94153/552726/657379/657474/670503/65(7060/B1-13_-_Appendix_4_-_2015-2035_LNG_Market_Assessment_Outlook_for_the_Kitimat_LNG_Terminal_-_A1W6T5_.

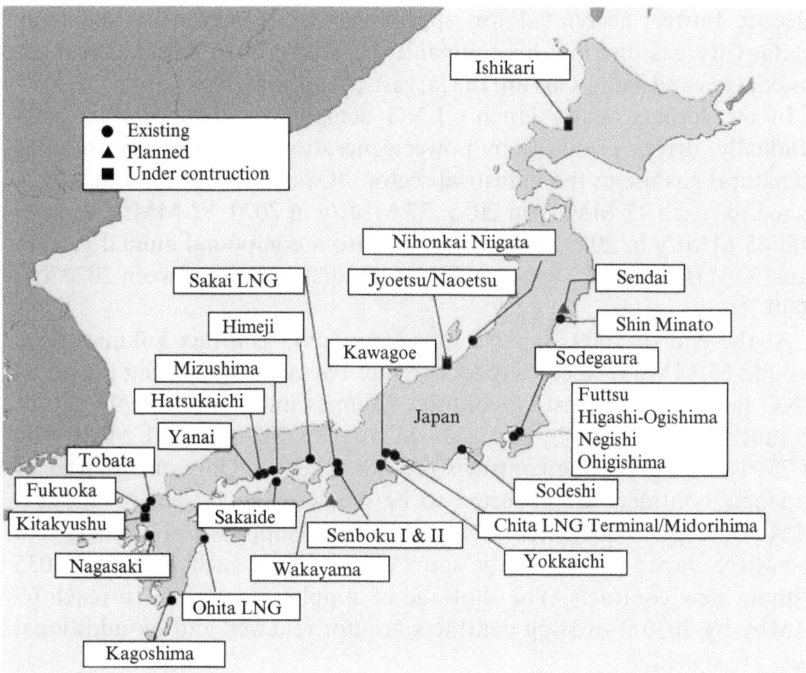

Source: Poten & Partners

Figure 3.1 Japan's LNG import terminals

Indonesia, Malaysia, Australia, Qatar, Brunei, Abu Dhabi, Oman and the US.[24] Other shipments come from various Atlantic Basin exporters on a spot and short-term contract basis. Japan has 28 existing LNG import terminals, with one currently under construction and seven planned to commence construction by 2018 (Figure 3.1). Existing send-out capacity is approximately 25 bcf/d.[25]

In 2009, seven electric utilities, Tokyo Electric Power Company (TEPCO), Chubu Electric Power, Kansai Electric Power, Tohoku Electric Power, Kyushu Electric Power, Chugoku Electric Power and Shikoku

---

pdf?nodeid=656825&vernum=0 (hereinafter Poten & Partners Kitimat Market Assessment).

24 While today Kenai LNG shipments to Japan represent only about 1% of the total LNG imported by Japan, back in 1969 the tiny Kenai LNG facility was Japan's only supplier.

25 Poten & Partners Kitimat Market Assessment.

Electric Power, accounted for approximately 65 percent of Japanese LNG. City gas distribution companies also buy LNG and Tokyo Gas, Osaka Gas and Toho Gas are the largest city utilities.[26]

In the coming years, Japan's LNG demand is projected to expand gradually, driven primarily by power generation and switching from oil to natural gas use in the industrial sector. "Overall LNG demand is projected to reach 72 MMt/y in 2015, 77 MMt/y in 2020, 82 MMt/y in 2025 and 85 MMt/y in 2030. This translates into a compound annual growth rate ('CAGR') of 1.4% between 2010 and 2020 and 1% between 2020 and 2030."[27]

At the end of 2009, Japan's long-term LNG contract volumes were around 67 MMt/y. According to Poten & Partners, comparing projected LNG demand with existing contract volumes indicates a supply gap of as much as 9 MMt/y by 2015, 27 MMt/y by 2020, and 51 MMt/y by 2025, assuming existing contracts are not renewed. Only one quarter of Japanese contracts are expected to be renewed (equivalent to about 3 MMt/y), largely because of declining feed gas supplies in Indonesia and elsewhere. Japan is likely to be short of supply for each year until 2035 without new contracts. The shortage of supply is projected to reach 65 MMt/y by 2030 if existing contracts are not renewed and no additional ones are signed.[28]

Japanese LNG buyers are aware of the potential of a supply tight market starting in around 2013 and are now seeking additional supplies to fill the gap. Driven by security of supply concerns, Japan will be active in securing reliable LNG supplies from Asia-Pacific and the Middle East over the coming years and may also occasionally source supplies from the Atlantic Basin on a spot to short-term contract basis.[29]

Having experienced volatile prices and fluctuating end use demand in the past several years – typified by the shift from a buyers' market in the early 2000s to a sellers' market in 2006 and then back to a buyers' market in 2010 – traditional Asian buyers are now seeking greater volume flexibility in new contracts beyond traditional take-or-pay obligations (for example, through downward quantity tolerance (DQT) of up to 10 percent per year and equity LNG volumes) to accommodate fluctuations in base load demand.[30]

With very limited domestic natural gas reserves or other energy

[26] Ibid.
[27] Ibid.
[28] Ibid.
[29] Ibid.
[30] Ibid.

resources, Japan will remain reliant on LNG imports and will always place a premium on security of supply. While economic growth is one key driver of LNG growth, in Japan, perhaps the more important driver is the operation and development of existing and future nuclear power plants.[31] The reason for this is that the share of nuclear and LNG in Japan's total electricity generation mix are roughly the same – 30 percent. Thus, LNG demand for power generation can suddenly increase if there is a nuclear power facility shutdown, as demonstrated by two separate incidents in recent years.

In July 2007, TEPCO's Kashiwazaki-Kariwa nuclear facility, with a total 8.2GW of capacity, was shut down due to an earthquake in northeast Japan.[32] As a result, TEPCO had to increase imports of LNG to record amounts of approximately 21.4 million tons, exceeding the record 19.2 million tons purchased in fiscal 2003, when all 17 of its reactors were shut down after TEPCO admitted hiding nuclear faults for a decade.[33]

On March 11, 2011, Japanese authorities informed the International Atomic Energy Agency (IAEA) that an earthquake and tsunami had struck Japan, resulting in damage to Japan's Fukushima Daiichi nuclear power plant. Flooding caused by the tsunami disabled diesel generators intended to provide back-up electricity to the plant's cooling system and Japanese officials declared a nuclear emergency situation.[34]

As a result of the Fukushima Daiichi twin disasters, Japan had to import record amounts of LNG in 2011. In September 2011, the IEA indicated that Japan is likely to import an additional 11 billion cubic metres (bcm) of gas (8.4 million tonnes of LNG) in 2011 to offset the loss of nuclear power generation. The IEA also projected that Japan's LNG demand could increase by nearly a third in 2012 if the country fails to restart any nuclear reactors and that Japan could struggle to secure enough LNG. The worst case is that Japan might need an additional 30 billion cm (22.8 million tonnes) of LNG with Qatar as a potential supplier, although it is unclear whether Qatar can supply this additional amount.[35]

---

31 Ibid.

32 Megumi Yamanaka, *Tokyo Electric Plans Record LNG Imports After Quake Shuts Plant*, Bloomberg, Oct. 24, 2007, http://www.bloomberg.com/apps/news?pid=newsarchive&refer=japan&sid=a765.KHQ.BRQ.

33 Ibid.

34 IAEA, *Fukushima Nuclear Accident Update Log*, March 11, 2011, http://www.iaea.org/newscenter/news/2011/fukushima110311.html.

35 Kwok W. Wan, *IEA Forecasts Japan LNG Worst-case*, Petroleum Economist, Sept. 28, 2011, http://www.petroleum-economist.com/Article/2908167/IEA-forecasts-Japan-LNG-worst-case.html.

According to the IEA, Japan's historical suppliers such as Qatar, Indonesia and Malaysia, as well as Russia, are working to make additional cargoes available for Japan. Neighboring countries including Taiwan and Korea have offered LNG cargoes based on time swaps.[36] Additional cargoes could come from other markets such as Europe depending on shipping availability and cost.[37] However, the increased demand for LNG coming from Japan has already impacted the daily rates of LNG tankers, which have increased from $40,000 in the summer of 2010 to $60,000 in the winter of 2010/11 and jumping to $80,000 in April 2011.[38]

While it is unclear at the time this book goes to print how long Japan's extra demand for LNG will be sustained and what the impact will be on global LNG markets, some experts have extended their projections of Japan's increased demand out to 2020.[39]

3.2.2 South Korea

Like Japan, South Korea has very limited domestic natural gas production and thus relies heavily on LNG imports for its natural gas needs. South Korea began importing LNG in 1986 and has maintained its position as the world's second-largest LNG import market ever since.[40]

Korea's natural gas market is highly seasonal, with winter peaks driven by space heating requirements, especially during periods of particularly cold weather. Large seasonal swings have led to a high dependence on spot market purchases and medium-term contracts to meet peak winter needs in the past. In order to manage this issue of seasonality, KOGAS has conducted an aggressive terminal upgrade program to boost storage

---

36 IEA Medium-Term Oil & Gas Markets (MTOGM) 2011, http://www.iea.org/textbase/nppdf/stud/11/mtogm2011.pdf.

37 It is also possible that the Kenai LNG plant might offer additional cargoes to Japan, although the plan to mothball that facility remains unchanged. Eileen Moustakis and Edward McAllister, *Conoco Says Kenai LNG Plant to Operate into October*, Reuters, Aug. 9, 2011, http://www.reuters.com/article/2011/08/09/energy-lng-kenai-idUSN1E7781JR20110809.

38 IEA MTOGM 2011.

39 Pat Roberts, *CWC Dynamic Insights Report*, World LNG Series Asia Pacific Summit, October 2011. The CWC Group organizes numerous quality gas and LNG events that involve the industry's leading players to address the key issues facing the global energy sector (http://www.thecwcgroup.com/gasandlng). The CWC's Dynamic Insights Report provides the latest insights from key players in the global gas industry and is written by industry experts. http://www.thecwcgroup.com/cdi/.

40 Poten & Partners Kitimat Market Assessment.

capacity at its three existing import terminals. As of 2009, three terminals have 46 storage tanks with total capacity of 2.79 MMt.[41]

State-owned Korea Gas (KOGAS)[42] imports over 95 percent of the LNG in Korea. KOGAS is the world's largest corporate buyer of LNG and South Korea's sole natural gas wholesaler, supplying South Korea's city gas users (63 percent) and power companies (37).[43] KOGAS operates three LNG import terminals at Pyeongtaek, Incheon, and Tongyeong, with a fourth under construction at Samchuk (Figure 3.2).

Korean steelmaker POSCO and power generator K-Power started LNG imports in 2005 and 2006, respectively, having secured supply contracts with Tangguh LNG (Indonesia). POSCO constructed the first and only private import terminal at Gwangyang in the south of the country and it shares capacity at the terminal with K-Power.[44]

Korean imports represent over 13 percent of the global trade, with LNG sourced from Qatar, Malaysia, Oman, Indonesia, Brunei, Australia, Abu Dhabi, Yemen, and Russia on a long-term basis and from Atlantic Basin exporters on a spot and short-term contract basis.

Due to the global financial crisis, in 2009 Korea's LNG imports fell by 12 percent[45] but in 2010 LNG demand increased from 33 bcm to 44 bcm, driven by heating demand and growth in power generation. In December 2010, the Korean government released its long-term supply/demand forecast projecting 3.5 percent growth over the 2008–13 period.[46] To meet the expected growth in LNG demand in the near term, KOGAS entered into short-term deals with Tangguh LNG in 2010 for 1.2 bcm (14 cargoes) through 2012; with Repsol for 1.9 bcm (21 cargoes) for 15 months from January 2011; and with GDF Suez to purchase 3.4 bcm (41 cargoes) from its global portfolio from the fourth quarter of 2010 until the end of 2013. KOGAS is also in negotiations with Russia to increase LNG supplies to Korea.[47]

In the longer term, overall LNG demand is projected to reach 34 MMt/y

---

41 Ibid.

42 Korea Gas Corporation (KOGAS) was incorporated by the Korean government in 1983 and has since grown to become the world's largest LNG importer. KOGAS is the nation's sole LNG provider and currently operates three LNG terminals and a nationwide pipeline network spanning over 2,739 km. KOGAS imports LNG from around the world and supplies it to power generation plants, gas utility companies and city gas companies throughout South Korea. KOGAS, *Who We Are*, http://www.kogas.or.kr/kogas_eng/html/main/main.jsp.

43 Poten & Partners Kitimat Market Assessment.

44 Ibid.

45 Ibid.

46 Ibid.

47 Ibid.

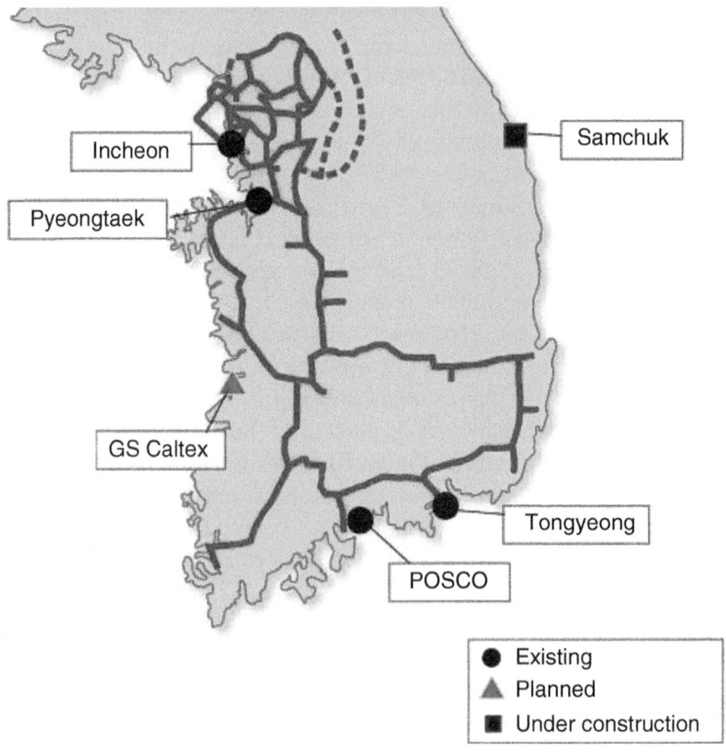

Source: Poten & Partners

Figure 3.2 South Korea's LNG import terminals

in 2015, 41 MMt/y in 2020, 45 MMt/y in 2025 and 49 MMt/y in 2030. South Korea's total long-term contract volume as of 2009 was 27 MMt/y and over the long term Korea will need to secure more long-term supply. Although Korea will receive additional LNG volumes from the new Gorgon, Wheatstone and Gladstone LNG projects in Australia starting in around 2015, Korea still needs to secure more long-term supply. According to some experts, the potential shortfall Korea is facing could be quite serious – as much as 10 MMt/y by 2015, 16 MMt/y by 2020, and 29 MMt/y by 2025 (assuming existing contracts will not be renewed). Like Japan, Korea is likely to be short of supply every year until 2035 without new contracts.[48]

---

[48] Ibid.

Although the primary driver for LNG demand in South Korea is economic growth, two other factors stand out – nuclear power plant development and the possibility of Russian pipeline gas imports.[49] In terms of nuclear power plant development, nuclear currently comprises approximately 12 percent of Korea's primary energy consumption and this share is expected to grow to 28 percent by 2030. It is unclear at this time whether Korea will revisit its planned energy mix in light of the Japanese Fukishima disaster and whether this will result in an increase in LNG in Korea's energy supply mix.[50]

As in the case of Japan, with limited domestic natural gas reserves and other energy resources, Korea's reliance on LNG will remain and a premium will always be placed on security of supply. Of the traditional Asia-Pacific importers, Korea has the highest growth rate and also the strongest seasonal fluctuations due to high winter heating loads.[51]

3.2.3 Taiwan

As with Japan and South Korea, Taiwan has limited domestic natural gas production with no pipeline imports, and therefore relies almost exclusively on LNG imports for power generation and city gas supplies. Taiwan began importing LNG in 1990 and by 2009 was the seventh largest LNG market, consuming 8.8 MMt of LNG or around 5 percent of the global LNG trade. Due to its lack of indigenous energy resources, Taiwan, like Japan and South Korea, is expected to continue to rely on LNG for energy security reasons.[52]

Taiwan imports LNG primarily from Indonesia, Malaysia, Qatar, Oman and Australia under long-term contracts and some spot shipments and from Atlantic Basin exporters on a spot and short-term contract basis. State-owned Chinese Petroleum Corporation (CPC) is the sole importer of LNG in Taiwan and controls all aspects of natural gas supply in the country, including exploration and production (E&P), imports, domestic pipeline transportation and gas wholesaling. CPC operates two existing LNG import terminals at Yung-An (1990) and Taichung (2009). The two terminals have a combined send-out capacity of 1.7 bcf/d. CPC's customers consist of city gas companies and power companies. Taiwan

---

[49] Ibid.

[50] Pat Roberts, *CWC Dynamic Insights Report*, World LNG Series Asia Pacific Summit, October 2011.

[51] Poten & Partners Kitimat Market Assessment.

[52] Ibid.

Power Company (Taipower) is the main end-user of CPC-sourced LNG.[53]

With 5 percent GDP growth projected in the medium term and an expected increase in gas for power, Taiwan's LNG demand is projected to grow steadily. Overall LNG demand is projected to reach 12 MMt/y in 2015, 15 MMt/y in 2020, 17 MMt/y in 2025 and 18 MMt/y in 2030. Power generation accounts for over 80 percent of LNG use in Taiwan but Taiwan is also considering adding additional nuclear capacity.[54]

A shortfall in Taiwan's LNG supply portfolio started to emerge in 2004 and, as a result, Taiwan has relied on spot and short-term cargoes as necessary to fill the gap. This imbalance is projected to continue into the future with insufficient long-term contract volumes to fill the projected shortfall. Taiwan's total existing long-term contracted volume was 6.8 MMt/y in 2009 but additional term volumes from Papua New Guinea starting in approximately 2015 will partially help address the imbalance.[55]

According to some estimates, Taiwan needs much more LNG supply with the projected supply/demand gap being as much as 6.5 MMt/y by 2015 and 10 MMt/y by 2020, and remaining at that level until 2025 (assuming existing contracts are renewed). As with other traditional Asian markets, Taiwan is likely to be short of supply for each year until 2035 without new term contracts.[56]

3.2.4 China

Historically, China has been far less reliant on gas imports than other Asian countries and, unlike Japan, Korea or Taiwan, China has large domestic fossil fuel reserves – particularly coal, which supplies a large share of its energy and electricity needs, in addition to significant reserves of oil and gas.[57] Nonetheless, China started importing LNG in 2006 and as of 2011 has four LNG terminals with a total regasification capacity of approximately 21 bcm.[58] The existing LNG terminals are located in the eastern coastal region of China in Guangdong, Fujian and Shanghai, where rapid economic growth is underpinned by strong increases in energy consumption (Figure 3.3). This region lacks significant indigenous energy

---

[53] Ibid.
[54] Ibid.
[55] Ibid.
[56] Ibid.
[57] Ibid.
[58] IEA MTOGM 2011.

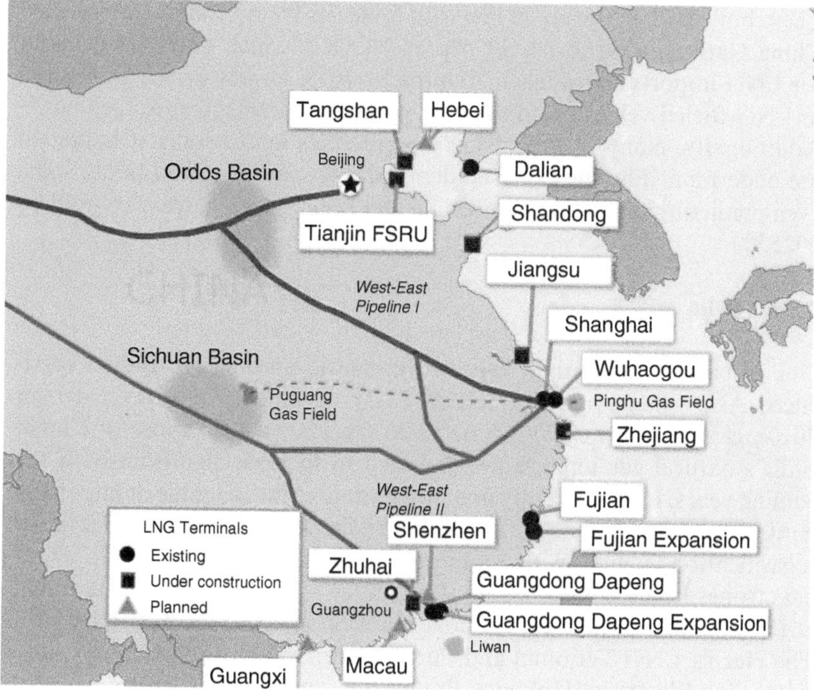

Source: Poten & Partners

Figure 3.3 China's LNG import terminals

resources and therefore relies heavily on imports of energy from other parts of China and overseas.[59]

In recent years, China has been rapidly expanding its portfolio of LNG regasification terminals and LNG import contracts, as well as expanding pipeline imports. This reflects the government's concerns that energy shortages could stifle the country's economic development and potentially trigger domestic unrest.[60] In 2011, China imported approximately 13.1 mtpa of LNG, representing the world's second highest growth rate (36.1 percent) in LNG demand.[61] In the coming decades, China's demand for LNG is expected to increase even more significantly, with some experts predicting that China's demand will surge five-fold to 44 mtpa by 2020.

---

59 Poten & Partners Kitimat Market Assessment.

60 Ibid.

61 GIIGNL, The LNG Industry 2011.

These bullish forecasts are in line with forecasts from China's state-owned China National Offshore Oil Corp. (CNOOC), which estimates demand for LNG imports could reach 30 mtpa by 2015, largely driven by a surge in LNG-fueled vehicles and the development of LNG storage facilities.[62] Additionally, many of the LNG mega projects in Australia (Chapter 8) are underpinned by the expected demand spike in China and India, some even predicting that demand in China and India could increase six-fold by 2025.[63]

3.2.5 India

Due in part to lower than expected domestic production, India experienced Asia's fastest growth rate in LNG demand (+37.4 percent over 2010), closely followed by China (+36.1 percent). Like those of China, India's natural gas imports are expected to increase significantly in the coming years. India currently does not have pipelines capable of importing natural gas from outside the region but India does have two LNG import/regasification facilities with a total capacity of 20 bcm.[64]

Petronet LNG Limited set up the country's first LNG import terminal at Dahej, Gujarat, in 2004 with an expansion in 2009 (see Appendix C).[65] The Hazira LNG Terminal and Port is partnered by Shell Gas B.V. and Total Gaz Electricité Holdings France and entered into service in April 2005 (see Appendix C). The Hazira LNG terminal operates and supplies LNG to a large number of customers in different industrial segments ranging from power and fertilizer to ceramic, steel, small and medium industries. According to Shell, "LNG has proved to be a cheaper, clean

---

62 Fayen Wong, *China's LNG Demand to Surge Five-Fold by End of Decade – GDF*, Reuters, Oct. 26, 2011, http://www.reuters.com/article/2011/10/26/china-lng-idUSL3E7LQ13320111026.

63 Stephen Bell, *Woodside Tips LNG Demand Spike in China and India*, The Australian, October 26, 2011, http://www.theaustralian.com.au/business/mining-energy/woodside-tips-lng-demand-spike-in-china-and-india/story-e6frg9df-1226177328349. "WOODSIDE Petroleum chief executive Peter Coleman said today demand for liquefied natural gas in China and India may rise six-fold by 2025."

64 IEA MTOGM 2011.

65 Petronet LNG was formed as a Joint Venture by the Government of India to import LNG and set up LNG terminals in the country. Petronet LNG also involves India's leading oil and natural gas industry players including GAIL (India) Limited (GAIL), Oil & Natural Gas Corporation Limited (ONGC), Indian Oil Corporation Limited (IOCL) and Bharat Petroleum Corporation Limited (BPCL). Petronet LNG, http://www.petronetlng.com.

Source: IEA MTOGM 2010, Gas Supp. p. 88

Figure 3.4 India's natural gas infrastructure and LNG terminals

and efficient replacement to liquid fuels being deployed by many industries for their power requirements."[66]

India does not have enough import capacity to meet expected future demand and is currently expanding both pipeline capacity and LNG regasification capacity (Figure 3.4). Two LNG terminals are under construction in India.[67] Petronet is in the process of building a terminal at Kochi, Kerala, that will have a capacity of 5 MMTPA.[68] In addition, Indian Oil is building a 5 million tonnes per year LNG terminal at Ennore in southern India, which is expected to come online by 2014.[69] New LNG import projects in India are challenged by how much the biggest consumers – power and fertilizer sectors – are willing for pay for imported LNG since the Indian government currently has price controls on what these two sectors can charge consumers. While many expect the price controls to be removed or

---

[66] Shell, *Hazira LNG*, http://www.shell.com/home/content/ind/aboutshell/shell_businesses/india_business_structure/hazira_lng.

[67] IEA MTOGM 2011.

[68] Petronet LNG, http://www.petronetlng.com.

[69] *Indian Oil in Talks with BG, Others to Secure LNG Supplies*, Reuters, March 28, 2012, http://www.reuters.com/article/2012/03/28/indian-oil-lng-idUSL3E8ES3B720120328.

modified, the ability to pay market prices for LNG has been limited since the costs often cannot be passed on to the consumer.[70]

3.2.6 Supplying the Asia-Pacific Region

Large amounts of incremental LNG supply are needed to feed growing Asian demand and new export projects may need to be developed in order to keep pace with the expected increase in consumption.[71] There is currently a significant number of competing LNG supply projects being planned in Australia that target the Asia-Pacific market, but only about 30 percent are likely to be completed within the needed time period due to economic, political, and engineering, procurement and construction (EPC) cost constraints.[72]

Some experts have projected that the Asia-Pacific region's need for new export capacity would start in 2014 and grow to reach 47 MMt in 2020, 85 MMt in 2025, and 130 MMt in 2035. These projects are beyond the more than 35 MMt/y of Asian-bound capacity currently under construction in Australia, Papua New Guinea and the Middle East.[73] As a result, some of the Asian buyers may be looking to LNG export projects being planned in Canada and the US (discussed in detail in Chapters 11 and 12). At least one expert has opined that the Canadian Kitimat LNG project (KM LNG) "is a natural choice to provide additional supply to the Asian markets to cover part of the emerging supply gap" due to Canada's proximity to the Asia-Pacific market, the large resource base, and Canada's political stability and regulatory certainty.[74]

3.2.7 Contracts and Pricing in Asia-Pacific

The Asian LNG market is predominantly based on long-term supply contracts. In 2009, more than 95 percent of total LNG deliveries were under such long-term contracts. Long-term contracts are the norm in order to address both buyer concerns with security of supply risks, and sellers' insistence on firm commercial arrangements to underpin construction

---

[70] Ibid. Other major LNG buyers in India, such as petrochemical plants and refineries, can afford to pay more for LNG and these customers are paying an average of \$13–15 per mmBtu and sometimes as much as \$16 per mmBtu for imported LNG. Ibid.

[71] Poten & Partners Kitimat Market Assessment.

[72] Ibid.

[73] Ibid.

[74] Ibid.

and operation of LNG liquefaction facility economics, including financing arrangements. Unlike in the Atlantic Basin, there are no liquid gas markets to provide a ready fall-back market for buyers or sellers in the Asia-Pacific region.

3.3 THE EUROPEAN LNG MARKET

In 1964, Algeria began delivering LNG to the UK, thus establishing Europe as the world's first LNG market. Deliveries to France, Spain and Italy followed shortly after. The discovery of natural gas in the North Sea in the late 1960s allowed the United Kingdom to cease LNG imports until relatively recently when declines of North Sea reserves necessitated the resumption of LNG imports into the UK. Other European countries, such as Turkey, Greece and Portugal, began importing LNG in the 1990s and early 2000s.

The growth of LNG in Europe has been more gradual than that in Asia-Pacific primarily because LNG has had to compete with pipeline gas, both domestically produced and imported from Russia. Currently, there are eight European countries that import LNG – the UK, France, Spain, Italy, Belgium, Turkey, Greece and Portugal. Historically, only Spain and Portugal have received more than 50 percent of their total gas supply from LNG imports.[75]

In 2008, LNG accounted for 10.4 percent of Europe's natural gas supply. Domestic production from Norway, the UK, the Netherlands, Germany and Italy provided over half of Europe's supply. Pipeline imports from North Africa, Algeria and Libya into Spain and Italy accounted for 8.5 percent of Europe's supply. However, the main source of European imports of natural gas is the former Soviet Union, which supplies about 30 percent of Europe's natural gas needs. Most of the gas comes from Russia but Turkmenistan, Azerbaijan and others are also suppliers.[76]

The role of LNG in Europe's energy supply mix is expected to grow in the coming decades due primarily to increased concerns about security of supply and declining domestic production.[77] In particular, the UK is increasingly looking to LNG to supply its energy needs,[78] and in 2011 the

[75] Andy Flower, *LNG in Europe*, LNG Industry, Winter 2009, at pp. 10–15, www.lngndustry.com.

[76] Andy Flower, *LNG in Europe*, LNG Industry, Winter 2009.

[77] Ibid.

[78] IEA MTOGM 2011.

UK overtook Spain as the world's third-largest LNG importer behind Japan and South Korea.[79]

In terms of energy security, according to the IEA, about 80 percent of Russian exports to Europe transit through Ukraine, which at times has caused significant problems for Europe.[80] For example, in January 2009, a dispute arose between Russia and Ukraine over the prices Ukraine pays for natural gas imports. While the dispute did not involve Russia's other European customers, it led to supply disruptions throughout Europe as gas supplies that transit through Ukraine were interrupted until the dispute was resolved. This was the third such Russian–Ukraine dispute since 2005 and arguably the worst ever since it occurred during the middle of an unusually cold winter in Europe.[81] For these reasons, diversifying supply is of particular importance for Europe, with LNG potentially playing a key role.

According to Gas Infrastructure Europe (GIE), as of 2010 there are 20 existing LNG terminals in Europe and six expansions under construction.[82] There are six new LNG terminals committed or under construction and another 32 LNG terminals are under study or planned.[83] (See Figure 3.5.)

The following subsections present an overview of the European LNG market, country by country.

---

[79] In 2011, the UK imported a record 19.6 million metric tons of LNG, up 32% from 14.8 million in 2010. Spanish LNG imports fell by about 18% to 17.4 million tons, while French imports rose 16% to 11.6 million tons. Dinakar Sethuraman, *U.K. Overtaking Spain to Become World's Third-Biggest LNG Buyer*, Bloomberg, Jan. 4, 2012, http://www.bloomberg.com/news/2012-01-04/u-k-overtaking-spain-to-become-world-s-third-biggest-lng-buyer.html.

[80] IEA WEO-2006, www.iea.org.

[81] Andy Flower, *LNG in Europe*, LNG Industry, Winter 2009 at pp. 10–15.

[82] Gas Infrastructure Europe (GIE) was established on March 10, 2005, as a non-profit organization representing 70 member companies from 25 countries and includes gathering operators of gas infrastructures across Europe including transmission pipelines, storage facilities and LNG terminals. GIE is the umbrella organization for its three subdivisions: GTE (Gas Transmission Europe) representing the Transmission System Operators, GSE (Gas Storage Europe) representing the Storage System Operators, and GLE (Gas LNG Europe) representing the LNG Terminal Operators. GIE, http://www.gie.eu.

[83] GLE (Gas LNG Europe), *LNG Terminals in Europe*, GLE presentation to CEER LNG Workshop – Session I, 6/09/2011. http://www.energy-regulators.eu/portal/page/portal/EER_HOME/EER_WORKSHOP/CEER-ERGEG%20EVENTS/GAS/1st%20workshop%20on%20access%20to%20European%20LNG%20terminals/GLE_LNG_Infrastructure_in_the_EU_CEER%20Workshop%20on%20LNG_20110906_final_0.pdf.

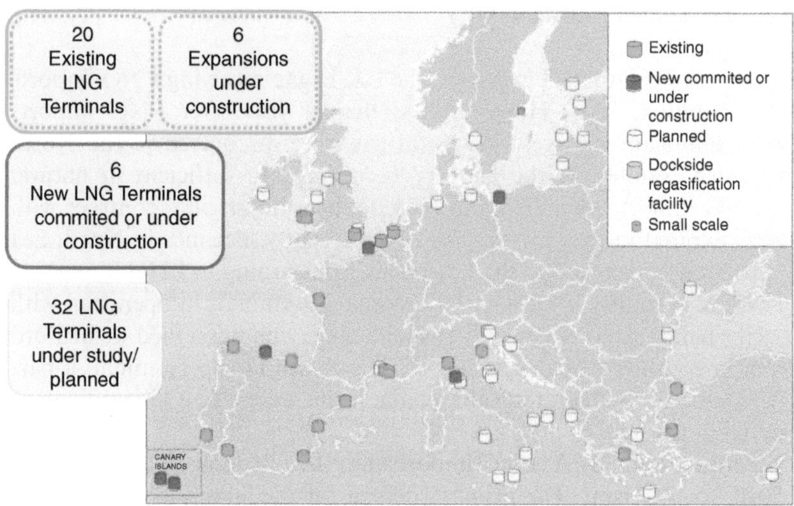

Source: Gas Infrastructure Europe

Figure 3.5 LNG terminals in Europe

3.3.1 Spain

Spain has historically been Europe's largest LNG importer and the third largest importer worldwide after Japan and Korea. Spain's 2008 LNG imports were 22.1 million tonnes, representing just over 50 percent of the total for Europe. Spain relies on LNG to meet 73 percent of its natural gas demand, making it the European country most dependent on LNG. The primary reason for Spain's dependence on LNG is due to geography since the Pyrenees Mountains are a barrier to the construction of pipelines connecting Spain to the European gas grid. There are only two small pipelines connecting Spain to Europe – one linking Spain to France and one used to carry gas contracted by Spanish gas company, Gas Natural, from Norway. There is also the Pedro Duran Farrell pipeline, which passes through Morocco and across the Straits of Gibraltar and allows Spain to import gas from Algeria. Due to the lack of pipeline gas, Spain has more LNG terminals than any other country in Europe and as of 2009 six facilities were in operation with a seventh, the El Musel terminal in the north of the country, under construction.[84]

[84] Andy Flower, *LNG in Europe*, LNG Industry, Winter 2009, at pp. 10–15.

3.3.2 The United Kingdom (UK)

As the world's first LNG importer, the UK began receiving LNG imports from Algeria in 1964. However, less than a year after LNG imports began, significant reserves of natural gas were discovered in the North Sea, which resulted in the country becoming self-sufficient in natural gas. LNG imports dwindled and ceased when the original contract with Algeria expired in the early 1980s. More recently, declines in North Sea gas production have led to the UK once again turning to LNG imports.[85]

The UK currently has four LNG receiving terminals in operation with capacity being expanded. Two terminals were commissioned at Milford Haven in southwest Wales in 2009.[86] The South Hook Terminal, a part of the Qatargas 2 integrated value chain, is the largest LNG regasification terminal in Europe.[87]

According to the IEA, LNG imports into the UK jumped by 85 percent in 2010, to 18.8 bcm. The primary reason for the large increase was the UK's taking of many Qatari cargoes throughout 2010 that would otherwise have gone to the US but for the shale gas revolution underway in the US (see Chapter 10). Since the UK's NBP prices were higher than HH prices in 2010, many producers, such as Qatar, sought buyers in the UK. Qatar in particular used the UK's South Hook import terminal "as a strategic outlet for the LNG from its mega-trains."[88]

The UK's Dragon LNG terminal also received a cargo from Qatar for the first time in August 2010. Notably, the Isle of Grain terminal received a cargo re-exported from Cheniere's Sabine Pass LNG import terminal – the first US cargo since the historic *Methane Pioneer*'s delivery to the UK's Canvey Island over 50 years ago. In early 2011, Centrica reached an agreement with Qatargas to supply 3.3 bcm (2.4 mtpa) of LNG for three years to the Isle of Grain terminal. Given the UK's increase in LNG imports, it now appears that the UK is set to play an "important role as an intake point of LNG to Continental Europe."[89] In fact, according to preliminary data, in 2011 the UK overtook Spain as the world's third-largest LNG importer behind Japan and South Korea.[90]

---

85 Ibid.

86 Ibid.

87 Southhook LNG, http://www.southhooklng.co.uk/cds-web/view.do.

88 IEA MTOGM 2011.

89 Ibid.

90 See note 79. The IGU's most recent LNG report confirms that in 2011 the UK was the world's third-largest LNG importer with imports of 18.6 mtpa. Spain

3.3.3 France

France has historically been Europe's second-largest market for LNG and imported 9.5 million tonnes of LNG in 2008. France is well connected to Europe's pipeline grid and also has capacity to expand pipeline imports of natural gas.[91] The French gas market is dominated by GDF Suez, which is 35 percent state owned and has the largest gas transport network in Europe. Over 75 percent of GDF Suez's natural gas requirements are covered over the next decade by long-term contracts.[92]

France currently has three LNG terminals in operation. Fos-Tonkin located on the Mediterranean Sea and Montoir de Bretagne on the Atlantic Ocean are both owned by Elengy, a subsidiary of GDF Suez created in early 2009. The third terminal, Fos-Cavaou, received a test cargo in October 2009 and started commercial operations in April 2010. As of 2010, the following three new terminals were under consideration, with FID expected sometime in 2010, with projected start-up dates of 2013, 2014, and 2015 respectively: Le Havre-Antifer (9 bcm project sponsored by Gaz de Normandie (73 percent) and Compagnie Industrielle Maritime (27 percent)); Dunkerque (10–13 bcm project sponsored by EDF, the Port of Dunkerque and possibly Total); and Fos-sur-Mer (8 bcm project undertaken by Shell and Vopack).[93]

Although France is Europe's second-largest market for LNG, the share of natural gas used in France is actually relatively low (15 percent) compared to Europe's average of 25 percent. France's lower use of natural gas is primarily due to the large use of nuclear for power generation in France.[94]

3.3.4 Italy

Italy is Europe's third-largest natural gas user, behind the UK and Germany, with ENI as the dominant gas company in Italy. Natural gas demand in Italy has more than doubled in the past 25 years and represents approximately 40 percent of Italy's total energy demand.[95] After nuclear energy was rejected in the late 1980s, natural gas became the primary fuel

---

was the world's fourth-largest importer with imports of 17.1 mtpa. IGU World LNG Report 2011 at p. 11.

91 Andy Flower, *LNG in Europe*, LNG Industry, Winter 2009, at pp. 10–15.

92 IEA MTOGM 2010.

93 Ibid.

94 Ibid.

95 Ibid.

for power generation in Italy and accounted for over 54 percent of the power output in 2008.[96]

Italy's LNG imports tripled in 2010 to 9 bcm, with LNG entering through two LNG terminals, Panigaglia LNG and Adriatic LNG. The Adriatic LNG terminal commenced operations in 2009 and in 2010 it received 7 bcm of LNG, primarily from Qatar. Like South Hook in the UK, the Adriatic LNG terminal appears to be one of the LNG terminals used by Qatar as an outlet for its LNG production.[97]

In addition to the two LNG terminals, Italy has five pipelines through which imported gas enters the country. In recent years, Italy has been faced with winter gas shortages due to the lack of import and storage capacity. In order to meet expected increased demand and improve energy security, a number of projects have been planned since the early 2000s. To date, however, Italy has made limited progress in expanding its LNG, pipeline and storage infrastructure.[98]

3.3.5 Belgium

The Zeebrugge terminal in Belgium was built to receive LNG for the country's main gas company, Distrigas. However, a long-term contract between Distrigas and Algeria was not renewed and expired in 2006. Since then, the Zeebrugge terminal has primarily been used to import LNG from Qatar under a long-term contract between Distrigas and RasGas. In recent years, the terminal has also received spot cargoes from Egypt, Nigeria, Norway, Trinidad and Tobago, and Malaysia.[99] The Zeebrugge LNG terminal is a gateway to supply LNG into Northwestern Europe and is operated by Fluxys, an independent operator of both the natural gas transmission grid and storage infrastructure in Belgium. Zeebrugge is also a hub where any LNG can be redelivered for consumption on the Belgian market, or traded on the Zeebrugge Hub for onward transmission to supply other end consumer markets in any direction including the UK, the Netherlands, Germany, Luxembourg, France and Southern Europe.[100]

About 1,200 LNG carriers have docked at the Zeebrugge terminal since it was commissioned in 1987. A fourth storage tank and additional send-out capacity were added to the terminal in 2004–08 and this enhancement

---

96 Ibid.

97 IEA MTOGM 2011.

98 IEA MTOGM 2010.

99 Andy Flower, *LNG in Europe*, LNG Industry, Winter 2009, at pp. 10–15.

100 Zeebrugge LNG Terminal, http://www.fluxys.com/en/About%20Fluxys/Infrastructure/LNGTerminal/LNGTerminal.aspx.

doubled the terminal's "throughput capacity to 9 billion cubic metres of natural gas per year, allowing reception of 110 ships per year instead of 66 ships per year previously." An additional enhancement is under consideration pending a review of demand interest.[101]

3.3.6 The Netherlands – Rotterdam

In 2005, N.V. Nederlandse Gasunie and Royal Vopak decided to jointly develop Gate ("Gas Access To Europe") terminal, the first independent LNG import terminal in the Netherlands. Construction began in 2008, and in June and July 2011 Gate terminal received three LNG cargoes, which were used for commissioning and testing.[102] The long-awaited Gate LNG terminal opened in September 2011[103] and received its first commercial LNG cargo on September 1, 2011, when the Q-Max LNG carrier *Bu Samra* arrived from Qatar.[104]

The terminal will have an initial annual throughput capacity of 12 billion cubic meters per year (bcma) and can be increased to 16 bcma in the future. Five European energy companies (DONG Energy from Denmark, EconGas from Austria, E.ON Ruhrgas from Germany, RWE Supply & Trading from Switzerland and Eneco from the Netherlands) have signed long-term throughput agreements with Gate terminal.[105] Gate LNG terminal will be another LNG hub in Europe similar to the Zeebrugge Hub in Belgium.[106]

3.3.7 Turkey

Turkey began importing LNG in 1994 to reduce its dependence on pipeline imports primarily from Russia but also from Iran and Azerbaijian.

---

[101] Ibid.

[102] Gate Terminal Press Release, Sept. 1, 2011, *First Commercial LNG Cargo at Gate Terminal*, http://www.gate.nl/en/news-media/news/news-items/first-commercial-lng-cargo-at-gate-terminal.html.

[103] LNG World News, *The Netherlands: Queen to Open Gate LNG Terminal September 23*, Sept. 2, 2011, http://www.lngworldnews.com/the-netherlands-queen-to-open-gate-lng-terminal-september-23/ (noting that Her Majesty Queen Beatrix of the Netherlands will open Gate terminal, located on the Maasvlakte in Rotterdam, on 23 September 2011).

[104] Gate Terminal Press Release, Sept. 1, 2011, *First Commercial LNG Cargo at Gate Terminal*, http://www.gate.nl/en/news-media/news/news-items/first-commercial-lng-cargo-at-gate-terminal.html.

[105] Ibid.

[106] IEA MTOGM 2011.

Turkey has two LNG terminals in operation. The Marmara Eregelisi terminal near Istanbul is owned and operated by the country's main gas company, Botas. The second terminal, the Aliaga terminal near Izmir, was built by an entrepreneur and sat idle for several years after its completion but was activated in December 2006 to import cargoes for Botas. In 2009, the terminal's operator, Egegaz,[107] imported two cargoes from Qatar.[108]

3.3.8 Portugal and Greece

Portugal and Greece each have one LNG terminal and both were built to diversify natural gas supply from pipeline imports from Algeria and Russia. Portugal's main LNG supplier is Nigeria and Greece receives most of its LNG from Algeria.[109]

3.3.9 Emerging European Importers

In addition to the traditional European LNG importers, a number of other European countries are also considering building import terminals including Poland, Croatia, Albania, Cyprus, Germany, Bulgaria, Romania, Lithuania, Estonia, Sweden and Eire.[110]

Polskie LNG[111] recently received an award of over 200 million Euros towards the construction of a new terminal at Świnoujście in northwest Poland. The terminal is a major energy project for the Polish government and offers Poland the opportunity to diversify its gas supply and improve the country's energy security. Like much of Europe, Poland is heavily dependent on Russian imports. Construction of the terminal began in March 2011 and it is expected to begin operations in July 2014.[112]

---

107 Egegez, http://www.egegaz.com.tr/en/default.aspx.

108 Andy Flower, *LNG in Europe*, LNG Industry, Winter 2009, at pp. 10–15.

109 Ibid.

110 Ibid.

111 Polskie LNG was established in 2007 by Polish Gas and Oil Company PGNiG (PGNiG SA). Gas Transmission Operator GAZ-SYSTEM S.A. (GAZ-SYSTEM S.A.), a company owned by the State Treasury and responsible for the security of natural gas supplies via transmission networks, became the owner of Polskie LNG. GAZ-SYSTEM S.A. will supervise the construction of the LNG terminal and the Polish Oil and Gas Company, PGNiG, is responsible for the supply and transport of LNG to the terminal in Świnoujście. Polskie LNG, http://en.polskielng.pl/nc/hidden/the-company.html.

112 The News.Pl, *News from Poland, European Commission Backs Poland's LNG Terminal*, Oct. 7, 2011, http://www.thenews.pl/1/12/Artykul/56439,European-Commission-backs-Polands-LNG-Terminal.

3.4 NORTH AMERICAN/ATLANTIC BASIN LNG MARKET

3.4.1 North America – the US, Canada and Mexico

In North America, the United States, Canada and Mexico have strong pipeline connections and exports to these countries are typically regarded as part of an integrated supply mix. With large supplies of natural gas and an extensive pipeline system connecting the US, Canada, and Mexico, North America has historically been able to supply almost all of its natural gas requirements from indigenous sources.

During the supply-constrained 1970s, however, the United States began importing LNG from Algeria. Four LNG import terminals were built in the US between 1971 and 1980: Lake Charles, LA, Everett, MA, Elba Island, GA, and Cove Point, MD. The US received a peak volume of 253 billion cubic feet (BCF) of LNG in 1979 (which represented 1.3 percent of US gas demand), after which LNG imports in the United States declined for a number of reasons. The first was deregulation, which led to an increase in North American domestic natural gas production. Another reason was price disputes with Algeria, then the sole LNG exporter to the United States. As a result, the LNG terminals at Elba Island and Cove Point were mothballed in 1980 and those at Lake Charles and Everett suffered from very low utilization.[113]

3.4.2 The Late 1990s Bring Renewed Interest in LNG in the US

In 1999, the first Atlantic Basin LNG liquefaction plant came online in Trinidad and Tobago. This event, combined with increasing US natural gas demand, particularly for electric power generation, and increasing natural gas prices, resulted in renewed interest in LNG for the American market.[114]

In the early 2000s, Trinidad and Tobago provided a full 66 percent of the US's LNG imports. According to the EIA, in 2002 the US imported 151 Bcf (3.2 million tons) from Trinidad and Tobago. In addition to imports from Trinidad and Tobago and Algeria, the US also received LNG cargoes from Brunei Darussalam, Malaysia, Nigeria, Oman, and Qatar.[115]

---

[113] Dominion, http://www.dom.com/business/gas-transmission/cove-point/history-of-lng.jsp.

[114] Ibid.

[115] US EIA Report no. 0637, *The Global Liquefied Natural Gas Market: Status and Outlook*, Dec. 2003, http://www.eia.gov/oiaf/analysispaper/global/uslng.html.

As a result of increased demand, the two mothballed US LNG import terminals were reactivated, Elba Island in 2001 and Dominion Cove Point in 2003. In 2003, there were four LNG import terminals in the continental United States with a total send-out capacity of about 1.2 Tcf (26.0 million tons) per year and an estimated baseload capacity of 880 Bcf (18.5 million tons) per year. The four terminals were:[116]

Cove Point, MD Cove Point received final permission to re-open from the Federal Energy Regulatory Commission in July 2003. Dominion, the terminal owner, began commercial operations in August and had received 18 Bcf (0.4 million tons) as of the end of September 2003, all from Trinidad and Tobago.

Elba Island, GA This terminal is the smallest US terminal and was reactivated in 2001. It received ten cargoes in 2002 and as of the end of September 2003 it had received 41 Bcf (0.9 million tons) from 18 shipments, all originating in Trinidad and Tobago.

Everett, MA This terminal is owned by Distrigas, which completed an expansion in early 2003 to serve a nearby power plant. As of September 2003, Everett had received 52 shipments carrying 117 Bcf (2.5 million tons), all from Trinidad and Tobago.

Lake Charles, LA This facility is owned by Southern Union and in 2003 operated above baseload capacity, having received 186 Bcf (3.9 million tons) from 81 cargoes through September 2003. Shipments to Lake Charles came from Trinidad and Tobago, Algeria, Malaysia, Nigeria, Oman, and Qatar.

At the time, all four terminals had either completed an expansion or were planning to expand their regasification capacity to meet rising demand. In addition, there were at least two dozen proposals to build new LNG regasification terminals in North America over the next several years. The EIA's *Annual Energy Outlook 2004* (AEO2004) projected that four new LNG regasification terminals would be constructed on the Atlantic and Gulf Coasts from 2007 through 2010 to meet an expected 58 percent increase in LNG imports that was projected for that time-frame.[117]

---

116 Ibid.
117 Ibid.

3.4.3 The Early 2000s – the US is Expected to Become a Large LNG Importer

Based on EIA long-term forecasts in the early 2000s, US natural gas consumption was projected to increase from 22.5 Tcf in 2002 to 26.2 Tcf in 2010 and 31.4 Tcf by 2025. Domestic gas production was expected to increase more slowly than consumption over the forecast period, rising from 19.0 Tcf in 2002 to 20.5 Tcf in 2010 and 24.0 Tcf by 2025. The difference between consumption and production was to be made up by imports, which were projected to rise from net imports of 3.5 Tcf in 2002 to 7.2 Tcf by 2025.[118]

Nearly all the increase in net US natural gas imports from 2002 to 2010 was expected to come from LNG, with an almost 2 Tcf (42 million ton) increase expected over 2002 levels. Net US LNG imports were expected to rise from 5 percent of net US natural gas imports in 2002 to 39 percent in 2010.[119] Net pipeline imports from Canada were expected to reach 3.7 Tcf in 2010, and then decline as Canadian fields matured and Canadian demand increased. It was projected that LNG would become the largest source of net US imports by 2015, as Canadian imports declined. Mexico, currently a net importer of US natural gas, was expected to remain so throughout the period, mainly to supply industry located on the United States–Mexican border. Exports to Mexico were forecast to decline after 2005 as terminals in Baja California, Mexico, came online to supply both the US and the Mexican markets.

While forecasts varied, many analysts expected "LNG to account for 12% to 21% of total U.S. gas supply by 2025, up from approximately 3% in 2005."[120] If these forecasts proved correct, the US would become increasingly dependent on LNG imports to supplement decreasing North American natural gas production. Moreover, the US would also need to significantly increase its natural gas infrastructure to support the additional imports, including building more LNG import terminals and adding more pipelines.[121]

Policy makers and US government officials began to encourage greater imports of LNG and the related infrastructure that would be needed to handle additional LNG imports. For example, in April 2005, US President

---

118 Ibid.

119 Ibid.

120 Paul W. Parfomak, Cong. Research Serv., *Liquefied Natural Gas (LNG) in U.S. Energy Policy: Infrastructure and Market Issues*, Jan. 31, 2006, http://www.cnie.org/NLE/CRSreports/06feb/RL32386.pdf.

121 Ibid.

Bush stated, "One of the great sources of energy for the future is liquefied natural gas . . . We need more terminals to receive liquefied natural gas."[122]

In June 2005, Department of Energy Secretary Samuel Bodman commented, "LNG seems to offer a solution to . . . the growing demand for natural gas that we will see all around the globe."[123] In November 2005, Federal Reserve Chairman Alan Greenspan testified before Congress that "severe reaction of natural gas prices to the production setbacks that have occurred in the Gulf highlights again the need to . . . import large quantities of far cheaper, liquefied natural gas (LNG) from other parts of the world."[124]

The US Congress, the Federal Energy Regulatory Commission, the Department of Energy, and other US federal agencies also began to promote greater LNG supplies by changing regulations, clarifying regulatory authorities, and streamlining the approval process for new LNG import terminals.[125] For example, the Energy Policy Act of 2005 (P.L. 109-58) included various incentives for domestic natural gas producers (Title III, Subtitle E). The act also amended Section 3 of the Natural Gas Act of 1938, granting the Federal Energy Regulatory Commission (FERC) explicit and "exclusive" authority to approve onshore LNG terminal siting applications (Sec. 311c) among other provisions.[126]

Federal officials and Members of Congress debated the merits and risks of increased US LNG imports. Some questioned the implications of such a policy, drawing analogies with the consequences of US dependency on foreign oil and citing potential instability among foreign LNG suppliers.[127] Others expressed concern about LNG safety and vulnerability to terrorism.[128]

---

122 Ibid., citing President George W. Bush, Press conference dated April 29, 2005.

123 Ibid., citing Samuel Bodman, US Energy Secretary, *Remarks to the USEA/Center for LNG Conference*, National Press Club, Washington, DC, June 16, 2005.

124 Ibid., citing Alan Greenspan, Chairman, US Federal Reserve Board, Testimony Before the Joint Economic Committee, US Congress. Nov. 3, 2005.

125 Ibid.

126 Ibid.

127 Hon. Peter Domenici, *U.S. Must Build LNG Ports to Avoid Spiraling Natural Gas Prices, Sen. Domenici Says*, Press Release, Feb. 15, 2005; Hon. John E. Peterson, Remarks at the Hearing of the House Resources Committee, Energy and Mineral Resources Subcommittee on "U.S. Energy and Mineral Needs, Security and Policy," March 16, 2005.

128 Hon. Edward Markey, "Democratic Reaction to the 9/11 Commission's Final Report and its Security Recommendations for Preventing Further Attacks," Press conference, Dec. 5, 2005.

Meanwhile, in the marketplace, a race was on to build LNG terminals, with LNG developers proposing over 30 new terminals throughout North America (Figure 3.6).

While federal actions sought to facilitate greater LNG imports, public concerns about the safety and siting of proposed terminals began to mount. Choosing acceptable sites for new LNG terminals would prove to be extremely difficult and controversial in the US.[129] Many developers sought to build terminals near major consuming markets such as California and the Northeast (Figure 3.6) in order to reduce delivery costs to these major markets.

Community opposition to proposed LNG projects in these markets, however, was particularly intense. For example, in California, several proposed LNG terminals were ultimately blocked due to community opposition, including a proposed terminal in Northern California's Humboldt Bay,[130] Sound Energy Solution's proposed terminal in Long Beach, California,[131] and BHP Billiton's proposed floating LNG facility off the coast of Malibu, California.[132]

Proposed LNG terminals on the East Coast of the United States did not fare much better and in April 2008 the New York Secretary of State rejected a proposal by Broadwater Energy to construct a floating storage and regasification unit for imported LNG in Long Island Sound.[133] The proposed facility had been approved by FERC subject to more than 80 mitigation measures to enhance safety and security and minimize environmental impacts.[134]

---

[129] For a summary of the controversy surrounding proposed LNG terminals in the US, see Virginia L. Thorndike, LNG: A LEVEL-HEADED LOOK AT THE LIQUEFIED NATURAL GAS CONTROVERSY (2007).

[130] Sierra Club, *Liquefied Natural Gas Threatens California's Coastal Communities*, available at http://www.sierraclub.org/ca/coasts/lng.

[131] California Energy Commission, Liquefied Natural Gas Projects, Long Beach LNG Facility, http://www.energy.ca.gov/lng/projects.html#long; see Letter from FERC suspending review of project due to Board of Harbor Commissioner's declining to enter into a lease with SES for the LNG site, http://www.energy.ca.gov/lng/documents/long_beach/2008-03-10_FERC_TO_SES_TERMINATING_REVIEW.PDF.

[132] Sierra Club, *Huge Victory Against Offshore LNG Terminal*, available at http://www.sierraclub.org/ca/coasts/victories/victory2007-04-19.asp.

[133] *New York Secretary of State Determines Broadwater's LNG Facility Not Consistent with the Long Island Sound Coastal Management Program*, New York Dept. of State Press Release, Apr. 10, 2008, available at http://www.dos.state.ny.us/pres/pr2008/41008.htm.

[134] *FERC Approves Broadwater LNG Project Subject to Safety, Environmental Measures*, FERC News Release, March 20, 2008, available at http://www.ferc.gov/news/news-releases/2008/2008-1/03-20-08-C-1.asp. FERC's review of the project took 38 months and 25,000 staff hours and produced a final environmental impact statement (EIS) exceeding 2,200 pages. Ibid.

Source: FERC, Office of Energy Projects

Figure 3.6 Existing and proposed North American LNG import terminals 2006

At one point, leading energy expert Daniel Yergin predicted that, due to environmental concerns in some regions of the United States, import terminals would probably need to be built in neighboring countries such as Mexico and Canada to supply the United States.[135] Mr Yergin's predictions were correct and to date the only West Coast terminal to have been built is Sempra's LNG import terminal, Energia Costa Azul, located in Baja California, Mexico (at the border of Southern California and Mexico).[136]

Opposition to LNG terminals was not just limited to the West and East Coasts. For example, in Alabama, a state assumed to be friendly to energy infrastructure, community groups effectively blocked two onshore terminal proposals and called for LNG import terminals to be built only offshore.[137]

Moreover, US opposition to LNG import terminals did not seem to dissipate over time, as evidenced by the long-standing opposition to Hess LNG's proposed Weaver's Cove terminal, a proposed East Coast LNG terminal at the north end of Fall River in Massachusetts. In June 2011, Hess LNG announced they were abandoning their plans to develop the controversial LNG terminal in Weaver's Cove, claiming that changing economics due to the significant increase in shale gas production was the reason for their decision to withdraw applications for the facility proposed back in 2003. Environmental groups had long opposed the facility and were supported by a host of public officials and federal and state lawmakers, some of whom had worked to pass laws complicating the ability of LNG tankers to reach Mount Hope Bay. While Hess said the opposition played no role in the decision to abandon the project, environmentalists hailed the decision, stating that the "project was never appropriate for Narragansett Bay, so its official demise is long overdue."[138]

In addition to public opposition, in some cases state and local agencies were at odds with federal agencies over LNG terminal siting approval. For example, Delaware's environmental secretary blocked the development

---

135 Daniel Yergin and Michael Stoppard, *The Next Prize*, Foreign Affairs, Vol. 82, No. 6 (2003) at pp. 109–111.

136 *Sempra Energy's New Baja California LNG Terminal Ready for Commercial Operations*, May 15, 2008, CNNMoney.com, available at http://money.cnn.com/news/newsfeeds/articles/marketwire/0397377.htm.

137 Ibid., citing Editorial, *Move ExxonMobil's LNG Plant Offshore*, Mobile Register, Nov. 30, 2003.

138 Chris Barrett, *Hess Abandons Weaver's Cove LNG Project*, Providence Business News, published online June 13, 2011, http://www.pbn.com/Hess-abandons-Weavers-Cove-LNG-project,58983.

of an LNG terminal on the Delaware–New Jersey border after ruling that part of the terminal would extend into Delaware's waters and violate Delaware's Coastal Zone Act.[139] In 2004, the California Public Utilities Commission (CPUC) sued FERC in federal court over FERC's assertion of sole jurisdiction over the siting of the Sound Energy Solutions (SES) LNG terminal in Long Beach. The CPUC later dropped its suit, however, after the passage of P.L. 109-58 mooted its arguments. In January 2005, Massachusetts and Rhode Island filed petitions in federal court to reverse FERC's approval of an LNG import terminal to be sited in Fall River, MA.[140]

Despite the controversy over terminal siting in the US, a number of new terminals ultimately did get built in the US. In 2005, an offshore facility, Gulf Gateway Energy Bridge, was added in the Gulf of Mexico to allow for additional imports. Another offshore facility, Northeast Gateway, was opened offshore from Boston. Several terminals were added in the Gulf of Mexico including Cheniere's Sabine Pass import terminal, which is discussed in detail in Chapters 11 and 12. However, the map of existing North American LNG import terminals (Figure 3.7) stands in stark contrast to the map of existing and proposed terminals shown above in Figure 3.6!

As it turns out, however, the US did not end up needing all of the proposed LNG import terminals anyway. By 2008, shortly after many of the new terminals came online, the dynamic LNG market had shifted, with very little LNG coming to the US as sellers sought higher-paying markets elsewhere. Many of the new US LNG import terminals remained underutilized. In an amazing turn of events, largely driven by US shale gas production, almost all US LNG import terminals are now looking to add liquefaction capability so they can export LNG as well. While this dramatic shift is discussed in detail in Chapters 10–12, the map of possible US LNG export sites (Figure 3.8) is somewhat ironically reminiscent of the maps of proposed import terminals in the early 2000s.

A little more irony can be found further north in Alaska where, on February 10, 2011, ConocoPhillips and Marathon Oil Corp. announced that the only existing LNG export terminal – Kenai LNG – would be closed due to "market changes" that precluded the company from securing the sales and supply agreements needed to keep the plant operating

---

139 Ibid., citing A. Fifield, *Del. Hands BP a Setback on Pier*, Philadelphia Enquirer, Feb. 4, 2005.

140 Ibid., citing *Rhode Island, Massachusetts Officials Ask Court to Block Weaver's Cove LNG Project*, Natural Gas Intelligence. Jan. 30, 2006.

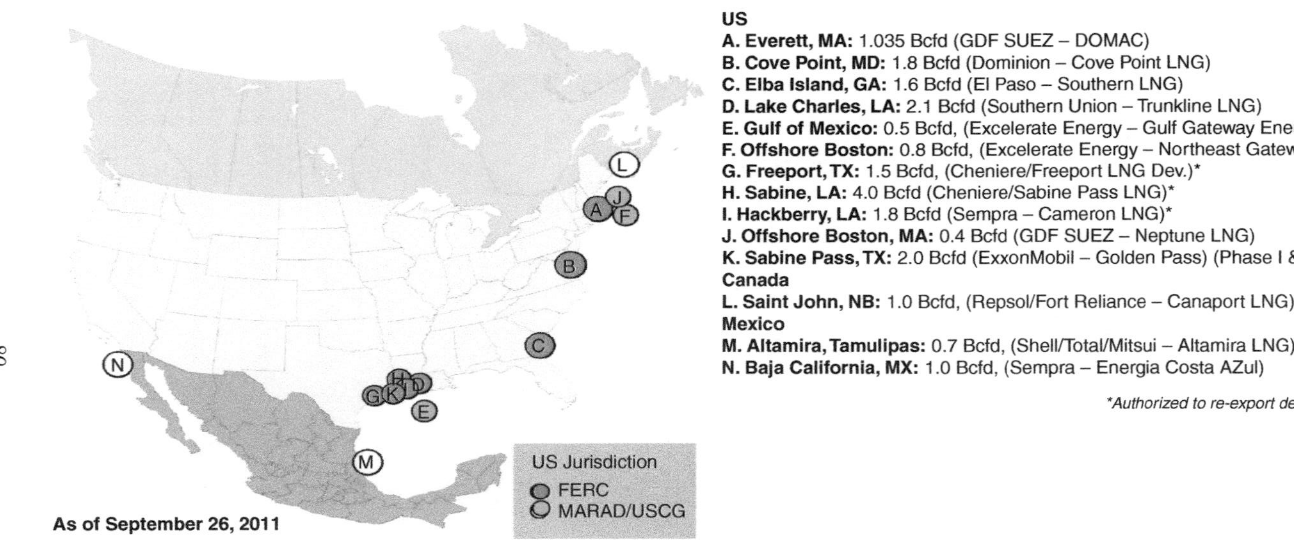

Note: There is an existing import terminal in Peñuelas, PR. It does not appear on this map since it can not serve or affect deliveries in the Lower 48 U.S. states.

Source: FERC, Office of Energy Projects, http://ferc.gov/industries/gas/indus-act/lng/LNG-existing.pdf

Figure 3.7 Existing North American LNG import terminals 2011

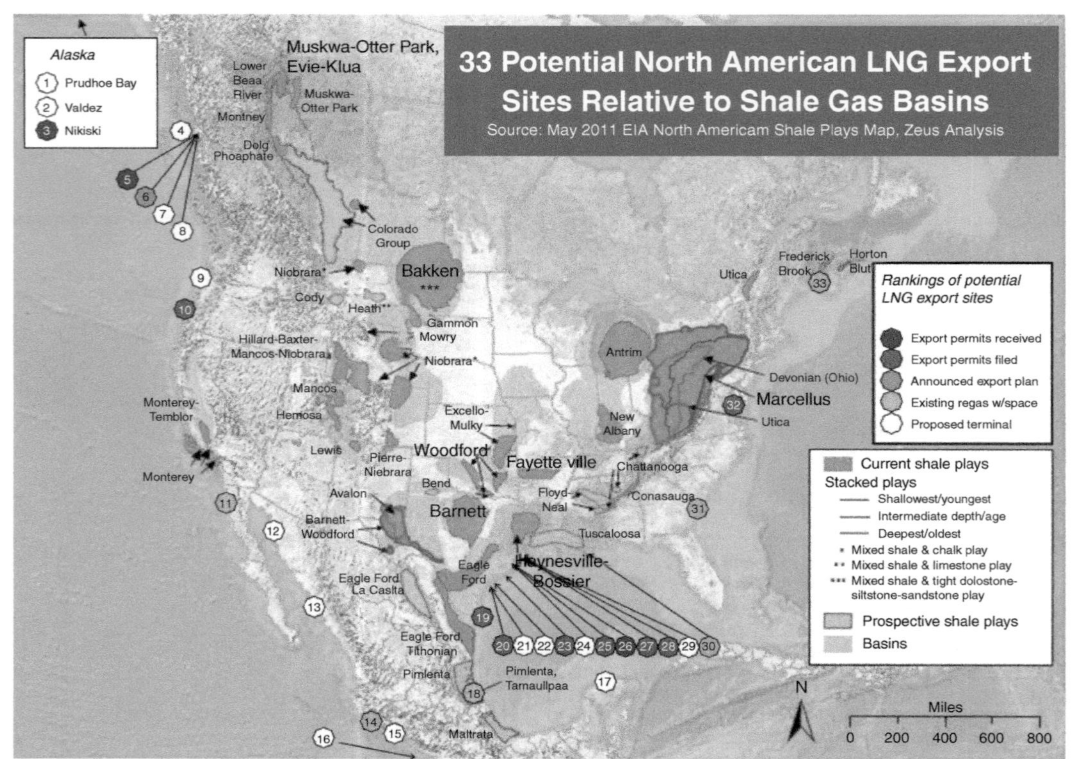

Source: Zeus Intelligence

Figure 3.8 33 potential North American LNG export sites

beyond April or May 2011.[141] Spokespeople for the companies explained that current market conditions and declines in production rates of natural gas drilling in Alaska's Cook Inlet made exports from Kenai LNG unfeasible and simply not economically viable.[142] Interestingly, the possibility of using the Kenai terminal to *import* LNG is also being explored to meet the local demand for natural gas.[143] The closure of the iconic Kenai plant marks a watershed of sorts since Kenai was the first dedicated LNG export terminal in North America and has been in operation for more than 40 years, supplying the world's largest LNG importer – Japan.[144]

---

141 Warren R. True, *Alaska LNG Plant to Close after 40 Years*, Oil&Gas Journal, Feb. 10, 2011, http://www.ogj.com/articles/2011/02/alaska-lng-plant-to.html. "Also, and with not a little [more] irony, the announcement of the Kenai plant closing exactly coincided with another announcement, from Canada's National Energy Board, that it would hold hearings next month on an application to build a new LNG plant in North America at Kitimat, BC." Ibid.

142 Isabel Ordonez, *Conoco to Stop LNG exports from Kenai Plant in Alaska*, Feb. 10, 2011, http://online.wsj.com/article/BT-CO-20110210-718575.html. According to one analyst, the closure of Kenai LNG "shows that the small facility didn't have the economy of scale to compete with a growing number of large suppliers in Asia" and that "although natural gas in the U.S. is now abundant it's still more expensive to produce than in other areas of the world [and that] we cannot really be competitive selling LNG from the U.S." Ibid., quoting Oppenheimer & Co. analyst Fadel Gheit.

143 Isabel Ordonez, *Conoco to Stop LNG Exports from Kenai Plant in Alaska*, Feb. 10, 2011, available at http://online.wsj.com/article/BT-CO-20110210-718575.html.

144 Another interesting fact about Kenai LNG is that Kenai supplied the conditioning cargoes to the Russian Sakhalin LNG Project. Confirming email from Sakhalin LNG on file with author. On April 10, 2000 (DOE/FE Order No. 1580), DOE/FE granted ConocoPhillips and Marathon blanket authorization to export the equivalent of 10 Bcf of natural gas from the Kenai LNG facility to international markets over a two-year period beginning on the date of the first export. Kenai LNG sent the commissioning cargoes to Sakhalin pursuant to this Order.

4. Global LNG supply

4.1 OVERVIEW OF NATURAL GAS RESOURCES AND GLOBAL LNG SUPPLY

Although worldwide natural gas resources are sufficient to meet projected increases in demand,[1] it should be noted that almost half of the world's proved natural gas reserves are found in just three countries: Russia, Iran and Qatar (Table 4.1).[2]

With the world's largest proved natural gas reserves, the Middle East and Africa are expected to account for 72 percent of the increase in natural gas exports by 2030, mainly to supply Europe and North America,[3] although Australia is also emerging as a key LNG exporter (see Chapter 8) and also potentially the US and Canada (see Chapters 11 and 12).

Recent analysis of supply expectations predicted that global LNG supply would rise 44 percent from 2007 to 2010 but would be backend loaded due to project delays. LNG supply was expected to rise from 22.3 bcf/d in 2007 to 32.2 bcf/d by 2010 but project delays in Norway, Nigeria, and Qatar shifted the expected supply growth to later years with only approximately 1 bcf/d incremental supplies added in 2008 and 3 bcf/d in 2009. As discussed in detail below, the expected "first wave" of LNG eventually hit between 2009 and mid-2011 with an anticipated "second wave" of LNG capacity due to come on-stream starting in 2014.

---

1 BP Statistical Review of World Energy 2012 ("World proved natural gas reserves at end-2011 were sufficient to meet 63.6 years of production").

2 According to BP, the Middle East still holds the world's largest proved reserves of 38.4%. The full chart of the world's proved natural gas reserves is available at http://www.bp.com. According to BP, proved reserves of natural gas are generally taken to be those quantities that geological and engineering information indicates with reasonable certainty can be recovered in the future from known reservoirs under existing economic and operating conditions. The reserves-to-production (R/P) ratio is determined by taking the reserves remaining at the end of any year and dividing by the production for that year. The resulting ratio is the length of time that those remaining reserves would last if production were to continue at that rate.

3 IEA WEO-2006.

Table 4.1 Worldwide natural gas proved reserves – select countries

Country	at end 2010 Trillion cubic meters	Trillion cubic feet	at end 2011 Trillion cubic meters	Share of total (%)	R/P ratio
US	8.2	299.8	8.5	4.1	13.0
Canada	1.8	70.0	2.0	1.0	12.4
Mexico	0.3	12.5	0.4	0.2	6.7
Total North America	10.3	382.3	10.8	5.2	12.5
Russian Federation	44.4	1575.0	44.6	21.4	73.5
Turkmenistan	13.4	858.8	24.3	11.7	*
Total Europe and Eurasia	68.0	2778.8	78.7	37.8	75.9
Bahrain	0.2	12.3	0.3	0.2	26.8
Iran	33.1	1168.6	33.1	15.9	*
Iraq	3.2	126.7	3.6	1.7	*
Qatar	25.0	884.5	25.0	12.0	*
Saudi Arabia	8.0	287.8	8.2	3.9	82.1
Total Middle East	79.4	2826.3	80.0	38.4	*
Algeria	4.5	159.1	4.5	2.2	57.7
Egypt	2.2	77.3	2.2	1.1	35.7
Nigeria	5.1	180.5	5.1	2.5	*
Total Africa	14.5	513.2	14.5	7.0	71.7
Australia	3.7	132.8	3.8	1.8	83.6
China	2.9	107.7	3.1	1.5	29.8
Indonesia	3.0	104.7	3.0	1.4	39.2
Malaysia	2.4	86.0	2.4	1.2	39.4
Total Asia-Pacific	16.5	592.5	16.8	8.0	35.0
Total World	196.1	7360.9	208.4	100.0	63.6
of which: OECD	18.1	660.2	18.7	9.0	16.0
Non-OECD	178.0	6700.7	189.7	91.0	90.0
European Union	2.3	64.4	1.8	0.9	11.8
Former Soviet Union	63.5	2638.5	74.7	35.8	96.3

Source: *BP Statistical Review of World Energy 2012*

4.2 OVERVIEW OF CURRENT LNG SUPPLY: THE "FIRST WAVE"

The Pacific area – including the Middle East – has historically been the largest LNG producing region in the world, although the ranking of suppliers has shifted in recent years. In 2002, Indonesia was the largest LNG exporter, followed by Algeria and Trinidad. In the early

Table 4.2 LNG production capacity added 2010–11

Country	Project	Capacity (bcm)	Online date
Qatar	Qatargas III Train 6	10.6	Feb. 2010
Qatar	Rasgas III Train 7	10.6	Nov. 2010
Yemen	Yemen LNG Train 2	4.6	Apr. 2010
Malaysia	Malaysia LNG Dua (II)	2.0	Apr. 2010
Peru	Peru LNG	6.1	June 2010
Norway	Nordic LNG	0.4	Nov. 2010
Qatar	Qatargas IV Train 7	10.6	Feb. 2011
Total		44.9	

Source: IEA MTOGM 2011

2000s, several new LNG projects were started in anticipation of rising demand. The "first wave" of LNG production to hit the market resulted in over 100 bcm of LNG production capacity added between early 2009 and mid-2011. In 2009 alone, three mega-trains in Qatar, Sakhalin II, Tangguh, Indonesia and Yemen LNG Train 1 were completed and added significantly to global LNG production capacity. In 2010, another 34 bcm of new LNG production capacity was added. (See Table 4.2.)

According to the IEA, total global liquefaction capacity stood at 373 bcm as of mid-2011, compared to 268 bcm as of the end of 2008. The rapid expansion of LNG capacity in recent years stands in stark contrast to the limited LNG production capacity added between 2005 and 2008, which averaged only 18 bcm/y.[4]

4.3 OVERVIEW OF FUTURE LNG SUPPLY: THE "SECOND WAVE"

Based on ten LNG projects that are under construction, are close to completion, or have reached a final investment decision (FID) recently, a "second wave" of LNG production capacity is on the way in the next few years. According to the IEA, these new projects will add 85 bcm of new LNG capacity so that total liquefaction capacity will be at least 458 bcm by 2017. Angola LNG is the first LNG project in Angola and will be completed by early 2012, with capacity of 7.1 bcm. Two projects in

4 IEA Medium-Term Oil and Gas Markets (MTOGM) 2011.

Table 4.3 LNG production capacity added 2011–17

Country	Project	Capacity (bcm)	Major stakeholders	Online date
Australia	Pluto LNG	6.5	Woodside, Kansai Electric, Tokyo Gas	Aug. 2011
Angola	Angola LNG	7.1	Chevron, Sonangol, Eni, Total, BP	Q1 2012
Algeria	Gassi Touil LNG	6.4	Sonatrach	2013
Algeria	Skikda new train	6.1	Sonatrach	2013
Australia	Gorgon LNG	6.8 (20.4)	Chevron, Shell, Exxon Mobil	2014 (2015)
Papua New Guinea	PNG LNG	9.0	Exxon Mobil, Oil Search	2014
Australia	Queensland Curtis LNG	11.6	BG, CNOOC, Tokyo Gas	2014
Indonesia	Donggi Senoro LNG	2.7	Mitsubishi, Pertamina, KOGAS	2014
Australia	Gladstone LNG	10.6	Santos, Petronas, Total, KOGAS	2015
Australia	Prelude LNG	4.9	Shell	2017
Total		85.3		

Source: IEA MTOGM 2011

Algeria are expected to be completed by 2013 and bring a total capacity of 12.5 bcm.[5]

Approximately 59 bcm of capacity is expected to be added globally over 2014–17 if all projects are completed on time. While half of the first wave of LNG came from Qatar (63 bcm), Australia dominates the second wave with 80 percent of the production capacity coming from Australia. The Gorgon mega project in Australia has three trains with 6.8 bcm (5 mtpa) each with a targeted completion date of 2014–15 for all three trains. Papua New Guinea (PNG LNG), Queensland Curtis LNG (QCLNG) and Donggi Senoro LNG are also expected to be completed in 2014. The Gladstone LNG and Gorgon projects have an expected completion date of 2015. If all of these projects are completed as planned, the global LNG production capacity will increase significantly in the coming years. (See Table 4.3.)

[5] Ibid. at p. 242.

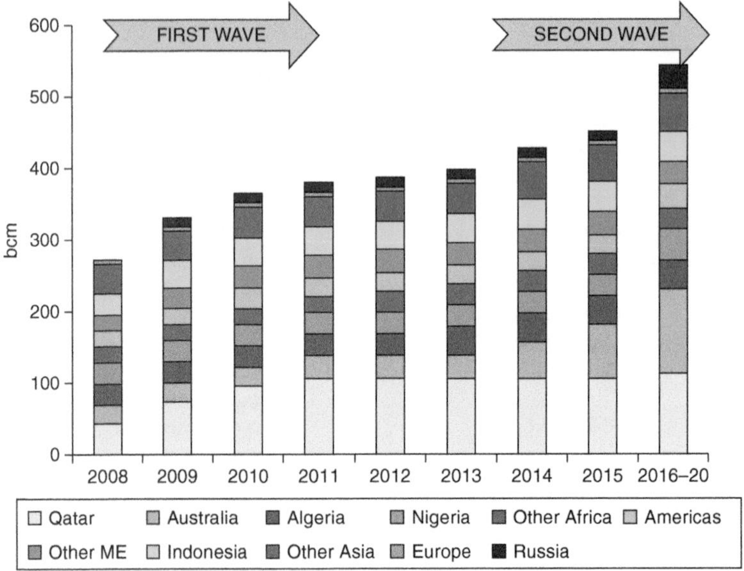

Source: IEA MTOGM 2011, ©OECD/IEA2011, p. 243

Figure 4.1 Future global LNG production capacities

The trend from 2014 onwards represents the start of the "second wave" of LNG production capacity additions. Based on currently proposed projects, global LNG production capacity is expected to reach 540 bcm by 2020 (Figure 4.1).

4.4 CHALLENGES FOR NEW PROJECTS

While there are many challenges involved in any large infrastructure project, the main challenges for new LNG projects involve cost and environmental issues. Project costs for new LNG projects vary depending on numerous factors including the design of the plant, environmental conditions of the project site, the location and technical or regulatory challenges of the particular location, availability of skilled labor, and length of construction. The timing of the project development is also an important factor and economies of scale do not always help to reduce project cost.[6]

---

6 Ibid. at p. 244.

Among the LNG projects recently completed, Sakhalin II in Russia had the highest cost of construction per tonne of annual LNG production, at over $2,000/tonne. Construction costs of Yemen LNG, Peru LNG and Qatargas IV were approximately $1,000/tonne. Projects currently under construction, including Pluto, Gorgon and PNG LNG, all have much higher costs ranging from $1,700 to $2,900/tonne. In addition, there are numerous environmental and technical challenges that vary project by project (see Chapters 7–9).

4.5 QATAR – THE WORLD'S LARGEST LNG EXPORTER

Today, the tiny Persian Gulf country of Qatar is by far the largest LNG exporter in the world with a combined capacity of 105 bcm (77 mtpa) of LNG, or 28 percent of global liquefaction capacity.[7] As of January 2011, Qatar's proven natural gas reserves stood at approximately 896 trillion cubic feet (Tcf).[8] Qatar holds almost 14 percent of total world natural gas reserves and has the world's third-largest reserves behind Russia and Iran.[9] The majority of Qatar's natural gas is located in the massive offshore North Field, which occupies an area roughly equivalent to Qatar itself. The North Field is part of the world's largest non-associated natural gas field and is a geological extension of Iran's South Pars field, which holds an additional 450 Tcf of recoverable natural gas reserves. In 2010, preliminary estimates from Qatar National Bank indicated that the oil and gas sectors accounted for over half of Qatar's 2010 GDP.[10]

As discussed in detail in Chapter 8, Qatar's LNG production comes from two LNG production projects at Ras Laffan – Qatargas and RasGas. Each project has seven LNG production trains with varying capacities. Although each train has different owners, the primary owner of all trains is Qatar Petroleum, the Qatari national oil company.[11] With the largest production capacity and a huge LNG tanker fleet, Qatar has

---

[7] IEA MTOGM 2011.

[8] EIA, Country Analysis Briefs, *Qatar*, http://www.eia.gov/countries/cab.cfm?fips=QA, citing Oil&Gas Journal.

[9] The IEA has noted that although worldwide natural gas resources are sufficient to meet projected increases in demand to 2030, more than half of the resources are found in just three countries: Russia, Iran and Qatar (IEA WEO-2006).

[10] EIA, Country Analysis Briefs, *Qatar*, http://www.eia.gov/countries/cab.cfm?fips=QA.

[11] IEA MTOGM 2011, at p. 185.

become a significant "swing producer" and is able to deliver LNG to all regions of the world.

4.6 AUSTRALIA – THE NEXT QATAR?

In recent years, Australia has emerged as the next Qatar in terms of expected LNG supply. By 2016, Australia's LNG production capacity could reach 75.8 bcm/y (55.7 mtpa), which would make it the second-largest LNG exporter after Qatar. Australia's projects are discussed in detail below in Chapter 8.

4.7 THE NEWEST LNG EXPORTERS – RUSSIA, YEMEN, PERU, ANGOLA AND PAPUA NEW GUINEA

In recent years, three new countries have emerged as LNG exporters – Russia, Yemen and Peru. Chapter 9 has a detailed discussion of the LNG projects in these countries. Angola and Papua New Guinea are expected to start LNG exports by 2012 and 2014, respectively.[12] A detailed discussion of the LNG project in Papua New Guinea is provided in Chapter 9.

4.8 INDONESIA

Indonesia has long been an important LNG producer, exporting its first cargo in 1977, and until being overtaken by Qatar was the largest LNG producer in the world. Indonesia has two LNG export terminals, Arun and Bontang, both of which are owned by Pertamina, the state oil and gas company. The Arun plants are aging and exports are declining and could cease altogether in 2014. As a result, the Bontang terminals are playing an increasingly important role in meeting long-term commitments to existing buyers and are developing new fields.[13]

In recent years, Indonesia took FID for the 2.7 bcm Donggi Senoro LNG project, which is set to come online in 2014. However, Indonesia's future as a reliable energy supplier is a bit uncertain as a policy dispute has

---

12 Ibid.

13 IEA MTOGM 2011, at p. 190.

raged over how much gas production should be reserved to meet increasing domestic demand.[14]

4.9 MALAYSIA

Malaysia has a long history as a reliable and stable LNG supplier. Of particular note is the fact that Malaysia is strategically located along one of the most important routes for the seaborne energy trade – the Strait of Malacca – which links the Indian and Pacific Oceans[15] (see Figure 4.2).

Malaysia's national oil and gas company, Petroliam Nasional Berhad (Petronas), holds exclusive ownership rights to all oil and gas exploration and production projects in Malaysia and is the single largest contributor of Malaysian government revenues. Since Malaysia's oil fields are maturing, the government is focused on enhancing output from existing fields and from new offshore developments of both oil and gas.[16]

In 2010, Malaysia was the world's tenth-largest holder of natural gas reserves. According to the *Oil and Gas Journal*, Malaysia held 83 trillion cubic feet (Tcf) of proven natural gas reserves as of January 2010 (Figure 4.3) with most of the country's natural gas reserves found in the eastern areas, predominantly offshore Sarawak.[17]

Malaysia also has one of the most extensive natural gas pipeline networks in Asia. The Peninsular Gas Utilization (PGU) project, which was completed in 1998, significantly expanded the natural gas transmission infrastructure on Peninsular Malaysia. The PGU system spans more than 880 miles and has the capacity to transport 2 billion cubic feet per day (Bcf/d) of natural gas.[18]

A number of pipelines link Sarawak's offshore gas fields to the Bintulu facility. In addition, Petronas is building the 310-mile Sabah–Sarawak Gas Pipeline between Kimanis, Sabah, and Bintulu, Sarawak, to transport gas from Sabah's offshore fields, such as Kota Kinabalu, to Bintulu for liquefaction and export. Some of the gas will be used for downstream projects in Sabah. This pipeline is expected to be completed by March 2011.[19]

---

[14] Ibid.

[15] EIA Country Analysis Brief, *Malaysia*, last updated, Dec. 2010, http://www.eia.gov/countries/cab.cfm?fips=MY.

[16] Ibid.

[17] Ibid.

[18] Ibid.

[19] Ibid.

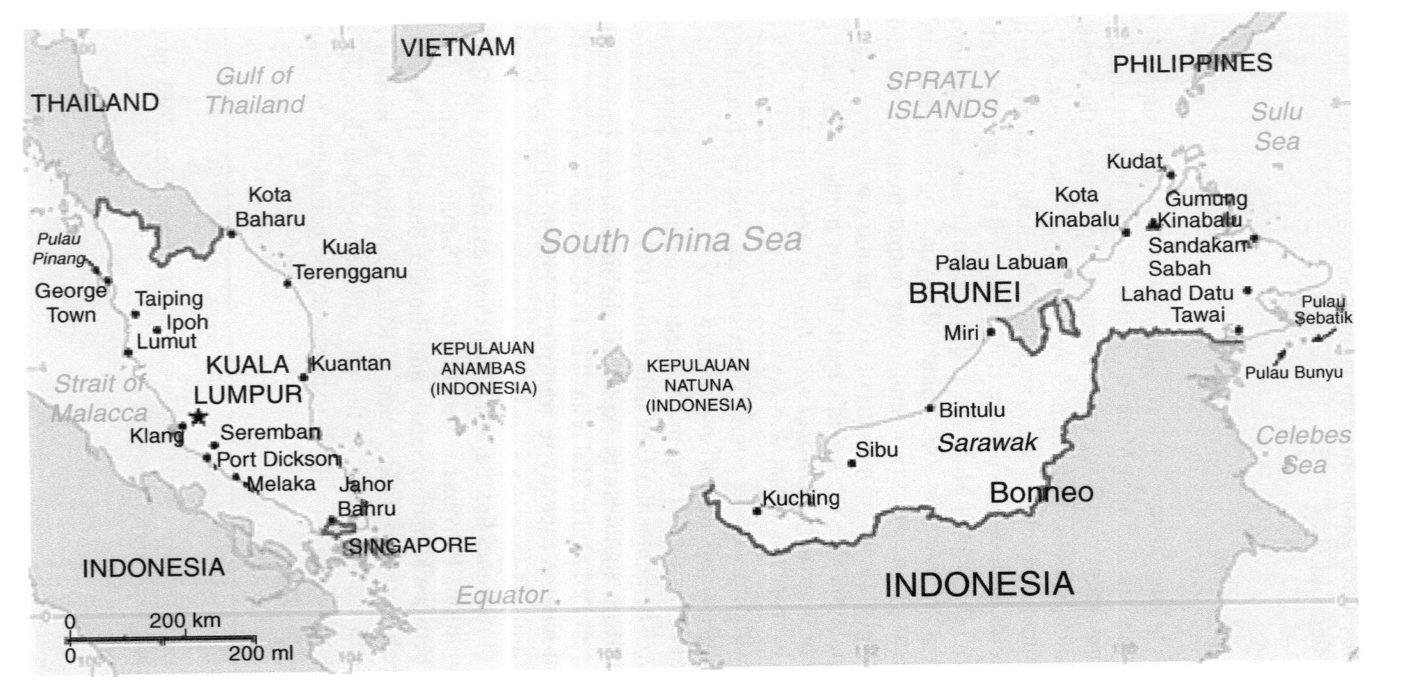

Source: US EIA

Figure 4.2 Map of Malaysia

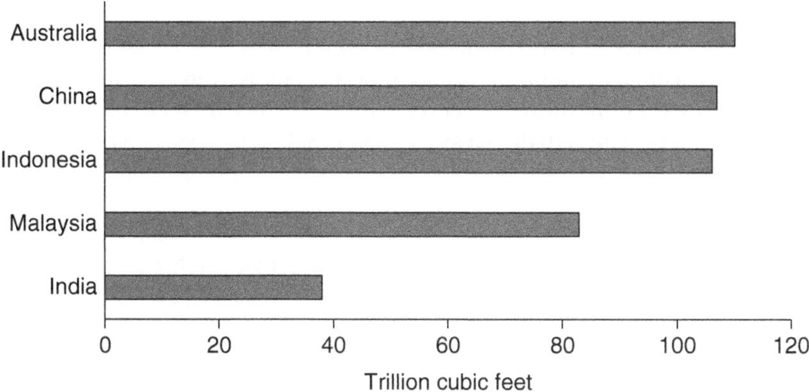

Source: US EIA

Figure 4.3 Top five Asia-Pacific proven natural gas reserve holders 2010

The Association of South East Asian Nations (ASEAN) is promoting the development of a trans-ASEAN gas pipeline system (TACP) aimed at linking 80 percent of ASEAN's major gas production and consumption centers. Because of Malaysia's extensive natural gas infrastructure and its location, the country is a natural candidate to serve as a hub in the ongoing TACP project.[20]

In 2009, Malaysia was the second-largest exporter of LNG in the world after Qatar, and exported over 1 Tcf of LNG, which accounted for 12 percent of total world LNG exports. Malaysia's LNG exports primarily went to Japan, South Korea, and Taiwan. LNG from Malaysia is transported by the Malaysia International Shipping Corporation (MISC), which owns and operates 27 LNG tankers, the single largest LNG tanker fleet in the world by volume of LNG carried. MISC is 62 percent owned by Petronas.[21]

The Bintulu LNG complex on Sarawak is the main hub for Malaysia's natural gas industry. Petronas owns majority interests in Malaysia's three LNG processing plants, all located at Bintulu, which are supplied by the offshore natural gas fields at Sarawak. The Bintulu facility is the largest LNG complex in the world, with eight production trains and a total liquefaction capacity of 1.1 Tcf per year.[22]

---

20 Ibid.
21 Ibid.
22 Ibid.

Construction began on Malaysia's newest terminal, the Sabah Oil and Gas Terminal (SOGT), in February 2007 and is expected to be completed by 2012. It will have handling capacity of 300,000 barrels of crude and 1 billion cubic feet of natural gas per day and will primarily serve Malaysia's export markets. The Sabah–Sarawak Gas Pipeline project is part of this development.

4.10 THE WORLD'S NORTHERNMOST LNG FACILITY – HAMMERFEST LNG, THE SNØHVIT PROJECT, NORWAY

The Snøhvit LNG project is the first offshore development in the Barents Sea and the world's northernmost LNG facility. Located outside of Hammerfest in Northern Norway, the Hammerfest LNG terminal is also Europe's first LNG export facility. (See Figure 4.4.)

The gas field for the project is called Snow White – Snøhvit in Norwegian (in Norway, energy projects are named after mythical characters). The field lies 340 miles north of the Arctic Circle in an area long considered out of reach for energy development due to the Barents Sea's shifting ice packs and hostile conditions, including powerful waves and extreme cold. Although no energy company knew how to operate in such a harsh environment, soaring global demand and high prices in the early to mid-2000s led to energy companies going to the "ends of the earth to find new supplies."[23] The Snøhvit LNG project was one of those "ends of the earth" projects undertaken mainly to supply natural gas to the eastern states of the United States. At the time, Dominion had even expanded its Cove Point, Maryland LNG import terminal to accommodate the Arctic gas to come from Snøhvit. While the United States no longer expects to become a major LNG importer, Snøhvit remains a pioneering project that opens up LNG trade in Arctic waters. This may be increasingly relevant as the warming Arctic opens up more trade routes, including the Northern Sea Route and the famed Northwest Passage.[24]

---

[23] Jad Mouawad, *A Quest for Energy in the Globe's Remote Places*, The New York Times, Oct. 9, 2007, www.nytimes.com.

[24] Thomas Hilsen, *Hammerfest LNG set sails for Japan via Arctic route*, The Barents Observer, Nov. 6, 2012, http://barentsobserver.com/en/arctic/hammerfest-lng-set-sails-japan-arctic-route-06-11, noting that "with rapid melting sea ice, Hammerfest LNG is located at the entrance gate for a possible new shipping route near the top of the globe to the energy hungry markets in Asia."

Source: Snøhvit LNG

Figure 4.4 Map of the Snøhvit LNG project location

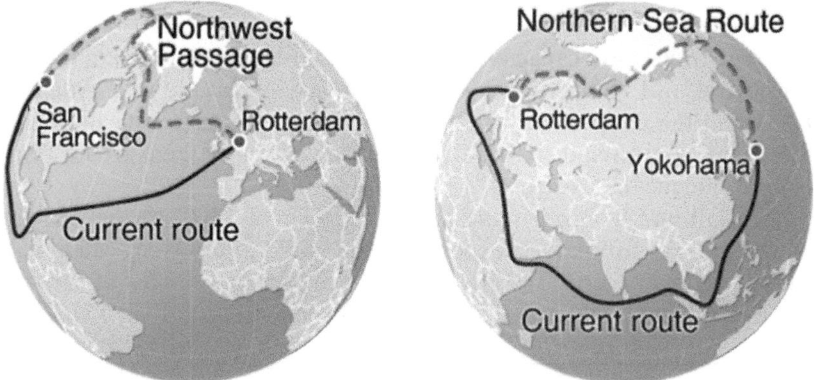

Source: Climate Action Programme, www.climateactionprogramme.org.

Figure 4.5 Northern Sea Route and Northwest Passage

According to benchmark data compiled for Chevron's Gorgon project in Australia (see Chapters 7 and 8), Snøhvit has the potential to be the most greenhouse gas efficient LNG plant in the world. This is in part due to the cold environment where Snøhvit is located since a lower ambient temperature results in both the gas turbines and the LNG process working more efficiently. In addition, and as with the Gorgon project, the Snøhvit development is based around a subsea gas production system which will significantly reduce greenhouse gas emissions by the subsurface injection of reservoir CO_2.[25]

The Snøhvit LNG project came on-stream in 2007 with a production capacity of approximately 4.3 mtpa.[26] According to Statoil, "at full capacity on Snøhvit, 700,000 tonnes of CO_2 will be stored per year, which equals the emission volume from 280,000 cars."[27]

4.11 TRINIDAD AND TOBAGO

Although the islands of the Caribbean Basin are predominantly net energy importers, Trinidad and Tobago is the exception and is a major exporter of LNG. In fact, the first new LNG plant built for the Atlantic Basin LNG trade in more than 20 years was the Trinidad Atlantic LNG plant, operated by partners BP, BG, GDF Suez, and Repsol-YPF.[28]

The Atlantic LNG Company operates four LNG trains at Point Fortin, on the southwestern coast of Trinidad. The first LNG train was completed in March 1999, with subsequent trains completed in 2002, 2003 and 2006. The four trains have capacity to produce a combined 14.8 million metric tons (Mmt) of LNG per year (775 Bcf of regasified natural gas). Atlantic LNG and the government of Trinidad and Tobago have indicated that there may be fifth and sixth trains, although currently there are no firm plans to pursue these projects.[29]

Trinidad and Tobago is the largest supplier of LNG to the United States, and the fifth-largest exporter in the world after Qatar, Malaysia, Indonesia, and Algeria. Trinidad and Tobago exported 236 Bcf of natural gas to the United States in 2009, about 50 percent of total US LNG net

---

[25] Chevron Australia, Gorgon, DRAFT ENVIRONMENTAL IMPACT STATEMENT/ENVIRONMENTAL REVIEW AND MANAGEMENT PROGRAMME FOR THE GORGON DEVELOPMENT, http://www.chevronaustralia.com/Libraries/Chevron_Documents/gorgon_ch13_LR.pdf.sflb.ashx.

[26] Snøhvit LNG, http://www.statoil.com/en/OurOperations/ExplorationProd/ncs/snoehvit/Pages/default.aspx.

[27] Statoil, *Carbon Storage Started on Snøhvit*, http://www.statoil.com/en/NewsAndMedia/News/2008/Pages/CarbonStorageStartedOnSnøhvit.aspx.

[28] EIA Country Analysis Briefs, *Trinidad and Tobago* (last updated March 2011), http://www.eia.gov/countries/cab.cfm?fips=TD.

[29] Ibid.

imports, but only 1 percent of total US natural gas supply. However, US LNG imports from Trinidad and Tobago in 2009 were almost half the amount received in 2005, due to the general decline in US LNG imports resulting from increased US shale gas production.[30]

4.12 ABU DHABI AND OMAN

Despite significant natural gas reserves, Middle Eastern countries (outside of Qatar) are not likely to add significant LNG export capacity, owing to growing domestic gas requirements. Abu Dhabi's Das Island plant, which has exported LNG since 1977, and Oman's Qalhat LNG, which has recently completed a third LNG train, are probably reaching the limits of their LNG capacities.

4.13 ALGERIA

Algeria has long been one of the world's largest LNG exporters and is also a major pipeline supplier of natural gas to Western Europe via Spain and Italy. In 1964, the world's first commercial LNG plant, the Arzew LNG plant in Algeria, came online and began deliveries to the United Kingdom under a 15-year contract. Algeria's Skikda plant came online in 1972. In 2004, an explosion and fire destroyed several of the trains at Skikda, which created an immediate need to replace lost natural gas export capacity. Sonatrach contracted with KBR to design a new LNG plant, which is supposed to offer more innovations and greater efficiencies.[31]

Algeria currently supplies about 20 bcm of LNG and over 30 bcm of pipeline gas exports to global gas markets. There are two trains under construction in Algeria that are expected to be completed by 2013; however, it is unclear whether these trains will add additional capacity or replace older plants.[32]

---

[30] Ibid.

[31] Kamel Bouzid, Pamela Roche, and David A. Coyle, *Skikda LNG Train Rebuild Project in Algeria Offers More Innovations*, LNG Journal, July/August 2010, pp. 35–9.

[32] IEA MTOGM 2011 at p. 191.

4.14 LIBYA

Although Libya has been exporting LNG since 1970, its plant at Marsa el Brega has been producing only a nominal amount of LNG (see Appendix A) and no plans have been announced for expansion of Libya's LNG capacity.

4.15 NIGERIA

The Nigerian LNG (NLNG) project at Bonny Island has been constantly expanded since it began commercial operation in 1999 and is a joint venture of Shell, Total, Eni, and the Nigerian National Petroleum Corporation (NNPC) (see Appendix A). Nigerian LNG currently has six trains in operation with an overall capacity of approximately 22 mtpa of LNG[33] and plans are currently underway to build Train 7, which will lift the total production capacity to over 30 mtpa of LNG.[34]

In 2009, Nigeria's LNG exports plummeted as a result of political unrest and sabotage. The situation improved in 2010 and LNG exports came back to 2008 levels although two trains remain shut down. However, a power supply problem in December 2010 reduced LNG production significantly. Nigeria has several LNG projects under consideration, including Brass LNG, Olokola LNG and Nigeria LNG Train 7. Although there is great potential for Nigeria to become an even larger LNG supplier, progress has been slow on recent projects due to the lack of a stable financial and tax regime and political uncertainty. At this point, it seems unlikely that any new projects will start producing before 2016.[35]

4.16 EGYPT

After Algeria, Egypt is the second-largest natural gas producer in Africa. Egypt's natural gas sector has expanded rapidly in recent years, with natural gas production more than tripling from 646 Bcf in 2000 to 2.2 Tcf in 2010. Natural gas consumption in Egypt has also been growing, with 2010 consumption of over 1.6 Tcf.[36] Egypt's natural gas

---

33 Bonny Island, hydrocarbons-technologies.com, http://www.hydrocarbons-technology.com/projects/bonny/.

34 Nigerian LNG Limited, http://www.nlng.com.

35 IEA MTOGM 2011 at p. 191.

36 EIA Country Analysis Briefs, *Egypt* (last updated July 2012), http://www.eia.gov/cabs/Egypt/pdf.pdf.

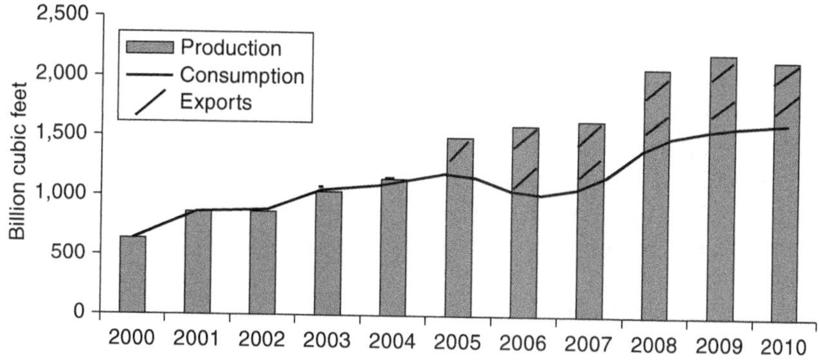

Source: US EIA

Figure 4.6 Natural gas production, consumption, and exports in Egypt, 2000–2010

production is expected to continue to grow to satisfy growing domestic demand and export commitments through the Arab Gas Pipeline and LNG exports. Accordingly, Egypt is expected to continue to be an important natural gas provider to Europe and the Mediterranean region, although exports are starting to compete with rising domestic demand.[37] (See Figure 4.6.)

In terms of Egypt's LNG capacity, two new LNG projects (a total of three trains) have come online during the past few years – SEGAS Damietta LNG Train 1 (Damietta LNG) and Egyptian LNG Trains 1 and 2 in Idku (Egyptian LNG).[38] The Spanish Egyptian Gas Company (SEGAS) LNG[39] complex in Damietta, Egypt, is situated on the Mediterranean coast 60 km west of Port Said (Figure 4.7). The project came on-stream during the final quarter of 2004 to export LNG to the Spanish market. This LNG project was the first facility of its type in Egypt and is one of

---

[37] Ibid.

[38] EIA Country Analysis Briefs, *Egypt* (update from Feb. 2011), http://www.eia.gov/cabs/Egypt/pdf.pdf.

[39] The operating company, SEGAS, is controlled by Union Fenosa Gas in conjunction with ENI of Italy (80%) and two state-owned Egyptian companies – Egyptian Natural Gas Holding Company (EGAS – 10%) and Egyptian General Petroleum Corporation (EGPC – 10%). Union Fenosa Gas is owned in a 50/50 partnership by Union Fenosa of Spain and Eni of Italy. Hydrocarbons-technology.com, *SEGAS Liquefied Natural Gas Complex, Damietta*, http://www.hydrocarbons-technology.com/projects/seagas/.

the world's largest capacity single-train facilities, producing 5.5 million t/yr (7.5 billion m^3 per year) of LNG.[40]

The project sponsors for the Egyptian LNG project (two 3.6 MTA trains) are BG Group,[41] Petronas, and Gas de France. The Egyptian LNG facilities are located at the Idku LNG Port, which is on Egypt's Mediterranean coast approximately 50 km east of Alexandria (Figure 4.7). The Idku LNG port is the largest specialized LNG export facility in Egypt and is capable of accommodating large LNG vessels up to 160,000 m^3.[42]

Egyptian LNG began deliveries in late 2005 to supply LNG to France, the United States, and Italy. The Idku site has sufficient space for an additional four LNG trains and expansion plans have been considered over the years.[43]

The combined LNG export capacity of Egypt is close to 600 Bcf per year with plans to expand in the near future pending export policy changes and legislation. In 2009, LNG exports were approximately 450 Bcf, with the largest recipient being the United States, which imported around 160 Bcf, representing 35 percent of Egyptian LNG exports for the year and also 35 percent of US LNG imports.[44]

In 2010, half of Egypt's LNG exports were sent to Europe, which imported about 180 Bcf. Over half of this amount went to Spain (110 Bcf). The US was the second largest importer of Egyptian LNG and imported just over 71 Bcf in 2011. Other importers included Korea (36 Bcf), Japan (21 Bcf), and Chile (18 Bcf). (See Figure 4.8.)

---

40 Ibid.

41 According to its website, "BG Group has played a leading role in the development of Egypt's natural gas industry and is responsible for around a third of all gas produced in Egypt. The Group's activities in Egypt span the gas chain from exploration, through development and production, to LNG." BG Group, Egypt, http://www.bg-group.com/OurBusiness/WhereWeOperate/Pages/pgEgypt.aspx.

42 Egyptian LNG, http://www.egyptianlng.com.

43 The Egyptian LNG Company owns both the Egyptian LNG site and common facilities. Its sister company, Egyptian Operating Company for Natural Gas Liquefaction Projects (Opco) (BG Group 35.5%), undertakes the operation of all trains. El Beheira Natural Gas Liquefaction Company (Train 1 Co.) (BG Group 35.5%) owns Train 1 and the Idku Natural Gas Liquefaction Company (Train 2 Co.) (BG Group 38%) owns Train 2. BG Group, Egypt, http://www.bggroup.com/OurBusiness/WhereWeOperate/Pages/pgEgypt.aspx.

44 EIA Country Analysis Briefs, *Egypt* (update from Feb. 2011), http://www.eia.gov/cabs/Egypt/pdf.pdf. In 2010, LNG exports fell to about 354 Bcf due to increased domestic demand for natural gas. EIA Country Analysis Brief, *Egypt* (last updated July 2012).

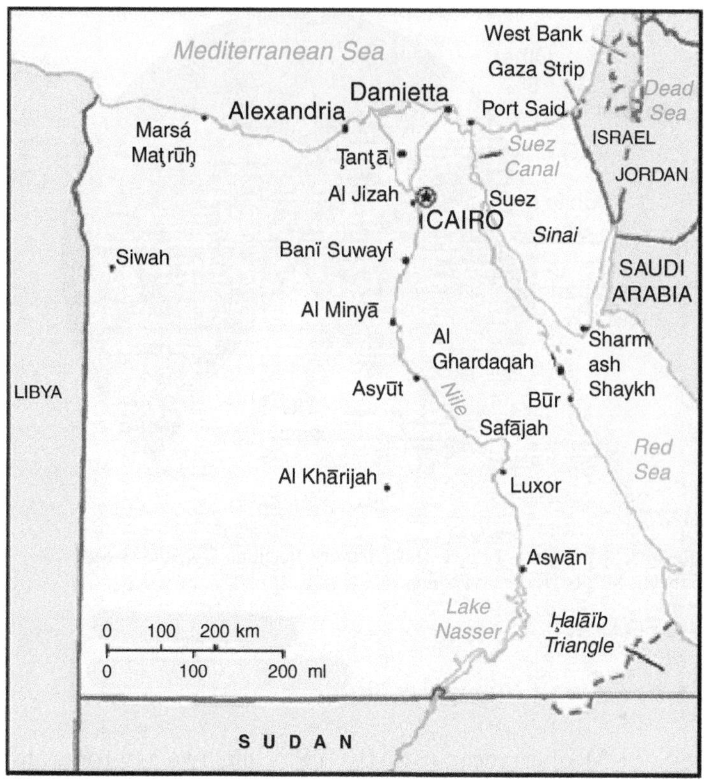

Source: US EIA

Figure 4.7 Map of Egypt

In addition to exporting LNG, Egypt exports gas through the Arab Gas Pipeline (AGP) to Lebanon, Jordan and Syria, with further additions being planned. The Arish–Ashkelon pipeline addition, which branches away from the AGP in the Sinai Peninsula and connects to Ashkelon, Israel, began operations in 2008. In recent years, domestic pressure over contracts, pricing for exports to Israel, and technical problems have caused interruptions in exports to Israel.[45]

In addition to oil and gas production, Egypt plays an important role in international energy markets through the operation of the Suez Canal

---

45 EIA Country Analysis Briefs, *Egypt* (update from Feb. 2011), http://www.eia.gov/cabs/Egypt/pdf.pdf.

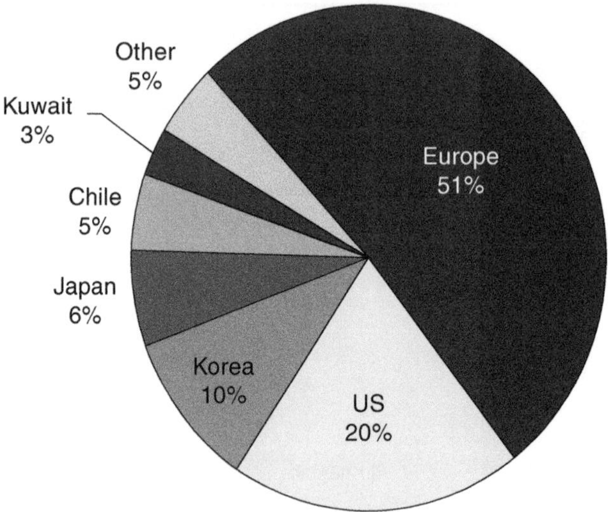

Note: Europe: Spain (60%), France, Italy, Turkey, Belgium, UK, and Greece
Other: Taiwan, Mexico, India and China

Source: US EIA

Figure 4.8 Egypt's LNG exports 2010

and the Suez–Mediterranean (SUMED) Pipeline, two key routes for the export of Persian Gulf oil and LNG[46] (Figure 4.9).

The Suez Canal connects the Red Sea and the Gulf of Suez with the Mediterranean Sea, spanning 120 miles. In 2011, 17,799 ships transited through the Suez Canal, of which 20 percent were petroleum tankers and 6 percent were LNG tankers.[47] With a width of only 1,000 feet at its narrowest point, the Suez Canal is unable to handle the VLCC (very large crude carrier) and ULCC (ultra large crude carrier) class crude oil tankers. The Suez Canal Authority is continuing enhancement and enlargement projects on the canal, and extended the depth to 66 ft in 2010 to allow over 60 percent of all tankers to use the canal.[48]

The 200-mile long SUMED Pipeline, or Suez–Mediterranean Pipeline,

---

[46] In addition, fees collected from operation of these two transit points are a significant source of revenue for the Egyptian government. EIA Country Analysis Briefs, *Egypt* (last updated July 2012), http://www.eia.gov/cabs/Egypt/pdf.pdf.

[47] Ibid.

[48] EIA Country Analysis Briefs, *Egypt* (update from Feb. 2011), http://www.eia.gov/cabs/Egypt/pdf.pdf.

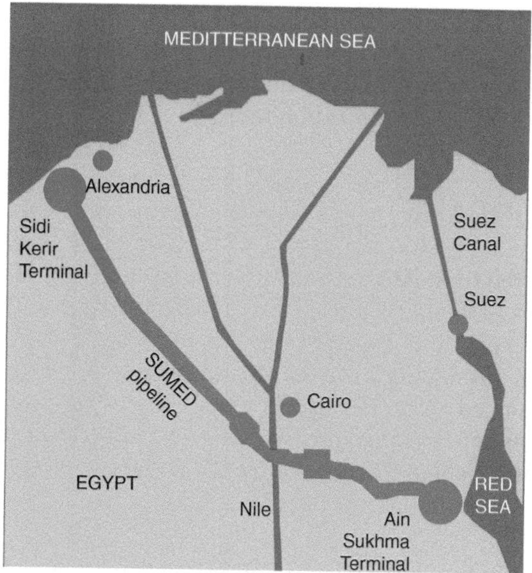

Sources: US EIA; Oil Capital Ltd.

Figure 4.9 Suez Canal and SUMED Pipeline

provides an alternative to the Suez Canal for those cargoes too large to transit the canal (laden VLCCs and larger). The pipeline has a capacity of 2.3 million bbl/d and flows north from Ain Sukhna on the Red Sea coast to Sidi Kerir on the Mediterranean. The SUMED is owned by Arab Petroleum Pipeline Co., a joint venture between the Egyptian General Petroleum Corporation (EGPC), Saudi Aramco, Abu Dhabi's National Oil Company (ADNOC), and Kuwaiti companies.[49]

According to the US EIA and the IEA, closure of the Suez Canal and the SUMED Pipeline would divert ships around the southern tip of Africa, the Cape of Good Hope, adding approximately 6,000 miles to transit, increasing both costs and shipping time. Shipping around Africa would add 15 days of transit to Europe and 8–10 days to the United States.[50]

According to the IEA, the political unrest of 2011 that toppled the

49 EIA Country Analysis Briefs, *Egypt* (last updated July 2012), http://www.eia.gov/cabs/Egypt/pdf.pdf.

50 Ibid.

Mubarek regime had minimal impact on the operation of the Egyptian LNG Idku plant and the Damietta LNG plant. Nonetheless, Egyptian LNG exports were down in 2011 due to growing domestic demand and relatively low LNG plant utilization rates.[51]

4.17 ANGOLA

According to the EIA, Angola had 10.9 trillion cubic feet (Tcf) of natural gas reserves as of January 1, 2011 – a significant increase from the 2007 estimated reserves of 2 Tcf. Natural gas production in Angola is tied directly to oil production and is often vented or flared, with limited volumes consumed domestically. In 2009, Angola's gross natural gas production was approximately 357 billion cubic feet (Bcf). Of this, 244 Bcf (67 percent) was vented or flared, 81 Bcf (23 percent) was re-injected to aid in oil recovery and only 24 Bcf (7 percent) was marketed for domestic consumption.[52]

Developments are underway to capture and market this natural gas for domestic electricity generation and to export most of it in the form of LNG. Chevron and state-owned Sonangol, together with other shareholders including Total, BP and Eni, are building the country's first LNG plant near Soyo in northern Angola. The plant is expected to be operational by early 2012.[53]

The natural gas will come from several offshore fields and, according to the partners, the project will process 1.1 billion cubic feet of associated gas per day and will eventually produce 5.2 million tons per year of LNG and process up to 125 million cubic feet per day of gas for the domestic market.[54] Initially, the LNG was to be directed to the Gulf LNG regasification plant in Pascagoula, Mississippi, where Sonangol holds a 20 percent share. However, given the current natural gas market conditions in the United States (surplus production and lower prices), Angolan LNG exports will likely be destined for Asian and European markets where prices are higher.[55]

---

[51] IEA MTOGM 2011 at p. 191.

[52] EIA Country Analysis Briefs, *Angola* (last updated Aug. 2011), http://www.eia.gov/cabs/Angola/pdf.pdf.

[53] Ibid.

[54] Angola LNG, http://www.angolalng.com/project/aboutLNG.htm.

[55] EIA Country Analysis Briefs, *Angola*.

4.18 EQUATORIAL GUINEA

Equatorial Guinea's natural gas production has grown rapidly over the past decade but is currently starting to level off. According to the US EIA, Equatorial Guinea had 1.3 trillion cubic feet (Tcf) of proven natural gas reserves as of January 1, 2011, with the majority of the reserves located offshore Bioko Island. From 2001 to 2009, natural gas production increased rapidly from 1 billion cubic feet (Bcf) to 232 Bcf as new projects came online. Domestic consumption over the same period went from 1 to 55 Bcf, increasing alongside of production until 2007, when the completion of the Punta Europa liquefied natural gas facility on Bioko Island allowed for greater exports.[56]

Most of Equatorial Guinea's natural gas production is exported in the form of LNG. In May 2007, Marathon Oil Corporation and its partners completed Train 1 of the $1.4 billion Punta Europa LNG facility on Bioko Island. In 2009, Equatorial Guinea exported approximately 153 Bcf of LNG, almost all of which went to Asia, mainly Japan (41 percent), South Korea (36 percent) and Taiwan (16 percent). Smaller volumes that year also went to India, China, France and Portugal. Initial LNG exports were destined for the United States and in 2007 the US imported close to 18 Bcf of LNG from Equatorial Guinea. However, since then, growing Asian demand and weak US demand have shifted the dynamics of Equatorial Guinea's LNG markets. Plans for a second LNG train are underway with an estimated start-up date of 2016.[57]

4.19 CAMEROON

According to the IEA, GDF Suez is currently considering investing in a 4.8 bcm LNG plant near Kribi in Cameroon.[58] In June 2010, Foster Wheeler AG announced that it had been awarded the contract by GDZ Suez to carry out the pre-front-end engineering design (pre-FEED) for the development of the onshore LNG plant and the offshore gas gathering infrastructure. According to the announcement, the Cameroon LNG project[59] seeks to establish a national gas transportation network linking Cameroon's offshore gas resources with the state-sanctioned onshore

[56] EIA Country Analysis Briefs, *Equatorial Guinea* (last updated Jan. 2011), http://205.254.135.24/EMEU/cabs/Equatorial_Guinea/pdf.pdf.

[57] Ibid.

[58] IEA MTOGM 2011 at p.254.

[59] Cameroon LNG, www.gdfsuez.com/document/?f=files/en/cameroun-uk.pdf.

site near Kribi on the southern coastline of Cameroon. Foster Wheeler is working with GDF Suez and GDF Suez's partner, Cameroon's Société Nationale des Hydrocarbures (SNH).[60]

The Cameroon LNG project involves a single-train onshore LNG plant with a production capacity of up to 3.5 million tonnes per annum. The project also includes an upstream gas gathering system and subsea pipeline tie-ins to offshore production facilities. For the pre-FEED contract, Foster Wheeler will develop the preliminary design basis for the liquefaction facility, gas gathering and treatment facilities and LNG export facilities, and will prepare capital cost estimates and other documentation. The pre-FEED contract is scheduled to be completed in early 2011.[61] The IEA has a projected start date for Cameroon LNG of 2016.[62]

4.20 EAST AFRICA – AN EMERGING LNG SUPPLIER?

In 2012, several companies announced significant finds of natural gas off the East Coast of Africa that could make East Africa a potential LNG exporter in the coming decades. In May 2012, Italian oil company ENI announced it had discovered a vast new gas field at its exploration block in Mozambique. Separately, Anadarko Petroleum announced another find off the coast of Mozambique. In addition, BG Group and Ophir Energy announced in May 2012 that they had made a big find at the Mzia well in Tanzanian waters.[63]

Eni has estimated that, with the latest discovery made in Mozambique, its block could potentially hold up to 52 trillion cubic feet (tcf) of gas. Anadarko estimates its reserves off the coast of Mozambique are also about 50 tcf. Finds off Tanzania might be 20 tcf, with the US Geological Survey estimating that gas fields off the coast of Kenya, Tanzania and Mozambique might contain 253 tcf of gas.[64]

---

60 LNG World News, *Foster Wheeler Awarded Contract by GDF Suez for Cameroon LNG Project*, Posted on June 28, 2010, http://www.lngworldnews.com/foster-wheeler-awarded-contract-by-gdf-suez-for-cameroon-lng-project/.

61 Ibid.

62 IEA MTOGM 2011 at p. 254.

63 Sarah Young, *Huge Finds Make East Africa the Next Big Gas Source*, Reuters, May 16, 2012, http://uk.mobile.reuters.com/article/topNews/idUKBRE84F0O320120516.

64 Tom Bergin, *East Africa to Join World Gas Giants*, Reuters, May 21, 2012, http://uk.reuters.com/article/2012/05/21/africa-gas-idUKL5E8GL2B320120521.

Since around 10 trillion cubic feet would be enough to meet an entire year's gas consumption by France, Germany, Britain and Italy, the size of the recent discoveries plus the potential reserves should justify the billions of dollars it will cost to build LNG export facilities. At the time of writing, interest and enthusiasm are mounting on East Africa's potential to become a major LNG player in the coming decades, although it remains to be seen whether this materializes.[65]

65 Ibid., "We can help vault Mozambique into being one of the world's three largest LNG exporters," citing Anadarko CEO Jim Hackett. See also *Eastern El Dorado?*, The Economist, April 7, 2012, noting that "At long last east Africa is beginning to realize its energy potential."

5. Global LNG demand and emerging demand markets

5.1 OVERVIEW OF GLOBAL LNG DEMAND

Until about the late 1990s, LNG was a niche industry operating mostly in the Asia-Pacific region. As the world entered the 21st century, however, global demand for LNG surged in a "perfect storm" created by the industrial and commercial boom around the world that resulted in an ever-growing appetite for all energy resources.

Between 2000 and 2008, the LNG industry entered a period of rapid growth with huge increases in supply coming from a growing number of LNG producing countries. The number of LNG exporting countries grew from 8 in 1996 to 15 at the beginning of 2008. The number of importing countries also grew, with the United States expected to become one of the largest LNG importers. To supply the growing number of LNG importing and exporting countries, more players entered the LNG business and, as indicated by the list in Appendix D, today there are quite a few companies involved in the LNG business.

Between 2008 and 2009, the world endured the "worst recession since the Second World War"[1] with demand for all energy dropping significantly, including global natural gas demand, which dropped 2 percent, the biggest drop since the 1970s.[2] As a result of weak demand due to the recession, the "wave" of new LNG supply coming from recently completed projects, and the unexpected surge in production of US shale gas, in 2009 there was a sizable oversupply of LNG, with everyone in the natural gas industry wondering how long the "gas glut" would last.[3]

In 2010, as global economies appeared to be emerging from the recession, global natural gas demand resumed its long-term upward trajectory with the IEA projecting that natural gas will be the only fossil fuel for which demand is higher in 2035 than in 2008.[4] Most of the growth in gas

1 IEA WEO-2010 at p. 67.
2 Ibid. at p. 179.
3 Ibid.
4 Ibid.

trade takes the form of LNG, with trade in LNG being expected to double between 2008 and 2035.[5] In its WEO-2010, the IEA predicted that the "gas glut" would last into 2011 before starting to decline, with the excess supply disappearing only by 2020.[6]

However, in early 2011, another "perfect storm" occurred in the global LNG markets in the form of the unfortunate disaster at Japan's Fukushima nuclear plant, which led to Japan significantly increasing LNG imports to make up for lost nuclear power. As a result of Japan's nuclear disaster, some countries that were considering increasing nuclear power have indicated they will most likely switch from nuclear to natural gas to replace, or supplement, their remaining nuclear fleet. Another unforeseen event in the "perfect storm" of 2011 was the unrest in the Middle East, which included the disruption of natural gas supplies from Libya.[7]

As discussed in detail in Chapter 1, this "perfect storm" of conditions has led the IEA to predict that the coming decade may well be the "Golden Age of Gas," with the "gas glut" now expected to disappear much quicker and most likely by 2015.[8] While the ultimate wildcard for all natural gas demand is the pace and strength of the global economic recovery,[9] the IEA nonetheless remains optimistic about the longer-term strength of LNG markets with recent data suggesting that LNG demand remains strong.

Global gas demand reached an estimated 3,284 bcm in 2010, up almost 7.4 percent from its 2009 level, one of the highest growth rates recorded over the past 40 years. As the first wave of new LNG was brought online, global LNG trade increased by 25 percent in 2010, reaching 299 bcm, the largest percentage increase ever noted. Global LNG trade now represents 9 percent of global gas demand.[10]

Growth in LNG demand is expected to continue over 2011–16 as most regions of the world are expected to increase their use of natural gas and import more LNG for a variety of reasons including economic growth, environmental policies and fuel switching (Figure 5.1).

---

5 Ibid.

6 Ibid.

7 IEA Medium-Term Oil and Gas Markets (MTOGM) 2011 at p. 157, http://www.iea.org/textbase/nppdf/stud/11/mtogm2011.pdf.

8 Ibid.

9 As this book goes to print, the outlook for the global economy is anything but clear with concerns about another global economic recession mounting. *Be Afraid*, The Economist, Oct. 1, 2011 (suggesting that unless political leaders act more boldly, the world economy is likely to slip into a global recession. This article also comments that the real reason to "be afraid" is that "[a]t a time of enormous problems, the politicians seem Lilliputian").

10 IEA MTOGM 2011 at p. 142.

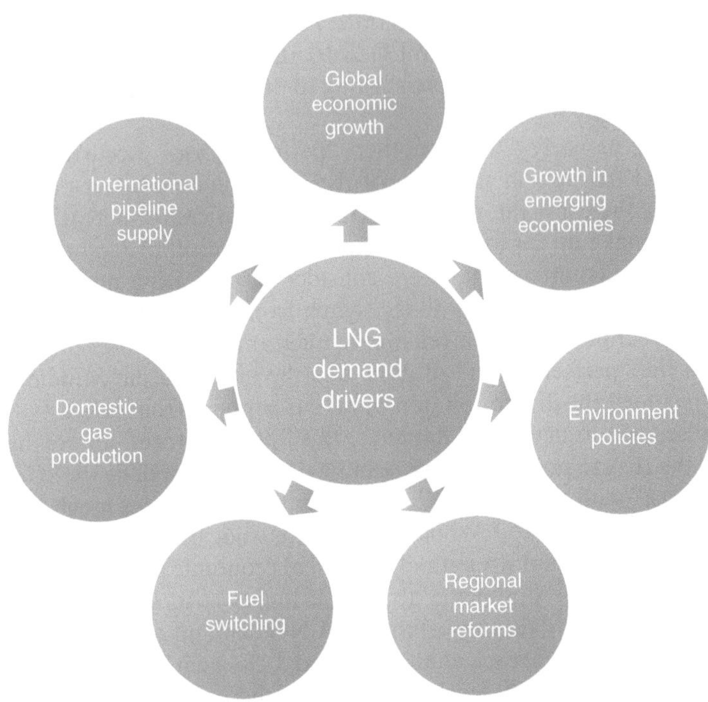

Source: Larry Persily, *Demand and LNG Production Growth, Competition, Pricing are the Variables*, February 27, 2012, Alaska Natural Gas Transportation Projects, Office of the Federal Coordinator, http://www.arcticgas.gov/Demand-and-LNG-production-growth-competition-pricing-are-the-variables.

Figure 5.1 Factors driving global LNG demand

As discussed in detail in Chapter 3, the Asia-Pacific region is currently the largest demand market for LNG and is expected to remain so going forward, accounting for an estimated 60 percent of the total global LNG imports. Europe is expected to account for approximately 29 percent of the LNG market going forward.

According to analysis by the Office of the Federal Coordinator for Alaska Natural Gas Transportation Projects (OFC), one of the key topics of discussion amongst LNG experts is whether the increased demand will cause the historically distinct LNG markets to converge at some point.[11]

11 The Office of the Federal Coordinator for Alaska Natural Gas Transportation Projects (OFC) was established by the US Congress in 2004 to

For the moment, the consensus seems to be that the markets are too distinct to converge anytime soon since the price of gas in Asia and most of Europe is still priced against the cost of substitute fuels in the region. Over time, and in light of the prospects of North American LNG exports (see Chapter 10), this could change.

Despite the dominance of the Asia-Pacific and European markets in terms of LNG demand, a number of other regions, such as Southeast Asia, Latin America and the Middle East and North Africa (MENA) region, are also emerging as LNG importers. While their share of LNG imports is expected to remain small (4 percent of total LNG imports), the development of additional demand markets is an interesting and important trend that is worthy of discussion and is the focus of this chapter.

5.2 EMERGING ASIA-PACIFIC IMPORTERS – SOUTHEAST ASIA

While demand from the three traditional Asia-Pacific LNG importers, Japan, Korea and Taiwan, is poised to grow in the coming years, demand from the countries in Southeast Asia is also growing fast (6 percent a year on average) and reached 160 bcm in 2009.[12] Due primarily to growing demand, more countries in this region are considering building LNG regasification terminals and new LNG importers are emerging including Thailand, Singapore and Malaysia.

In 2004, Thailand's national oil and gas company, PTT Public Company Limited, established the PTT LNG Company Limited (PTTLNG) in order to build and operate Thailand's, and Southeast Asia's, first LNG receiving/regasification terminal, Map Ta Phut LNG terminal.[13] Map Ta Phut is a deepwater port on Thailand's south coast (220 km southeast of Bangkok) and is becoming one of the most important industrial areas in the country.[14]

---

expedite and coordinate federal permitting and construction of a pipeline and enhance transparency and predictability of the federal regulatory system to deliver natural gas from the Arctic to American markets. The OFC coordinates with over 20 federal agencies, the Canadian federal government, the State of Alaska (which leases all the known natural gas reserves and owns portions of the right of way), tribal governments and other stakeholders. Alaska Natural Gas Transportation Projects Office of the Federal Coordinator, http://www.arcticgas.gov.

[12] IEA MTOGM 2011 at p. 262.

[13] PTTLNG, http://www.pttlng.com/en/.

[14] *Map Ta Phut LNG Regasification Terminal*, Thailand, hydrocarbons-technology.com, http://www.hydrocarbons-technology.com/projects/thailandptt/.

Construction on the $880 million, 5 mmpta/6.8 bcm/y Map Ta Phut LNG terminal began in 2008, the same year that Thailand's PTT entered into a contract with Qatar securing 1 mmpta/1.4 bcm/y of LNG for 15 years, while also pursuing other supply sources under both long-term contracts and spot market purchases. In June 2011, the Map Ta Phut LNG terminal received its first commissioning cargo.[15] PTT has also indicated that it is considering an additional $400 million investment to double the capacity of the terminal to 10 million tonnes by 2016.[16]

Singapore is also building a 4.8 bcm/y LNG receiving terminal, which will be operational by 2013 and has capacity to expand to 8.2 bcm/y in the future. To supply the terminal, Singapore has entered into a 20-year agreement to purchase 4.1 bcm of LNG from Queensland Curtis LNG, BG's CBM to LNG project starting in 2015. Singapore is in negotiations with other suppliers as well, such as Qatar and Indonesia.[17]

Interestingly, even Malaysia, a historic LNG exporter, is also constructing an LNG regasification terminal. PETRONAS, the national oil and gas company of Malaysia, is developing Malaysia's first LNG regasification terminal in Melaka, which is due for completion in July 2012. The new terminal will facilitate the importation of LNG by PETRONAS and third parties and ensure "security of gas supply for the nation in the future."[18]

Other Southeast Asian countries, such as Vietnam, are also looking to LNG to meet growing demand for energy, especially in the power sector. PetroVietnam Gas (PV Gas) was considering building an onshore LNG terminal but ultimately opted for a 1.4 bcm floating LNG terminal. In July 2010, PV Gas signed a memorandum of understanding (MOU) for supply of LNG to the power plant in Ho Chi Minh City for 20 years.[19]

Most of the planned terminals in the Southeast Asian region are floating storage and regasification units (FSRUs), which have a shorter building time than onshore terminals, making them a better option for countries facing more immediate natural gas shortages.[20]

With the "world price of energy increasing constantly, Map Ta Phut is an example of an LNG regasification facility being built by a country hedging its bets (not wanting to be left behind in the rush for natural gas)." Ibid.

15 IEA MTOGM 2011.

16 Randy Fabi, *UPDATE 1-Thai PTT has not Scrapped LNG Deals with Qatar-exec*, Reuters, Sept. 20, 2011, http://af.reuters.com/article/commoditiesNews/idAFL3E7KK1FI20110920.

17 IEA MTOGM 2011.

18 PETRONAS, http://www.petronas.com.my.

19 IEA MTOGM 2011.

20 Ibid. at p. 262.

5.3 WILL SOUTHEAST ASIA DEVELOP A REGIONAL LNG HUB?

There are currently nine intra-regional natural gas pipelines in Southeast Asia that connect the natural gas producing countries – Indonesia, Myanmar and Malaysia – with the demand countries – Singapore, Malaysia and Thailand.[21] Only one pipeline expansion is being planned and the outlook for pipeline interconnections looks doubtful due to the fact that the major exporters in the region are facing declining production.[22] Another limitation to increasing pipeline connections is the large differences in economic development throughout the region.

While there is currently a lack of potential for additional pipeline connections, there is the potential for the region to develop an intra-regional LNG trading hub based on small-scale LNG projects. Small-scale floating LNG (FLNG) projects are well suited to supply isolated and remote areas with small LNG vessels and gas could be used to complement variable power sources such as hydropower in regions not connected to existing electricity networks.[23]

Singapore has a vision of developing into an LNG hub for LNG and gas trading in Asia and is geographically well positioned to serve the LNG demand centers of Northeast Asia as well as the major supply regions of the Middle East and Australia. The Government of Singapore has introduced various tax incentives to help spur the development of an LNG trading hub in Singapore.[24] In addition, the Singapore terminal also wants to offer LNG storage to third parties and this might enable the development of a regional gas price, similar to spot prices in the US and the UK.[25] While it remains to be seen whether Singapore will emerge as a major LNG trading hub in the coming years, it is a development worth watching.

5.4 EMERGING EUROPEAN LNG IMPORTERS

In addition to the traditional European LNG importers discussed in detail in Chapter 3, a number of other European countries are also considering building import terminals, including Poland, Croatia, Albania, Cyprus,

---

21 Ibid. According to the IEA, Malaysia imports gas from Indonesia.

22 Ibid.

23 Ibid. at p. 263.

24 Singapore LNG Corporation, http://www.slng.com.sg/business-the-future.html.

25 IEA MTOGM 2011 at p. 263.

Germany, Bulgaria, Romania, Lithuania, Estonia, Sweden and Eire.[26] Polskie LNG[27] recently received an award of over 200 million Euros towards the construction of a new terminal at Świnoujście in northwest Poland. The terminal is a major energy project for the Polish government and offers Poland the opportunity to diversify its gas supply and improve the country's energy security, since, like most of Europe, Poland is heavily dependent on Russian imports. Construction of the terminal began in March 2011 and it is expected to begin operations in July 2014.[28]

5.5 LATIN AMERICAN LNG DEMAND

In recent years, gas demand in Latin America has been growing fast at 3.5 percent per year over 2000–2008 and represents approximately 20 percent of the total primary energy supply. Despite rapid growth, however, Latin American gas demand represents only about 4 percent of the total global gas demand.[29]

Due to LNG exports from Trinidad and Tobago, the Latin American region as a whole is a net gas exporter but several countries are looking to increase their natural gas use, in part to meet increased energy demand, to replace use of expensive oil products, and/or to reduce reliance on hydro power, which has experienced problems due to irregular rainfall patterns. Additionally, the region also suffers from natural gas supply shortages of pipeline gas primarily due to policy and regulatory failures that have limited investment and development of resources. As a result, a number of countries in Latin America already import LNG and may increasingly look to LNG to meet their natural gas needs.[30] (See Figure 5.2.)

---

26 Andy Flower, *LNG in Europe*, LNG Industry, Winter 2009, at pp. 10–15. www.lngndustry.com.

27 Polskie LNG was established in 2007 by Polish Gas and Oil Company PGNiG (PGNiG SA). Gas Transmission Operator GAZ-SYSTEM S.A. (GAZ-SYSTEM S.A.), a company owned by the State Treasury and responsible for the security of natural gas supplies via transmission networks, became the owner of Polskie LNG. GAZ-SYSTEM S.A. will supervise the construction of a LNG terminal and the Polish Oil and Gas Company, PGNiG, is responsible for the supply and transport of LNG to the terminal in Świnoujście. Polskie LNG, http://en.polskielng.pl/nc/hidden/the-company.html.

28 *European Commission Backs Poland's LNG Terminal*, The News. Pl, News from Poland, Oct. 7, 2011, http://www.thenews.pl/1/12/Artykul/56439,European-Commission-backs-Polands-LNG-Terminal.

29 IEA MTOGM 2010.

30 Ibid.

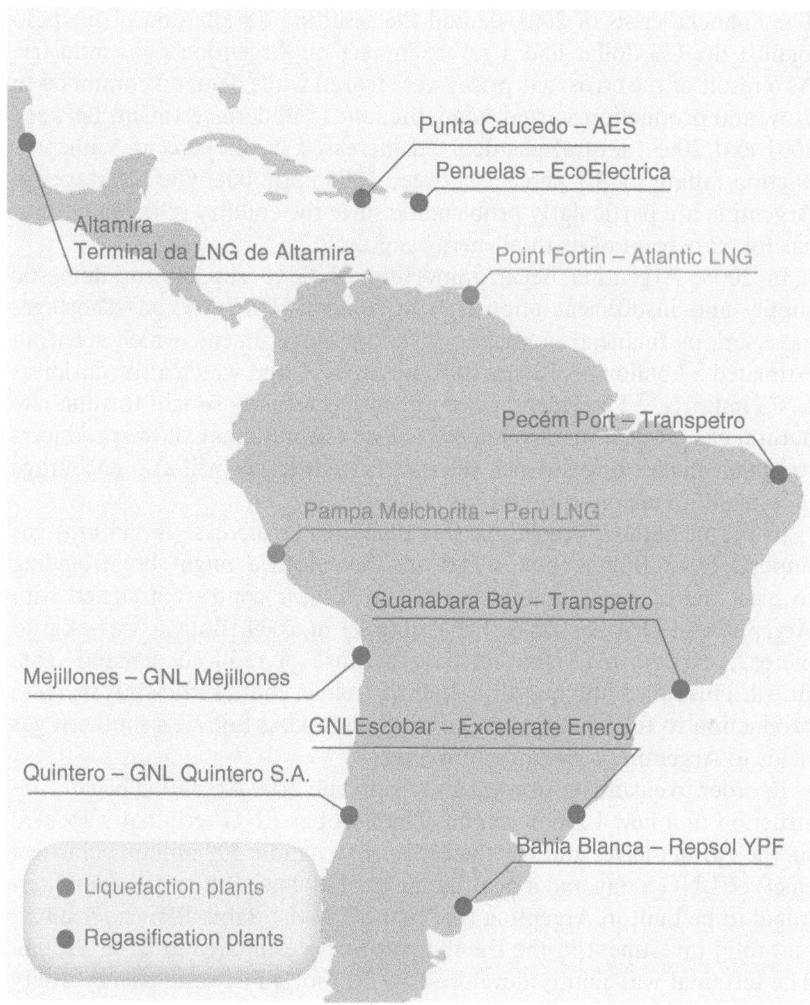

Source: GIIGNL, *The LNG Industry in 2011*

Figure 5.2 Latin American liquefaction and regasification plants

5.5.1 Argentina

Argentina has the largest gas market in Latin America with natural gas reserves estimated at 428 bcm as of 2009. Argentina's natural gas sector was opened up to private investment in the 1990s, which transformed the country from a natural gas importer to a major gas exporter in the region.

The financial crisis of 2001–02 and the resulting devaluation of the peso against the US dollar had a severe impact on Argentina's gas industry. As a result of the crisis, gas prices were frozen while demand continued to grow and production started stagnating due to underinvestment. Between 2001 and 2008, natural gas demand increased by 25 percent, with production falling by 3.1 percent between 2006 and 2008. Gas shortages in Argentina are particularly problematic since the country relies on natural gas for 52 percent of its total energy supply.[31]

In 2008, Argentina began importing LNG to supplement domestic supply and insufficient imports from Bolivia. Domestic gas shortages have serious financial implications for the government, which spent an estimated $7 billion on energy subsidies in 2008 and which also subsidizes LNG imports.[32] The federal government has recently sought to stimulate natural gas production through a Gas Plus Program that allows producers to obtain higher prices which the government hopes will also encourage investment in unconventional gas projects.[33]

In the meantime, Argentina was planning to increase its natural gas imports from Bolivia but it appears that Bolivia might be struggling to meet the output requirements of the 20-year contract it signed with Argentina in October 2006. For example, in 2009, Bolivia was able to increase exports to Argentina only because of reduced demand from Brazil. This called into question whether Bolivia could sufficiently increase production to fill a new proposed pipeline linking Bolivia's southern gas fields to Argentina's Northern provinces.[34]

In order to ensure natural gas supply, in late 2009 Argentina began construction of a new LNG terminal. The Escobar LNG terminal is located on the Paraná River and has the capacity to handle 500 million cubic feet (mcf) of LNG a day and a peak capacity of 600mcf.[35] It is the second terminal to be built in Argentina (the first being the Bahia Blanca terminal) and total investment in the Escobar project is estimated at $140–$150m. The terminal was jointly developed by Repsol YPF, state energy utility Enarsa and Excelerate Energy. Repsol YPF and Enarsa jointly own the facility.[36]

The facility features a 150,900m³ floating storage and regasification

---

31 Ibid.
32 Ibid.
33 Ibid.
34 Ibid.
35 *Escobar LNG Terminal*, Project Summary from hydrocarbons-technology.com, http://www.hydrocarbons-technology.com/projects/escobar-terminal/.
36 Ibid.

vessel, a new distribution pipeline, a high-pressure gas manifold, four docking dolphins and a berthing tower. The regasification vessel uses Excelerate Energy's ship-to-ship transfer technology to unload LNG through the loading arms. It vaporizes LNG onboard and unloads it through a high-pressure manifold connected to the LNG loading arms. The high-pressure manifold enables liquid transfer or high-pressure gas discharge to be carried out. From the manifold the LNG is transported directly to the Argentine natural gas system through the new distribution pipeline. The new pipeline is connected to the Transportadora Gas del Norte pipeline, which carries natural gas to northern Argentina.[37]

In April 2011, Morgan Stanley purchased 14 cargoes which will be sourced from Nigeria and Trinidad for the Escobar terminal. Morgan Stanley was awarded a contract to supply 70 percent of the LNG required by the terminal between June and December. Spain's Gas Natural will supply the remaining 30 percent of the LNG required by the terminal.[38]

Argentina is expected to pay $10.70 per mmBtu of natural gas supplied to Escobar. About 45 to 50 LNG cargoes are expected to be imported through both the Escobar and Bahia Blanca terminals in 2011.[39] As noted in Chapter 12, in early 2012, Argentina announced the planned seizure and nationalization of a 51 percent stake in YPF SA from majority stakeholder Repsol YPF SA. The potential impact of the nationalization on Argentina's natural gas markets is unclear as of the time this book goes to print.

5.5.2 Bolivia

Bolivia has the second-largest natural gas reserves in Latin America after Venezuela with an estimated 710 bcm at the end of 2009. Bolivia is also the largest exporter of natural gas in the region although production has been stagnating since the 2006 nationalization process with production currently at 15 bcm, while 2008 contracted demand was around 17.5 bcm. Bolivia's primary export markets are Argentina and Brazil, which generally command higher prices than the domestic market.[40]

In May 2010, President Evo Morales enacted the announced nationalization of four major electricity companies in Bolivia, some with foreign participation. In addition to the nationalization, President Morales

37 Ibid.
38 Ibid.
39 Ibid.
40 IEA MTOGM 2010.

ordered a 20 percent reduction in actual electricity prices. The nationalization is expected to make future investors much more cautious and may severely limit Bolivia's ability to increase natural gas production.[41]

5.5.3 Brazil

Brazil's natural gas imports have more than doubled over five years from 5.1 bcm in 2003 to 12 bcm in 2008. Gas demand is expected to continue to grow due to an expected growth in gas-fired power generation and overall growth in Brazil's economy. Brazil's gas production is also expected to grow and should reach 50 bcm by 2019, with most of the increase coming from recent pre-salt discoveries.

Petrobras is the largest producer of natural gas in Brazil and the company controls almost 90 percent of Brazil's natural gas reserves. State-controlled Petrobras[42] is looking at options to monetize associated natural gas when production from the Tupi pre-salt field starts after 2016, but so far many foreign companies have been reluctant to invest in the natural gas sector due to the perceived dominance of the state-controlled company and the lack of a clear regulatory framework for gas.[43]

Brazil imports natural gas from three primary sources – from Bolivia and Argentina and LNG primarily from Trinidad and Tobago. Bolivia is by far Brazil's biggest supplier and in 2008 99 percent of Brazil's total natural gas imports came from Bolivia via the Gasbol pipeline. Due to Bolivia's recent nationalizations, however, Brazil is eager to diversify supply and may look more to LNG imports to meet seasonal demand for natural gas.[44]

Brazil currently has two LNG regasification terminals – the 2.0 bcm Pecem terminal in the northeast, and the 4.8 bcm Guanabara Bay terminal in the southeast near Rio de Janeiro. Both facilities were installed in the past few years and both are floating regasification and storage units (FRSUs) provided by Golar LNG. The Pecem terminal received its first LNG cargo from Trinidad and Tobago in July 2008. The Guanabara Bay

---

[41] Ibid.

[42] Petrobras is a publicly traded corporation, the majority stockholder of which is the Government of Brazil. Petrobras is active in the following sectors: exploration and production, refining, oil and natural gas trade and transportation, petrochemicals, and derivatives, electric energy, biofuel and other renewable energy source distribution. Petrobras, http://www.petrobras.com.br.

[43] IEA MTOGM 2010.

[44] Ibid.

terminal came online in May 2009. In 2009. Brazil received 0.7 bcm of LNG, all from Trinidad and Tobago.[45]

Petrobras is planning a third offshore LNG terminal, the 2.2 bcm Tergas terminal, which is expected to come online by 2013. Petrobras is also considering building a floating LNG liquefaction plant in the offshore Jupiter field, which could process associated gas coming from the pre-salt fields. Were this to happen, Brazil could supply its own LNG import terminals as well as export to other countries in Latin America.[46]

5.5.4 Chile

In 2009, Chile became the third-largest LNG importer in Latin America and the only one on the Pacific Coast of South America. Most of Chile was entirely dependent on Argentina for its natural gas imports until its two LNG import terminals – Quintero and Mejillones – started operations in 2009 and 2010, respectively. The state-owned companies, the National Petroleum Company (ENAP) in the Central Region and the National Copper Corporation (CODELCO) in the north, are the main sponsors and developers of the two LNG projects, which were started in response to the Argentinian financial and gas crisis of the early to mid-2000s.[47]

The first phase of the GNL Quintero regasification terminal, located 114 km from Santiago, was inaugurated in July 2009 with an initial fast-track phase that used unloading vessels as storage tanks to store the LNG until it could be regasified and injected into pipelines linking the terminal to consumers in Central Chile. This allowed the project to be in operation about a year earlier than it would have been, since the two larger storage tanks would not be operational until late 2010. The estimated cost of the project is $940 million, with an initial send-out capacity of 2.3 bcm. GNL Chile has a contract with BG Group to buy up to 2.3 bcm/y of LNG from BG Group's global portfolio.[48]

Chile's second import terminal, GNL Mejillones, began operations during the second quarter of 2010. The project is a 50/50 joint venture (JV) between Suez Energy and the state-owned mining company, CODELCO. Suez is expected to provide 0.8 bcm of LNG for the first three years from its global LNG portfolio, including from Yemen LNG. GNL Mejillones is located in the northern area of Chile and is a priority for the mining sector

45 Ibid.
46 Ibid.
47 Ibid.
48 Ibid.

and the region, which is heavily dependent on gas and other fossil fuels. The terminal has a send-out capacity of 2 bcm and an estimated cost of $500 million. As of mid-2010, GNL Mejillones was considering building permanent storage tanks.[49]

While it would seem logical for Chile to import LNG from Peru, a long-standing border dispute between the countries has made this impossible. Chile also has a border dispute with its neighbor to the northwest, Bolivia, dating back to the 1879–83 War of the Pacific, that makes gas exports from Bolivia to Chile unlikely in the near future.[50]

5.5.5 Colombia

Compared to Argentina and Brazil, Colombia is a small natural gas market, consuming about 10 bcm. Colombia is self-sufficient in natural gas and has been exporting natural gas to Venezuela since 2008 via its one pipeline connection with that country. Gas production in Colombia is highly concentrated with just two fields accounting for around 90 percent of production: the Guajira field operated by Ecopetrol and Chevron and the Cusiana field operated by Ecopetrol, BP and Total.[51] The government has been encouraging greater use of natural gas with natural gas vehicles (NGVs) in particular being heavily promoted. As of January 2010, Colombia had 305,000 NGVs, representing close to 10 percent of the total vehicles.[52]

5.5.6 Venezuela

Venezuela has the largest natural gas reserves in Latin America with 4.8 tcm (2.7 percent of proven global gas reserves). Despite large reserves, however, gas production in Venezuela has been declining from a peak of 31 bcm in 1998 to 23 bcm in 2009. While the state-owned PDVSA has ambitious plans to increase production, it has not been able to reach its targets, in part due to a lack of foreign investment after President Chavez announced plans to nationalize various industries, including electricity and telecoms.[53]

In October 2009, the largest discovery in Venezuela's history was made by Eni and Repsol – the Perla field. This field is estimated to hold over 220 bcm, but it could hold more and might produce over 25 bcm/y. Although

---

[49] Ibid.
[50] Ibid.
[51] Ibid.
[52] Ibid.
[53] Ibid.

Eni and Repsol have a 50/50 JV on the field, PDVSA has plans to take 35 percent in the project.[54]

Venezuela has ambitious plans to develop three LNG liquefaction trains with each train having a different holding structure, with PDVSA having majority stakes in each, along with 11 different companies. This different ownership could be an obstacle for the projects to reach FID and it is unlikely that the projects will be operational in 2014 as planned.[55]

5.5.7 Peru

Despite relatively small gas reserves (415 bcm), Peru recently became the second LNG exporter in Latin America, after Trinidad and Tobago, and the first LNG exporter in South America. The Peruvian government is optimistic about Peru's ability to expand its reserves and recent discoveries may aid in that.[56] A detailed description of the Peru LNG project is found in Chapter 9.

5.6 MIDDLE EAST/NORTH AFRICA (MENA) – THE NATURAL GAS PUZZLE

Perhaps one of the most surprising recent trends in the global gas markets is that the Middle East/North Africa (MENA) region is currently faced with what some industry leaders have called "a natural gas puzzle."[57] The "puzzle" is that although the region as a whole holds 40 percent of the world's proven gas reserves, the reserves are not evenly distributed. This results in an abundance of gas in some countries but a shortage of gas in many countries in the MENA region.

While historically there has been a dearth of information about natural gas markets in the MENA region, the so-called "gas puzzle" has resulted in more analysis being done on these markets in recent years. According to the

---

54 Ibid.

55 Ibid.

56 Ibid.

57 Simon Webb and Eman Goma, *Shell Says Middle East Needs to Solve "Gas Puzzle,"* Reuters, April 26, 2010, http://www.reuters.com/article/2010/04/26/shell-mideast-gas-idUSLDE63O0BJ20100426?rpc=401&feedType=RSS&feedName=rbssEnergyNews&rpc=401. See also Malcolm Brinded, "Natural gas: changing the Middle East energy landscape," Speech given by Malcolm Brinded, Executive Director, Shell Upstream International, at the Middle East Petroleum & Gas Conference in Kuwait, April 26, 2010, http://www.shell.com/home/content/media/speeches_and_webcasts/archive/2010/brinded_kuwait_26042010.html.

IEA, natural gas demand growth in the MENA region has been at least 8 percent per year and often as much as 10 percent. Dubai and some areas of Saudi Arabia had demand growth of 13 to 15 percent in 2008. The increased demand is due in part to rapid economic growth, which has led to increased demand for power generation driven by natural gas. Pricing policies in the MENA region have also resulted in artificially low natural gas prices, which has spurred economic growth but also encouraged more demand.[58]

According to the IEA, the installed electricity generating capacity in the MENA region is about 200 Gigawatts (GW) and in 2008 the six Gulf Cooperation Council (GCC) states of Saudi Arabia, Kuwait, United Arab Emirates, Qatar, Bahrain and Oman accounted for 82 GW of installed electricity capacity. Iran had about 50 GW of installed capacity and Egypt had about 25 GW. Approximately 55 to 60 percent of the installed generating capacity in the MENA region is gas-fired. Small consumers such as Bahrain and Qatar are 100 percent gas-fired, while larger consumers such as Algeria and the United Arab Emirates (UAE) are 90 percent gas-fired. Egypt is 75 percent gas-fired and Saudi Arabia relies on gas for about 45 percent of its electricity generation with the balance being provided for by oil. Kuwait also generates about half of its power from oil.[59]

The Oxford Institute for Energy[60] has recently published the first academic book providing a comprehensive analysis of natural gas markets in the MENA region.[61] The book concludes that, despite large reserves, the majority of countries in the region are encountering serious difficulties meeting rising demand for natural gas, projected by the authors to be 6–7 percent per annum. The difficulties are due in part to very low domestic gas prices (one-third to one-sixth of the cost of new domestic production) that are just a fraction of the price of internationally traded gas. While the low gas prices have helped spur economic growth, they have also resulted in some countries having to import pipeline gas and LNG to meet the growing demand. The book suggests that MENA countries need to raise domestic gas prices to at least cost-based – and eventually to internationally traded – levels.[62]

---

58 IEA MTOGM 2010, "Focus on Middle East and North Africa," at p. 14.

59 Ibid.

60 The Oxford Institute for Energy specializes in advanced research into the economics and politics of international energy. http://www.oxfordenergy.org/.

61 Bassam Fattouh and Jonathan Stern (eds), NATURAL GAS MARKETS IN THE MIDDLE EAST AND NORTH AFRICA, The Oxford Institute for Energy Studies (2011), http://www.oxfordenergy.org/shop/natural-gas-markets-in-the-middle-east-and-north-africa/.

62 Ibid.

According to industry experts, the current analysis suggests that solving the MENA gas puzzle will require three crucial elements: (1) increasing natural gas supplies through exploration and development; (2) expanding the region's natural gas and LNG infrastructure; and (3) adjusting the region's gas pricing policies.[63]

While demand for natural gas in the MENA region is likely to remain high, the prospects of the region being able to increase production and export capacity remain doubtful. The only exception might be Qatar but even Qatar's ability to increase production beyond the current projects is uncertain. Algeria's exports will most likely continue to grow slowly and will probably peak before the end of this decade. Although Iraq and Israel might become modest exporters of natural gas, the general outlook for exports from the region is bleak.[64] Thus, to achieve gas supply security for all countries in the MENA region, something more will need to be done.

The second element of solving the gas puzzle is to continue investing in the infrastructure that enables countries with a gas deficit to import their needs. In that context, LNG is an interesting option. By investing in LNG regasification terminals, countries can tap into the fast-globalizing LNG market and diversify their gas supplies. Regasification terminals can be built relatively quickly (in three years or less) and, unlike cross-border pipelines, do not require the same bilateral agreements and cooperation. The MENA region is likely to be an attractive market for LNG exporters because of its anti-cyclical nature. Most countries in MENA require LNG in the summer months for air conditioning and cooling, while demand in Europe and Asia is relatively low during these months.

Perhaps the most challenging aspect of solving the MENA region's natural gas puzzle relates to pricing. Part of the reason for the gas supply challenge in this region is a history of low natural gas pricing. There have been times when regional gas prices were a fifth of the gas prices of the UK and the US, a tenth of gas prices in the Far East and less than one-twelfth of the energy equivalent price of oil. While the availability of cheap gas has helped drive industrial growth and keep inflation low, it has also

---

63 Malcolm Brinded, "Natural gas: changing the Middle East energy landscape," speech given by Malcolm Brinded, Executive Director, Shell Upstream International, at the Middle East Petroleum & Gas Conference in Kuwait, April 26, 2010, http://www.shell.com/home/content/media/speeches_and_webcasts/archive/2010/brinded_kuwait_26042010.html.

64 Bassam Fattouh and Jonathan Stern (eds), NATURAL GAS MARKETS IN THE MIDDLE EAST AND NORTH AFRICA, The Oxford Institute for Energy Studies (2011), http://www.oxfordenergy.org/shop/natural-gas-markets-in-the-middle-east-and-north-africa/.

encouraged relatively high per capita consumption of energy resources. While some countries – mainly net gas exporters – can subsidize domestic gas with the income from exports, other countries are only likely to attract imports at competitive export prices.

As a result, MENA countries are increasingly trying to develop energy policies that remove energy subsidies and promote energy efficiency and conservation. One potential approach might be to have industrial users pay higher, more market driven gas prices. For example, in May 2008, the Egyptian government announced an increase in the price of natural gas to energy intensive industries of nearly 60 percent. This resulted in significant cuts in the government's energy subsidy bill. It remains to be seen in the coming decades what new energy policies will arise from the MENA region and whether the region can solve its natural gas puzzle.

5.6.1 Kuwait – a Middle East Importer of LNG

All countries in the MENA region face the problem of growing summer peak power demand as more air conditioning is used during this time. To meet this increased seasonal demand for power, Kuwait has turned to LNG imports via a floating regasification facility with LNG supply coming from various producers. Interestingly, one of the very first LNG cargoes from the Sakhalin II project in eastern Russia was delivered to Kuwait.[65]

5.6.2 Dubai (UAE) and Bahrain

Like Kuwait, Dubai has started to import LNG in order to meet high summer seasonal demand for natural gas. Bahrain is also planning an offshore LNG terminal to meet seasonal demand. The LNG terminals in Dubai and Kuwait are both FSRUs and the shorter building times made the FSRU a practical option for both Kuwait and Dubai.[66]

5.6.3 Iran – a Future Exporter?

With one of the largest gas reserves in the world, Iran would have great potential to be a major natural gas producer were it not for the economic sanctions from the UN, the US and the EU over Iran's nuclear program. As a result of the sanctions, Iran has limited access to the advanced technologies needed to successfully develop an LNG project and has faced

---

[65] IEA MTOGM 2010 at p. 14.

[66] IEA MTOGM 2011 at p. 266.

serious difficulties developing three proposed LNG projects: Iran LNG, Persian LNG and Pars LNG. While these LNG projects have stalled, Iran still has the option to develop its gas fields through pipeline exports to neighboring countries or to use the gas for domestic consumption.[67]

5.6.4 Iraq – a Future Exporter?

Iraq has approximately 3.2 tcm of proven gas reserves and thus has the potential to become a significant gas producer and potentially an exporter.[68] However, Iraq has continued to struggle to boost its energy production, with one of the biggest challenges being the continued delay in passing an oil and gas law. Although the Iraqi Cabinet approved the Draft Hydrocarbon Law in February 2007, disagreement over various provisions and among various parties has prevented the Iraqi Parliament from approving the law.[69] Other challenges for the Iraqi energy sector include a lack of critical infrastructure such as roads, bridges, gas transmission networks and gas-fired power plants.[70]

Most of Iraq's natural gas resources are located in the South Eastern Basrah province although there are also fields in the North and the West. Iraq has never attempted to develop its natural gas resources, in part due to years of sanctions, wars, and political unrest. Currently, only 2 bcm of natural gas is used in the domestic market while an estimated 7 bcm is flared or vented due to the lack of transport infrastructure and lack of demand.[71]

According to the World Bank, Iraq loses $5 million a day by flaring off associated gas and is one of the world's biggest gas-flaring polluters. According to data collected in conjunction with the World Bank-led Global Gas Flaring Reduction Partnership,[72] Iraq's flared gas has risen to 10 bcm/year, up from 3 bcm/year in 1994, due to Iraq's increased oil

---

[67] IEA MTOGM 2010 at p. 252.

[68] IEA MTOGM 2011 at p. 238.

[69] See Susan L. Sakmar, *The Status of the Draft Iraq Oil and Gas Law*, Houston J. Int'l Law, Vol. 30, No. 2 (2008), available at http://ssrn.com/abstract=1931874. This article analyzes the main provisions of the Iraq Draft Hydrocarbon Law and highlights the issues that must be resolved in order for the Iraqi government to pass the Draft Hydrocarbon Law. The article provides critical insight into Iraq's oil and gas reserves, the background of Iraq's Draft Hydrocarbon Law, and the main issues of contention including foreign participation in Iraq's energy sector and revenue sharing.

[70] IEA MTOGM 2011 at p. 238.

[71] Ibid.

[72] World Bank, Global Gas Flaring Reduction Partnership, http://web.

production, putting it in fourth place behind Russia, Nigeria and Iran.[73] World Bank officials are currently working with officials in Iraq to come up with ways to reduce flaring, including implementing regulations and pricing systems that can provide incentives as well as penalties to help reduce gas flaring.[74]

Three different oil and gas licensing rounds organized by the Ministry of Oil have taken place since 2008, offering 11 oil fields and 3 gas fields. The gas fields only found bidders in the third round, reflecting uncertainty on the future of Iraq's gas industry.[75] Shell, Europe's largest energy producer, has been in talks with the Iraqi government since 2008 to set up a venture to capture flared gas. In June 2010, the Iraqi cabinet approved a $12 billion project to capture gas flared from the Rumaila, Zubair and West Qurna-1 oil fields and Shell agreed to operate the project through the Basrah Gas Company (BGC), including Iraq's South Gas Company (51 percent); Shell (44 percent) and Mitsubishi (5 percent). (See Table 5.1.) So far, Shell has struggled to get the final approvals and contract details.[76]

The captured gas from the three oilfields would mainly serve the domestic market but could be exported if the Iraqi government decides on that option.[77] Three export possibilities have been considered so far – the Nabucco pipeline, the Arab Gas Pipeline and an LNG project in the Basrah region.[78]

5.6.5 Israel – Importer, Exporter, Neither or Both?

Recent discoveries of large offshore natural gas reserves make Israel a particularly interesting case study on the role of natural gas and LNG in

---

worldbank.org/WBSITE/EXTERNAL/TOPICS/EXTOGMC/EXTGGFR/0,,menuPK:578075~pagePK:64168427~piPK:64168435~theSitePK:578069,00.html.

[73] *Estimated Flared Volumes from Satellite Data, 2006–2010*, World Bank, Global Gas Flaring Reduction Partnership, http://web.worldbank.org/WBSITE/EXTERNAL/TOPICS/EXTOGMC/EXTGGFR/0,,contentMDK:22137498~pagePK:64168445~piPK:64168309~theSitePK:578069,00.html.

[74] *Iraq Needs to Cut Flaring: World Bank*, Bloomberg, Sept. 29, 2011, http://www.iraqenergy.org/news/?detailof=3953&content=Iraq-Needs-to-Cut-Flaring:-World-Bank-.

[75] IEA MTOGM 2011 at p. 239.

[76] Ibid.

[77] *Iraq Needs to Cut Flaring: World Bank*, Bloomberg, Sept. 29, 2011, http://www.iraqenergy.org/news/?detailof=3953&content=Iraq-Needs-to-Cut-Flaring:-World-Bank-.

[78] IEA MTOGM 2011 at p. 239.

Table 5.1 Gas fields under development in Iraq

Licensing round	Field	Plateau (mcf/d)	Company
Ministry of Oil			
Round 1	Rumaila, Zubair, West Qurna	700 currently flared but up to 3000 later	South Gas Co. (51%); Shell (44%); Mitsubishi (5%)
Round 3	Akkas	400	KOGAS
Round 3	Mansouriya	320	Kuwait Energy/ TPAO/KOGAS
Round 3	Siba	100	Kuwait Energy/TPAO
Kurdistan Regional Government	Khor Mor Chemchemal	300 200	Crescent (40%); Dana (40%); MOL (10%); OMV (10%)

Source: IEA MTOGM 2011

a country's energy supply mix. Historically, Israel has been an "island" in terms of electricity supply since, for political reasons, its electrical grid is not connected to any of its Arab neighbors.[79] Faced with an electrical power system unable to meet the demands of a growing economy, in recent years Israel has been forced to make some difficult decisions about the future of its energy supply mix.

Fortunately for Israel, the decision making process has been aided by two significant discoveries of natural gas offshore. In 2009, the offshore Tamar field was discovered with reserves of approximately 240 bcm (8.4 Tcf). In 2010, the larger offshore field Leviathan was discovered with estimated reserves of 450 bcm (16 tcf). The IEA has indicated that the estimated reserves of the Tamar and Leviathan fields are sufficient to meet Israel's growing domestic gas demand and might also allow Israel to become an LNG exporter.[80]

While Israel was in the process of analyzing the potential of its newly discovered gas fields, a study was done to help the Israeli government exploit the use of natural gas while minimizing the potential drawbacks.[81] Although the study was conducted before the reserves of the two fields were

[79] Steven W. Popper, *Energy Resiliency: A New Way for Israel to Tap into the Future*, Rand Review, Spring 2010, www.rand.org.

[80] IEA MTOGM 2011 at p.253.

[81] Steven W. Popper, *Energy Resiliency: A New Way for Israel to Tap into the Future*, Rand Review, Spring 2010, www.rand.org.

firmly established, it found that Israel should increase its use of natural gas through pipeline imports and/or LNG imports. Since Israel did not have an LNG terminal, the study recommended that Israel should plan for an LNG terminal but delay construction until future demand and costs became clearer.[82]

In October 2011, state-owned Israel Natural Gas Lines announced it had signed a deal with Italian marine contractor Micoperi to build an offshore floating LNG (FLNG) terminal costing approximately $140 million.[83] The 2.5 bcm terminal will be built some 10 km out from the Mediterranean coastal city of Hadera and will be the unloading point for ships carrying the natural gas, which will then be fed directly into Israel's underwater gas pipeline. Construction is due to begin in the second half of 2012 and be completed by the end of the year. The proposed FLNG terminal is viewed as a quick solution and "of the utmost strategic importance for the country's ability to ensure a continuous energy supply to its power stations and to safeguard its energy security."[84]

In the meantime, however, Israel could face a natural gas shortage as early as the third quarter of 2012 since Israel's sole working gas field is nearly depleted and gas supplies from Egypt have been continuously disrupted due to chaos and sabotage in the Sinai Peninsula. Although natural gas production in Israel is set to soar in coming decades, production from the Tamar field is not expected to begin until 2013 and production from the even larger Leviathan field will not be online until about 2017.[85]

Thus, the offshore LNG terminal is being viewed as just one of Israel's stop-gap solutions. The government has also instructed a number of energy companies exploring its territorial waters to speed up operations and threatened to let their licenses expire if they do not meet their commitments.[86] Israel may also have to resort to burning more coal or diesel to power its power plants, although these two options are less desirable from a cost and environmental standpoint.[87]

---

[82] Ibid.

[83] Ari Rabinovitch, *Israel Signs Deal for $140 mln Off-shore LNG Terminal*, Reuters, Oct. 31, 2011, http://www.reuters.com/article/2011/10/31/israel-lng-idUSL5E7LV1N920111031

[84] Ibid.

[85] Ibid.

[86] Ibid.

[87] Ibid.

5.7 NORTH AMERICAN LNG DEMAND – FROM IMPORTER TO EXPORTER

In 2006, it was widely expected that natural gas supply from indigenous sources in the North American region would not keep pace with demand over the longer term and, as a result, North America, and in particular the US, was expected to be a major source of demand growth for LNG in the 21st century.

In its *Annual Energy Outlook 2007*, the US Energy Information Administration (EIA) predicted that US LNG imports would grow from 0.6 trillion cubic feet in 2005 to 4.5 trillion cubic feet in 2030.[88] According to the EIA, "the future direction of the global LNG market is one of the key uncertainties [going forward]. With many new international players entering LNG markets, competition for available supply is strong, and the supplies available to the U.S. market may vary considerably from year to year."[89]

This was consistent with some analysts who predicted that the US LNG market would be a "supply-push" not a "demand-pull" market. In this type of market, other regions will price LNG away from the US when they need it but, since other regions lack significant storage, the US will attract LNG supplies during times of low demand and will absorb the world's excess when there is nowhere else to put it.[90]

As discussed in detail in Chapters 10–12, with the recent increase in production of shale gas in the US, it now appears that, far from being a large importer of LNG, the US may ultimately become a large exporter of LNG!

---

88 Energy Information Administration, Annual Energy Outlook 2007 (EIA AEO 2007), http://www.eia.dow.gov/oiaf/archive/aeo07/pdf/trend_4.pdf.

89 Energy Information Administration, Annual Energy Outlook 2008, Early Release, (EIA AEO 2008), http://www.eia.dow.gov/oiaf/archive/earlyrelease08/production.html.

90 Credit Suisse Equity Research Report, *Natural Gas Group Sector Review: LNG*, Dec. 31, 2007.

6. The globalization of LNG: the evolution of LNG trade, pricing and contracts

6.1 THE EVOLUTION OF LNG TRADE: FROM REGIONAL TO GLOBAL TRADE

The LNG markets originally developed as a niche business under which a relatively small number of sellers supplied specific regional markets with LNG under long-term contracts. Historically, there was very little trade in terms of a spot or short-term market and likewise, there were very few cargo diversions from the originally intended destination. As discussed in detail in Chapter 3, LNG trade has historically been divided into the Asia-Pacific and Atlantic Basin/North American regions with very little trade occurring between the two.[1] For example, LNG trade data for the period 1995–2005 indicate that suppliers in both the Asia-Pacific and the Atlantic Basin regions dedicated over 99 percent of their supply to markets in the same region.[2]

In addition to trade being mostly regional, the vast majority of the cargoes to those regions were committed under long-term contracts that were generally required to underpin the financing and capital investment required for the capital-intensive LNG projects. Typical contracts normally specify delivery of gas to a particular location for a duration of 20–25 years. Historically, the contracts have usually been structured on a take-or-pay basis for specified volumes, with pricing linked either to crude oil or to a basket of crude oil and refined products. In this way, risk is shared between the LNG supplier and the buyer, where the supplier assumes the price risk and the buyer assumes the volume risk.[3]

---

[1] Energy Information Agency, *The Global Liquefied Natural Gas Market: Status and Outlook* (2003), http://www.eia.doe.gov/oiaf/analysispaper/global/lng-market.html.

[2] *Liquefied Natural Gas*, Topic Paper no. 13, Working Document of the National Petroleum Council (NPC) Global Oil & Gas Study (2007), http://www.npc.org/Study_Topic_Papers/13-STG-LiquefiedNaturalGas.pdf.

[3] Ibid.

In more recent years, global LNG trade patterns have shifted, with more LNG being traded between the historically distinct LNG regions and more players entering the market (see Appendix D for a list of companies with major holdings in LNG projects). The growth in LNG trade over the past few years has led many to question whether the LNG markets have become "globalized" and whether LNG could ever trade as a global commodity. In order for this to happen, a number of factors would have to converge including: (1) increased patterns and flow of international trade of LNG; (2) the establishment of a single price; (3) liquid trading; and (4) flexibility of supply.[4] Of these, the establishment of a single price is perhaps the most complicated and difficult to achieve due to differences in the global pricing of natural gas, which is discussed in detail below.

6.2 LNG TRADE IS SET TO GROW SIGNIFICANTLY IN THE 21st CENTURY

While identifying and developing new gas fields is critical to maintaining gas supplies, bringing gas to markets is equally important and identifying the market will determine the infrastructure to be built, the project's development costs, the economics and which partners are needed. While pipeline gas has historically dominated international trade in gas,[5] in recent years there has been a surge in LNG trade for a number of reasons. One reason is the limited flexibility of pipelines once the end-point is decided.[6] This is where LNG has the distinct advantage since LNG that was earmarked for a particular country can potentially be redirected if markets change. The flexibility of LNG has been demonstrated in recent years by the redirecting of Qatari LNG that was originally destined for the US as well as by the additional cargoes that went to Japan after the Fukushima disaster.

In addition to increased flexibility, the wave of emerging LNG importers continues to grow as more countries look to LNG to diversify supply options and also meet increasing demand for natural gas. For example,

---

[4] Michael Stoppard, *Towards a Global Gas Market*, Presentation to CeraWeek 2008, Houston, TX.

[5] Most notably supply to Western Europe from Russia, North Africa and Norway and supply to the US from Canada. *Liquefied Natural Gas*, Topic Paper no. 13, Working Document of the National Petroleum Council (NPC) Global Oil & Gas Study (2007), http://www.npc.org/Study_Topic_Papers/13-STG-LiquefiedNaturalGas.pdf.

[6] IEA Medium-Term Oil and Gas Markets (MTOGM) 2011.

some countries such as Malaysia and Indonesia are looking to import LNG because their gas reserves are located far from the demand centers, which historically were served by closer producing gas fields that are now maturing.[7]

Other countries are looking to increase LNG imports to compensate for the decline in domestic gas production, or to substitute LNG for imported oil and coal.[8] Substituting LNG for higher-priced fuels such as oil can have a significant impact for many countries. For example, Lebanon recently announced it was considering building an LNG import terminal to reduce its reliance on higher-priced gas oil. Switching to LNG would save Lebanon an estimated $1 billion on its energy bill, with a $2 billion impact on the economy as a whole.[9]

Underlying the need to import LNG in many countries are strong economic growth and large populations. For example, there are more than 400 million people in Indonesia, Malaysia and Brunei, Thailand and Singapore, and their combined GDP was nearly $1.47 trillion in 2010, on a par with India's economy.[10] In terms of gas consumption, these countries consumed 13.4 bcf/d in 2010, more than China, India or Japan consumed individually.[11]

Another reason why many countries are looking to build regasification/import capacity is to ensure security of supply. For example, Poland and the Baltic states launched their plans to develop LNG imports when Russia's Gazprom announced alternate supply routes to Europe.[12] Similarly, continued tension in the Middle East led Israel, Lebanon and Jordan to propose building LNG import capacity.[13] In Southeast Asia, Thailand and Singapore are looking to supplement and diversify away from an exclusive reliance on pipeline imports. Singapore currently imports pipeline gas from Malaysia and Indonesia but if additional supplies are insufficient, then Singapore is planning to supplement its existing LNG contract with BG. Thailand's PTT relies solely on pipeline imports from Myanmar but has always been uncomfortable with its reliance on only one gas supplier and is thus looking to LNG imports.[14]

---

7 International Gas Union (IGU) World LNG Report 2011 at p. 30.

8 Ibid.

9 *Lebanon to Base its Future on LNG*, LNG World News, July 4, 2012, http://www.lngworldnews.com/lebanon-to-base-its-future-on-lng/.

10 International Gas Union (IGU) World LNG Report 2011 at p. 31.

11 Ibid.

12 Ibid. at p. 32.

13 Ibid.

14 Ibid.

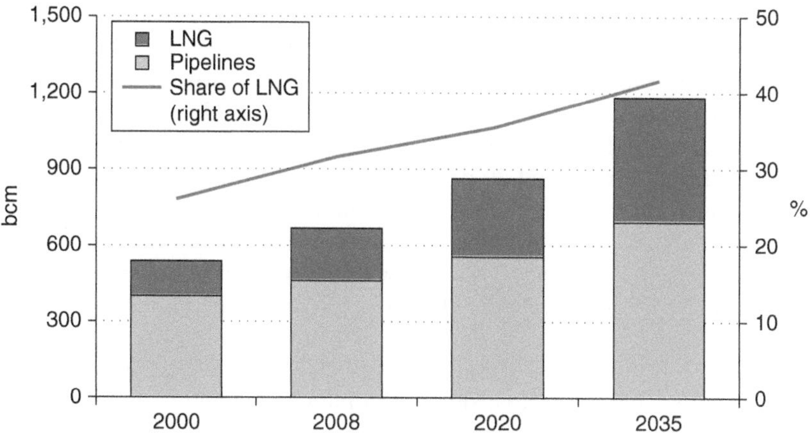

Source: IEA, *World Energy Outlook 2010*, ©OECD/IEA 2010, Figure 5.7, p. 195

Figure 6.1 World inter-regional natural gas trade by type

Due to the above factors, as well as others discussed throughout this book, international trade in natural gas is set to grow "rapidly" in the coming decades and more than half of the growth will be in the form of LNG.[15] According to the IEA, the share of LNG in total gas trade rises from 31 percent in 2008 to 35 percent in 2020 and 42 percent in 2035 (Figure 6.1).

Trade in LNG will account for almost 70 percent of the increase in inter-regional trade by 2030 and exports of LNG will grow from 90 bcm (8.7bcf/d) in 2004 to 470 bcm (45.5 bcf/d) in 2030.[16] In the IEA WEO-2010 New Policies Scenario, trade in LNG more than doubles between 2008 and 2035, reaching 500 bcm, or 11 percent of world

---

15 IEA WEO-2010 at p. 192.

16 IEA WEO-2006, www.iea.org. Between 2002 and 2007, global LNG trade expanded by around 50%. The global economic crisis in 2008–09 led to a sudden and pronounced drop in global gas demand. The drop in demand coincided with a huge increase of new LNG liquefaction capacity, which increased the supply of gas globally and resulted in a "supply glut" for most of 2008–09, with many in the industry wondering how long the supply glut would last. In 2010, the glut began to dissipate in response to a rebound in gas demand, especially from Asia and emerging economies, as well as cold weather across the Northern Hemisphere that called on more gas for heating. IEA World Energy Outlook 2011, Special Report, *Are We Entering a Golden Age of Gas?*, June 6, 2011, http://www.iea/org/weo/docs/weo2011/WEO2011_GoldenAgeofGasReport.pdf (hereinafter IEA Golden Age Report), at p. 31.

demand, with most of the incremental LNG supply going to Asia (Figure 6.2).

As discussed in detail in Chapter 1, in the IEA's Golden Age of Gas Report GAS Scenario, the volume of inter-regional trade of natural gas is expected to more than double by 2035, reaching more than 1 tcm by 2035 and accounting for around 20 percent of total gas trade.[17] Inter-regional trade in the form of LNG increases by around 290 bcm, and by 2035 trade in LNG accounts for about 50 percent of the overall trade in natural gas.[18] In the coming years, China, India and several countries in the Middle East and Latin America are expected to become increasingly reliant on LNG imports.[19]

To meet the anticipated growing demand for LNG, the LNG industry is in the midst of a rapid expansion of both liquefaction and regasification capacity. In terms of liquefaction capacity, 100 bcm per year of liquefaction capacity has come online since early 2009[20] (Figure 6.3). As discussed in detail in Chapter 8, most of this capacity is from Qatar but other countries, such as Australia, are expected to add significant capacity in the coming decades. This additional supply will facilitate even greater trade in LNG as both buyers and sellers seek to take advantage of the flexibility of supply.

In terms of global regasification capacity, the number of countries having regasification capacity has more than doubled since 2001 with eight countries (Argentina, Brazil, Canada, Chile, Kuwait, the Netherlands, Thailand, and Dubai, UAE) adding regasification capacity since 2006.[21] Global regasification capacity continued to grow in 2011 with an ever-larger number of LNG importing markets, many of which were not expected as recently as 2006.[22] As of the end of 2011, there were 89 regasification terminals around the world, representing 608 MTPA in regasification capacity. Of these, 29 started commercial operations between 2006 and 2011.

Regasification capacity is expected to continue to grow, and out of the 24 projects currently under construction 18 are completely new terminals and, once they are completed, five new countries will

---

17 IEA Golden Age Report at p. 32.

18 Ibid. The other 50% is composed of inter-regional trade via natural gas pipelines. While pipeline trade is important, a detailed discussion is beyond the scope of this book.

19 Ibid. at p. 67.

20 Ibid.

21 IGU World LNG Report 2011 at p. 35.

22 Ibid.

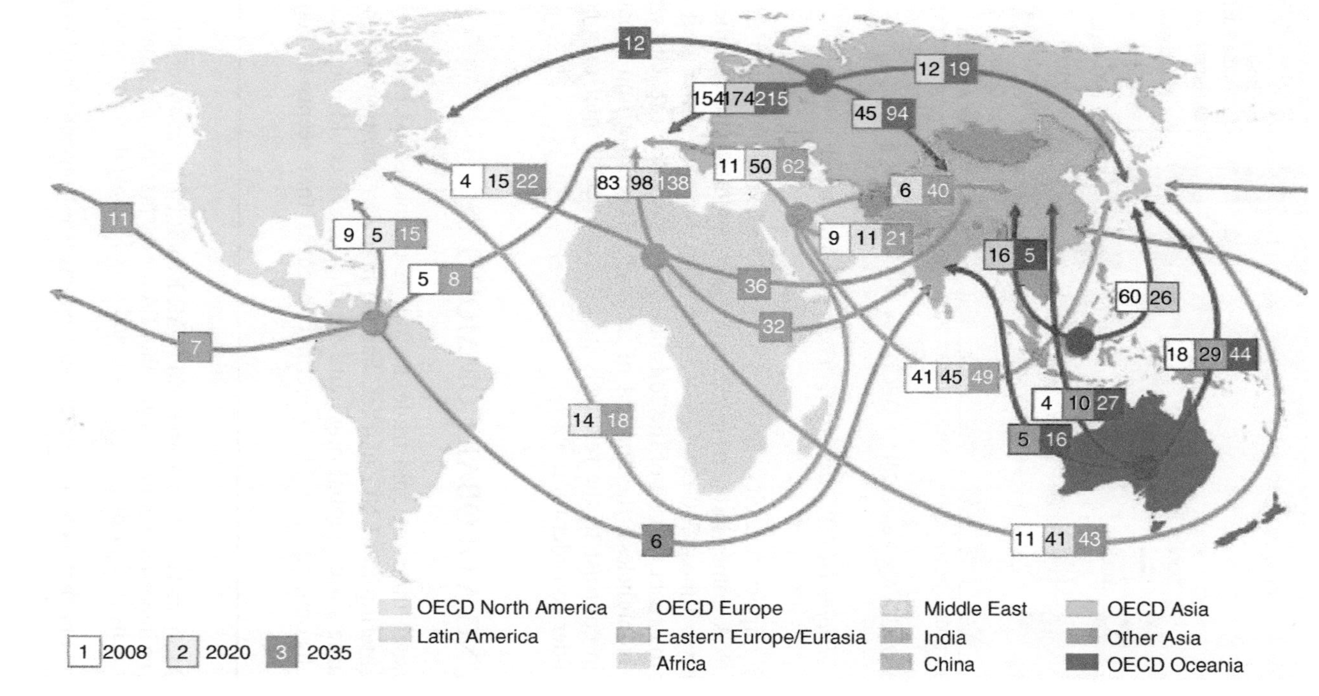

Source: IEA, *World Energy Outlook 2010*, ©OECD/IEA 2010, Figure 5.6, p. 194

Figure 6.2 Inter-regional natural gas net trade flows between major regions in the New Policies Scenario (bcm)

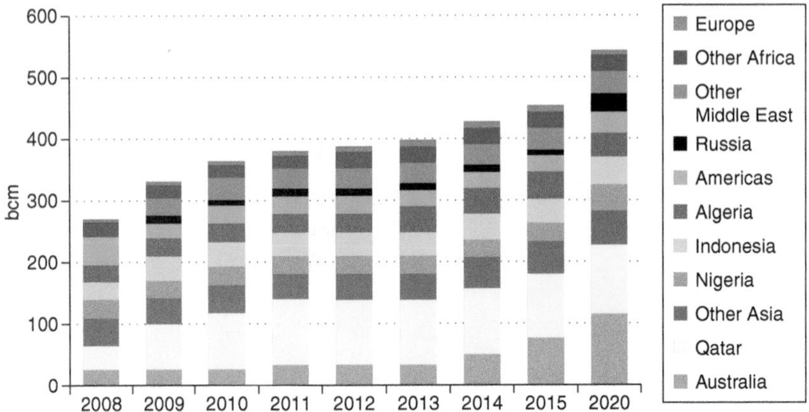

Source: IEA Golden Age Report 2011, ©OECD/IEA 2011, Figure 2.8, p. 71

Figure 6.3 Projected LNG liquefaction capacity by country

have LNG import capacity: Indonesia, Israel, Malaysia, Poland and Singapore.[23] A somewhat surprising amount of additional regasification capacity – almost 700 bcm/y – is under consideration although it is unrealistic to expect that all of the considered regasification capacity will be built.[24] Nonetheless, the amount of interest in regasification capacity stands in contrast to the relatively limited number of inter-regional pipelines under consideration.[25]

6.3 IS LNG A GLOBAL COMMODITY?

The increase in inter-regional trade, as well as the increased number of countries entering the LNG market, has led many to question whether the gas markets are "globalizing" and whether LNG could ever trade as a global commodity.[26] The general consensus that seems to have

---

[23] Ibid.

[24] IEA Medium-Term Gas Market (MTGM) Report 2012. As an indication of the growing importance of natural gas, the IEA for the first time has separated oil from gas in its 2012 medium-term report and issued a separate report focused entirely on gas.

[25] Ibid. noting that, in terms of pipelines, "not much is happening apart from a first decision on the much-awaited Southern Corridor."

[26] IEA MTOGM 2010 at pp. 158, 168.

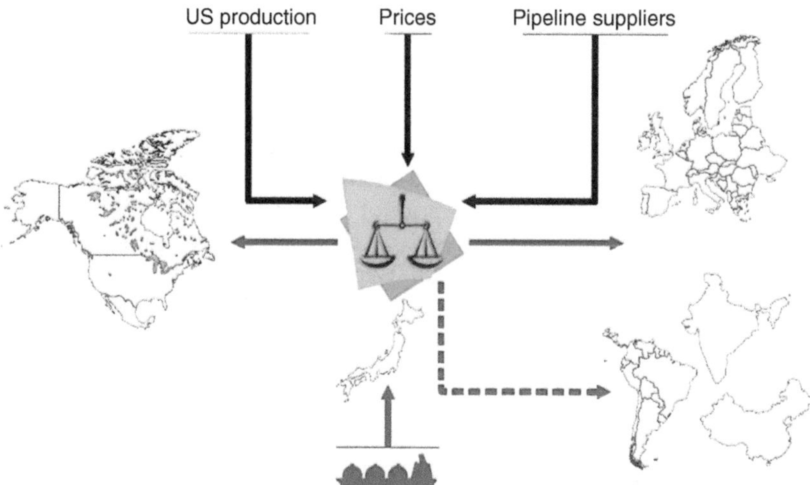

Source: IEA MTOGM 2010, ©OECD/IEA 2010, p. 159

Figure 6.4 Globalization of gas markets

emerged is that while the gas markets are "globalizing," they are not yet globalized since approximately two-thirds of global gas is still consumed in the country where it is produced.[27] Moreover, a truly global LNG market continues to be hindered by the lack of a single pricing structure for LNG, with natural gas remaining the most important commodity with no global market price yet.[28] As such, prices in the various LNG markets remain the balancing point between global gas markets (Figure 6.4).

As inter-regional trade in LNG continues to grow, however, the traditional pricing structures are being challenged and, as discussed below, there are a number of developments that might cause the traditional pricing structure to change over time, with the possibility that LNG will one day trade as a global commodity more similar to oil. For the near term, however, it is expected that the divergence among regional gas prices will decline but remain a feature of global gas markets.[29]

---

27 Ibid. at pp. 158–60.
28 IEA MTGM 2012.
29 Ibid.

6.4 TRADITIONAL PRICING STRUCTURE FOR LNG – OIL-LINKED PRICING

Approximately 70 percent of world LNG trade is priced using a competing fuels index, generally based on crude oil or fuel oil, and referred to as "oil price indexation" or "oil-linked pricing."[30] "The original rationale for oil-linked pricing was that the price of gas should be set at the level of the price of the best alternative to gas."[31] Historically, the best alternative was heavy fuel oil, crude oil or gas oil. While the substitutability of oil and gas has decreased over time, the traditional oil-linked pricing remains. As such, in most areas of the world, natural gas prices rise and fall broadly in line with oil prices.

This is true for most European markets with the exception of the United Kingdom, where the linkage between gas and oil prices is typically formalized by contract, so as oil prices move, gas prices automatically follow. In more recent years, the two largest gas markets in Europe – the UK and Germany – have set the two universally accepted reference points for natural gas prices: the UK National Balancing Point (NBP) and the German Border Price (GBP).[32]

In the United Kingdom, more than half the gas consumed is traded on spot markets with the virtual National Balancing Point (NBP) as the key trading point in the entry/exit based system. Recent long-term contracts supporting large infrastructure projects between the UK and Qatar were at NBP prices rather than oil prices. The other half of the UK gas market is delivered according to the terms of old North Sea prices, which incorporate many indices such as coal, inflation, electricity, fuel oil and gas oil.[33]

---

[30] Australian Government, Bureau of Resources and Energy Economics (BREE), Gas Market Report July 2012, www.bree.gov.au (hereinafter BREE Gas Market Report 2012). "This is the first of what is planned to be an annual report on the current state and projected developments in international and domestic gas markets." "The Bureau of Resources and Energy Economics (BREE) is a professionally independent, economic and statistical research unit within the Australian Government's Resources, Energy and Tourism (RET) portfolio. The Bureau was formed on 1 July 2011 and its creation reflects the importance placed on resources and energy by the Australian Government and the value of these sectors to the Australian economy." BREE, www.bree.gov.au.

[31] BREE Gas Market Report 2012 at p. 42.

[32] Anthony J. Melling, *Natural Gas Pricing and its Future: Europe as the Battleground* (2010), Carnegie Endowment for International Peace, www.CarnegieEndowment.org. This report is adapted from a 2009/10 study commissioned by Cheniere Energy Partners and submitted as part of Cheniere's application for approval to export LNG.

[33] International Energy Agency (IEA), Natural Gas Market Review 2006,

The German Border Price (GBP) is published in Germany by Bundesamt für Wirtschaft und Ausfuhrkontrolle (BAFA) each month. BAFA publishes the total value of gas imports into Germany during each month and the total quantity in energy units. The GBP price is determined by dividing the total value by the quantity to obtain the average gas price known as the GBP.[34]

In the Asia-Pacific region, LNG contracts are typically based on the historical linkage to the Japanese Customs-cleared Price for Crude Oil (JCC, or the "Japanese Crude Cocktail"). This is due to the fact that at the time that LNG trade began, Japanese power generation was heavily dependent on oil so early LNG contracts were linked to JCC in order to negate the risk of price competition with oil.[35] The formula used in most of the Asian LNG contracts that were developed in the late 1970s and early 1980s can be expressed by:[36]

$$P_{LNG} = \alpha \times P_{crude} + \beta$$

where

P_{LNG}	=	price of LNG in US$/mmBtu (US$/GJ × 1.055)
P_{crude}	=	price of crude oil in US$/barrel
α	=	crude linkage slope
β	=	constant in US$/mmBtu (US$GJ × 1.055).

Historically, there was little negotiation between parties over the slope of the LNG contracts, with most disagreements centered on the value of the constant β. Following the oil price declines of the 1980s, most new LNG contracts incorporated floor and ceiling prices that determined the range over which the contract formula could be applied.[37] Since suppliers had to make substantial investments in LNG liquefaction trains, a pricing model developed that provided a floor price. For suppliers, this floor limits the fall in the LNG price to a certain level even if the oil price were to continue falling. Conversely, buyers are protected by a price

---

Gas Prices and Market Design (hereinafter IEA Nat. Gas Mkt. Review 2006), http://www.iea.org/textbase/nppdf/free/2006/gasmarket2006.pdf.

34 Anthony J. Melling, *Natural Gas Pricing and its Future: Europe as the Battleground* (2010), Carnegie Endowment for International Peace, www.CarnegieEndowment.org.

35 James T. Jensen, *Asian Natural Gas Markets: Supply, Infrastructure and Pricing Issues*, a Background Paper prepared for the National Bureau of Asian Research, 2011 Pacific Energy Summit, available at http://www.jai-energy.com/pubs/nbr.pdf.

36 BREE Gas Market Report 2012 at p. 42.

37 BREE Gas Market Report 2012 at p. 42.

cap, which restricts LNG price rises when oil prices rise above a certain point.[38]

According to Poten & Partners' study in the Kitimat LNG export application, the Asia-Pacific LNG market is predominantly based on long-term supply contracts based on the JCC price. In 2009, more than 95 percent of total LNG deliveries were under such long-term contracts.[39] Long-term contracts are the norm in order to address both buyer concerns with security of supply risks and the seller's insistence on firm commercial arrangements to underpin construction and operation of LNG liquefaction facility economics, including financing arrangements.[40] Unlike in the Atlantic Basin, historically, there are no liquid gas markets in the Asia-Pacific region to provide a ready fall-back market for buyers or sellers.

In North America, the gas market originally operated in a similar manner to the European market with long-term, oil-based contracts. When the North American gas markets were liberalized in the 1980s and 1990s, a "hub" system developed whereby natural gas is now traded at over 40 principal centers, or hubs, spread across North America. The best known is Henry Hub in Louisiana, which is the reference point for the pricing of NYMEX gas futures contracts.[41] In many countries, including countries in the Middle East and North Africa (MENA), Latin America, the former Soviet Union, most of the rest of Africa, gas prices are set without any linkage to oil or costs and gas prices barely cover production costs.[42]

6.5 RECENT PRICING ISSUES

In terms of price, the last few years can be analyzed as three different stages. The first stage, from mid-2007 to mid-2008, represents the time when gas market prices, along with other fuel prices, were increasing and

---

38 IEA Nat. Gas Mkt. Review 2006.

39 Poten & Partners, *2015–2035 LNG Market Assessment Outlook for the Kitimat LNG Terminal*, Prepared for KM LNG Operating General Partnership, Oct. 2010, https://www.neb-one.gc.ca/ll-eng/livelink.exe/fetch/2000/90466/94153/552726/657379/657474/670503/65(7060/B1-13_-_Appendix_4_-_2015-2035_LNG_Market_Assessment_Outlook_for_the_Kitimat_LNG_Terminal_-_A1W6T5_.pdf?nodeid=656825&vernum=0. See Chapter 3, Section 3.2.

40 For a listing of the long-term and medium-term contracts in force in 2011, see GIIGNL, The LNG Industry 2011, at pp. 27–30.

41 IEA Nat. Gas Mkt. Review 2006.

42 IGU World LNG Report 2011 at p. 18.

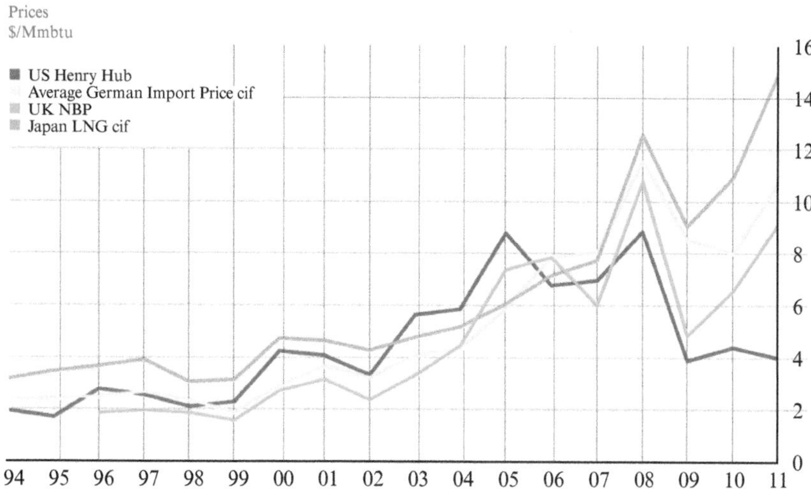

Source: *BP Statistical Review of World Energy 2012*

Figure 6.5 International gas prices, 1994–2011

regional prices converged, reflecting a tightening of the global gas markets.[43] (See Figure 6.5.)

In the second stage, mid-2008 to mid-2009, all regional prices decreased due to the global recession. By the end of 2009 and into the first half of 2010, spot gas prices weakened significantly relative to oil prices. This "decoupling" was primarily due to two things happening on the supply side – the surge in LNG capacity when new production came online and the unexpected boom in shale gas production in North America. The increase in supply coming at a time when demand was dropping in the face of recession led to a sizable "glut" of natural gas on the market.[44]

The significant imbalance in the gas markets led to a "large and unprecedented gap" between prices in the competitive markets of North America and the United Kingdom and prices in those markets where gas was indexed to oil under long-term contracts – namely continental Europe and Asia-Pacific.[45] For example, in 2009, the spot price for Henry Hub (HH)

---

[43] IEA MTOGM 2011 at p. 212.

[44] IEA WEO-2010 at p. 185.

[45] Because North America is a deregulated market for natural gas, gas is not traded based on an energy-equivalent value of gas vs. oil and the recent ratio of oil

in the US averaged $4/Mmbtu, the UK NBP averaged $5/Mmbtu, and Japan and continental Europe averaged $9/Mmbtu[46] (Figure 6.5).

The price decoupling caused buyers of oil-linked contracts in Europe to seek to renegotiate pricing terms with their suppliers. For example, Russia's Gazprom granted some important concessions on pricing, partially moving from oil to spot gas price indexation over a three-year period causing prices to fall in some key markets like Germany. Buyers were also granted some other concessions to provide them more flexibility when they were required to lift contracted volumes.[47]

The unprecedented decoupling of natural gas prices from oil caused some to question whether the historic link between oil and gas prices was a "divorce" or merely "a temporary separation."[48] At the time, the IEA predicted that as long as the gas supply glut existed, the pressure to move further away from oil indexation would remain, especially for long-term contracts.[49]

The third stage depicted in the graph (Figure 6.5 above) started in the second quarter of 2010 when HH and NBP spot prices started to diverge and NBP increased towards the GBP. The unexpected increase in NBP was due to a combination of factors. There were some LNG supply production problems from Norway and Qatar, which took some LNG off the market. There was also a tightening of global gas markets due to strong LNG demand in Asia and a switching from coal to gas in the European power sector.

By April 2011, the NBP monthly average day-ahead price was 85 percent higher than it was in April 2010, and more than twice as high as HH. In March 2011, the spread between the US HH and the UK NBP was an unprecedented difference of $5.8/MBtu. The spread between the Asian LNG price and the US HH was an even higher $8/MBtu.[50]

---

to gas prices reflects the huge discount in the value of gas versus oil. The following table illustrates the recent ratio between gas and oil prices:

US	Canada	Jan. 24, 2012 pricing
37:1 Heat Parity: 6:1 Discount: 84%	41:1	Henry Hub Spot: US $2.63/MMBtu AECO Spot: Cdn $2.40/Mcf WTI Spot: US $98.75/MMBtu

Source: Ziff Energy Group, http://www.ziffenergy.com

46 IEA WEO-2010 at p. 185.

47 Ibid.

48 Ibid. The IEA highlighted the separation with a special "spotlight" in the WEO-2010 – "Oil and gas prices: a temporary separation or a divorce?"

49 IEA WEO-2010 at p. 185.

50 Ibid.

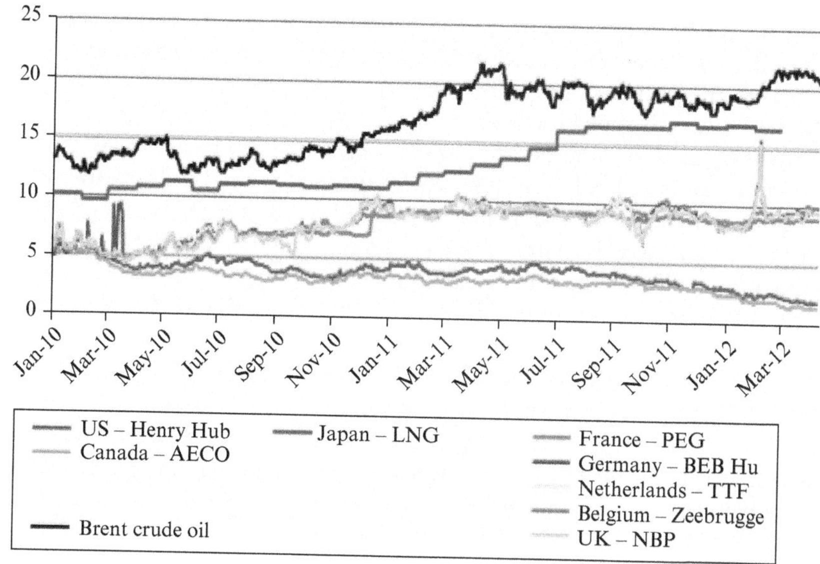

Source: US EIA

Figure 6.6 Global spot natural gas and crude oil prices with average monthly LNG prices in Japan, Jan. 2010–March 2012

Throughout most of 2011, the Asian and European markets faced rising natural gas prices, while the US HH price declined even further. However, by June 2011, European gas prices were back to the level of October 2008 while US HH prices continued to fall.[51] In 2011, US HH prices sunk to their lowest levels in ten years, while Europe's prices ultimately stabilized at $8-10/MBtu with Japan's LNG import price peaking at $17/MBtu by late 2011.[52] (see Figures 6.5 and 6.6.)

6.6 FROM CONVERGENCE TO DIVERGENCE: IS THE US MARKET MOVING AWAY?

Since LNG is the "glue" linking global gas markets, the IEA has recently analyzed how LNG arbitrage created a convergence between the markets in 2009–10 and more recently why the convergence appears to have

51 Ibid.
52 IEA MTGM 2012.

ended.[53] In 2007, the average US LNG import price was \$0.1/MBtu higher than the HH price. This increased to more than \$1/MBtu in 2008 as LNG went to the higher-priced NBP markets. Between 2007 and 2008, US LNG imports dropped more than 50 percent, from 22 bcm to 10 bcm. In mid-2009, the NBP and HH prices converged, the difference between the HH price and the US LNG import price increased, and US LNG imports increased slightly to 13 bcm. This continued up to the beginning of 2010 until the NBP and HH prices diverged again with the US LNG import price moving above the HH price again.[54]

In analyzing the different LNG import prices, the IEA noted that several countries (Trinidad and Tobago, and Egypt) export LNG to the US at prices linked to HH. Qatar and Nigeria account for 21 percent of US LNG imports with prices from those countries linked to NBP. As a result, the average US LNG import price is somewhere between HH and NBP, depending on import volumes from the respective supplier.[55] The delta between the LNG import price and the market price is an indication that US LNG imports of 12 bcm in 2010 were at a minimum and that any cargo that could be re-routed was. In other words, the only LNG that would come to the US was under contract and couldn't be rerouted or LNG that was necessary to avoid shutting down existing US terminals.[56]

In 2010, it became evident that LNG was no longer a price setter in the US and that the US market had become almost immune from global price developments with the US EIA noting that US net imports of natural gas have fallen for three consecutive years,[57] due largely to growing domestic

---

53 IEA MTOGM 2011 at p. 214.

54 Ibid.

55 Ibid.

56 LNG import terminals must be maintained in their cryogenic state in order for them to be able to perform the primary functions of the receipt, storage and regasification of LNG and send-out of natural gas. As such, a sufficient quantity of LNG must be retained in the LNG storage and piping systems regardless of whether or not a terminal is importing. This became a challenge for US import terminals that faced dwindling imports as market conditions changed due to shale gas production and higher priced LNG markets elsewhere. Donna J. Bailey, *US LNG Import Terminals the Perfect Storm*, LIQUEFIED NATURAL GAS: THE LAW AND BUSINESS OF LNG, 2nd edn (2012) (ed. Paul Griffin), at p. 166.

57 Between 2007 and 2010, annual net imports declined by about 1.2 trillion cubic feet, or nearly one-third. Net imports by pipeline account for over 80% of total net imports and come entirely from Canada (the United States is a net exporter to Mexico). Net pipeline imports fell by 28%. Net imports in the form of liquefied natural gas (LNG) were down nearly 50%. US EIA, Energy Today, April 1, 2011, *US Natural Gas Imports Fall for Third Year in a Row*, http://www.eia.gov/todayinenergy/detail.cfm?id=770.

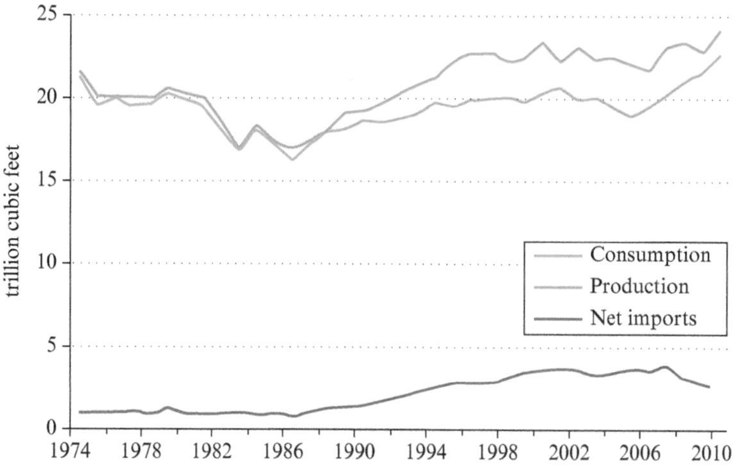

Source: US EIA, Energy Today, April 1, 2011, *US Natural Gas Imports Fall for Third Year in a Row*, http://www.eia.gov/todayinenergy/detail.cfm?id=770.

Figure 6.7 US natural gas consumption, production, and net imports, 1974–2010

production from shale gas formations, which more than tripled between 2007 and 2010 (Figure 6.7).

According to the IEA, whether the price divergence continues into the future depends on whether US production of shale gas is sustained and also the continued development of LNG re-exports (Section 6.8) or the development of new US export capacity (see Chapters 11 and 12).[58]

6.7 THE GROWING SPOT AND SHORT-TERM LNG MARKET

In recent years, the LNG markets have seen the emergence of a growing spot and short-term LNG market, which generally includes spot contracts and contracts of less than four years.[59] Short-term and spot trade allows divertible or uncommitted LNG to go to the highest value market in response to changing market conditions. The short-term and spot market

---

[58] IEA MTOGM 2011 at p. 214.

[59] Anthony Patten and Philip Thomson, *LNG Trading*, Liquefied Natural Gas: The Law and Business of LNG, 2nd edn (2012) (ed. Paul Griffin), at p. 55.

began to emerge in the late 1990s and early 2000s. The LNG spot and short-term market grew from virtually zero before 1990 to 1 percent in 1992, to 8 percent (400 Bcf or 8.4 million tons) in 2002.[60] In 2006, nine countries were active spot LNG exporters and 13 countries were spot LNG importers.[61]

Due to divergent prices between the markets in the past few years, the short-term LNG market has grown rapidly, and by 2009 the spot and short-term market had grown to 16.3 percent of the world LNG trade (491 cargoes and 65.1 10^6 m^3).[62] In 2010, the short-term and spot trade jumped up 40 percent from 2009 and accounted for 18.9 percent of the world LNG trade (727 cargoes and 91.3 10^6 m^3).[63] Europe in particular attracted a large number of spot cargoes due to the availability of uncommitted LNG supply from the Middle East.[64]

In 2011, the spot and short-term market again recorded strong growth, reaching 61.2 Mt (994 cargoes) and more than 25 percent of the total LNG trade.[65] Asia attracted almost 70 percent of the global spot and short-term volumes primarily due to Japan's increased LNG need following the March 2011 Fukushima disaster. Spot and short-term LNG imports into Korea almost doubled (10.7 Mt) and they almost tripled for China and India, with these two countries importing a combined 6.5 MT of LNG.[66]

By the end of 2011, 21 countries were active spot LNG exporters and 25 countries were spot LNG importers. The growing number of countries looking to participate in the spot market is indicative of the increased desire for flexibility to cope with market changes or unforeseen events such as Fukushima as well as the increased number of countries now participating in the LNG markets. Going forward, short-term trade is expected to continue to grow both in absolute terms and as a share of total trade[67] although the availability of future spot LNG cargoes is a source of much speculation.[68] In 2012, Angola LNG is expected to deliver its entire production of 7 bcm as spot cargoes. Although several Australian projects are expected to come online in the next few years, most of the capacity has

---

60 IEA.

61 IGU World LNG Report 2011 at p. 16.

62 GIIGNL, The LNG Industry 2010, www.giignl.org.

63 Ibid.

64 Ibid. at p. 3.

65 GIIGNL, The LNG Industry 2011, www.giignl.org.

66 Ibid.

67 Stephen Thompson, Poten & Partners, Inc., *The New LNG Trading Model Short-Term Market Developments and Prospects*, http://www.igu.org/html/wgc2009/papers/docs/wgcFinal00351.pdf.

68 IEA MTGM 2012 at p. 146.

been contracted for by Asian buyers and any available spot cargoes would most likely go to Asia.[69] For the near term, it appears that additional sources of flexible LNG are most likely to come from the US and Canada where several LNG export projects are planned.[70] In particular, the contracts for Cheniere's Sabine Pass Liquefaction project do not require a destination clause and are based on spot indexation.[71]

6.8 AN EMERGING TREND – LNG RE-EXPORTS

Re-exports of LNG occur when foreign LNG shipments are offloaded into the above-ground storage tanks located on-site at the imports terminals and then stored while waiting for price signals to dictate whether to deliver the LNG to the higher-paying markets in Asia, Europe, and South America.[72]

As an indication that re-exports may be a growing and profitable business opportunity, in 2011 there were over 40 cargoes reloaded, compared with 19 cargoes in 2010. The re-exported volumes went to 13 countries, with cargoes being re-exported from the Atlantic Basin to Europe, Asia and South America. (See Table 6.1.)

The ability to re-export LNG has allowed US LNG importers in particular to take advantage of the arbitrage opportunity between the US and other markets in recent years. For example, in 2010 the US LNG import price was around $5/Mbtu whereas the LNG price in Asia was around $10–11/Mbtu. With marginal operation costs of storage and reloading into LNG tankers, the main cost to re-export would be shipping to the new destination. Assuming freight to Asia is around $3/Mbtu, LNG re-exports would be a potential solution to under-utilized import terminals.[73]

At least in the US, the re-export business appears to be quite profitable even taking into account regasification and storage costs. For example, in 2011, the average landed price for US LNG imports was $7.5/MBtu, compared to an average FOB price for re-exports of $9.3/MBtu.[74] Re-exports

[69] Ibid. For a listing of the long- and medium-term contracts concluded in 2011 for the Australian projects see GIIGNL, The LNG Industry 2011, at p. 6.

[70] IEA MTGM 2012.

[71] Ibid.

[72] US EIA, Today in Energy, *Re-exports of Liquefied Natural Gas Rose Rapidly in Early Winter*, June 6, 2011, http://www.eia.gov/todayinenergy/detail.cfm?id=1670.

[73] IEA MTOGM 2011.

[74] IEA MTGM 2012.

Table 6.1 Re-exports of LNG cargoes, 2011

Export country/exporter	Import country	No. of cargoes
Belgium/Zeebrugge	Japan	3
Belgium/Zeebrugge	Netherlands	1
Belgium/Zeebrugge	South Korea	1
Belgium/Zeebrugge	Spain	4
Spain/Cartagena	Argentina	1
Spain/Cartagena	Italy	5
Spain/Huelva	Argentina	1
Spain/Huelva	Italy	3
Spain/Mugardos	Argentina	3
Spain/Mugardos	Italy	1
Spain/Mugardos	Kuwait	1
Spain/Mugardos	Taiwan	1
Mexico/ECA	Chile	1
USA/Sabine Pass	Brazil	3
USA/Freeport	Brazil	1
USA/Cheniere	Chile	1
USA/Sabine Pass	China	2
USA/Freeport	India	2
USA/Sabine Pass	India	2
USA/Cameron	Japan	1
USA/Freeport	South Korea	1
USA/Sabine Pass	South Korea	1
USA/Cameron	Spain	1
USA/Sabine Pass	Spain	1
USA/Sabine Pass	United Kingdom	1

Note: Duration: spot, in all cases.

Source: GIIGNL, *The LNG Industry 2011*, at p. 7

of foreign sourced LNG from two of the three US LNG terminals authorized to re-export (Freeport in Texas, Sabine Pass and Cameron in Louisiana) exceeded 12 billion cubic feet (Bcf) in January 2011 alone, equivalent to about 30 percent of US LNG import volumes during that month, and the highest volume since the start of re-export service in December 2009[75] (Figure 6.8).

[75] US EIA, Today in Energy, *Re-exports of Liquefied Natural Gas Rose Rapidly in Early Winter*, June 6, 2011, http://www.eia.gov/todayinenergy/detail.cfm?id=1670.

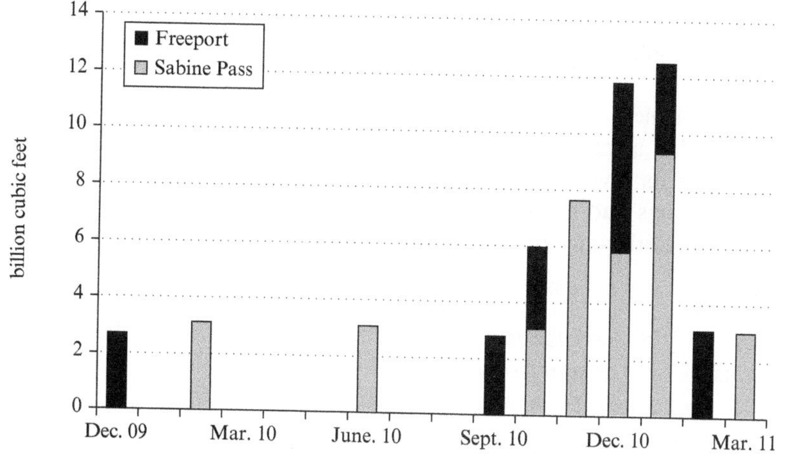

Note: No re-export volumes at Cameron as of March 2011.

Source: US EIA, Today in Energy, *Re-exports of Liquefied Natural Gas Rose rapidly in Early Winter*, June 6, 2011

Figure 6.8 US re-exports of LNG, Dec. 2009–March 2011

6.9 THE EVOLUTION OF LNG CONTRACT TERMS

Just as the LNG markets and trading have evolved over the past decade, so too have the contracts for LNG, with both buyers and sellers searching for more flexibility and shorter-term contracts becoming more common. Nonetheless, long-term (15–25 years) sale and purchase agreements (SPAs) have historically been the cornerstone of the commercial arrangements between LNG buyers and sellers and remain so today and for the foreseeable future.[76]

6.9.1 The Sale and Purchase Agreement (SPA)

The SPA is the key economic driver in an LNG project, and the agreement underpinning much of the LNG project financing. For example, SPAs typically provide most, if not all, of the revenues for an upstream LNG project and also provide anchor supplies for development of LNG regasification facilities.[77] In fact, new LNG trains are typically not launched

---

[76] Tusiani & Shearer, LNG: A Non-Technical Guide, at Chapter 12.
[77] Steven Miles and Thomas Holmberg, Baker Botts LLP, *LNG Master*

without at least some long-term contract coverage. SPAs also have firm take-or-pay obligations for buyers and firm deliver-or-pay obligations for sellers.

Although there is generally a common group of core issues included in every SPA, a model agreement does not exist, owing to the diversity of participants involved in supply and offtake and the variety of commercial models and gas markets.[78] Nonetheless, some of the key issues, such as quantities and price review, warrant a brief discussion here.

Quantities

The quantity clause sets forth the quantity obligations of the seller and the buyer as well as the terms of delivery. For example, the seller may be obligated to make the LNG available for delivery at its facilities with the buyer providing for the LNG shipping – namely, free on board (FOB) as per Incoterms®.[79] If the seller is providing the LNG shipping to the buyer's facilities, then the term is "delivery at terminal" (DAT) or "delivery at place" (DAP), as per Incoterms® 2010.[80] The primary obligations set out in the quantity clause include:[81]

- the base annual contract quantity (ACQ) obligations
- the amount of any downward quantity flexibility (usually for the buyer but sometimes for the seller)

Sales Agreements (MSAs) and their Value in a Destination Flexible LNG Market, Presentation to the CWC Group, LNG Contracting, Pricing and Trading Workshop B, San Antonio, TX, April 21, 2008.

78 Susan H. Farmer and Harry W. Sullivan, *LNG Sale and Purchase Agreements*, LIQUEFIED NATURAL GAS: THE LAW AND BUSINESS OF LNG, 2nd edn (2012) (ed. Paul Griffin), at p. 29. While contracts and project financing are critical elements of LNG projects, a detailed discussion of these issues is beyond the scope of this book. For a useful book on these issues, see Paul Griffin (ed.), LIQUEFIED NATURAL GAS: THE LAW AND BUSINESS OF LNG, Globe Business Publishing Ltd., 2012. Tusiani & Shearer, LNG: A NON-TECHNICAL GUIDE, Chapter 12, also provides an excellent discussion of the upstream gas supply agreements.

79 Susan H. Farmer and Harry W. Sullivan, *LNG Sale and Purchase Agreements*, LIQUEFIED NATURAL GAS: THE LAW AND BUSINESS OF LNG, 2nd edn (2012) (ed. Paul Griffin), at p. 32. Incoterms® is the International Chamber of Commerce's rules on the use of domestic and international trade terms and was originally created in 1936 with the most recent publication Incoterms® 2010. Ibid.

80 Ibid. Noting that the former "delivery ex-ship" or DES term was deleted in Incoterms® 2010 and that the current rule is either DAT or DAP. However, parties can still continue to use DES so long as the long-term SPA refers to Incoterms® 2000. Ibid.

81 Ibid. at pp. 32–3.

- any requirements for the buyer to take delivery ("make good") the quantities previously exercised as downward flexibility
- the buyer's obligation to take or pay for a minimum quantity of LNG
- the parties' obligations with respect to *force majeure* make-up quantities
- the parties' rights and obligations with respect to quantities in excess of the ACQ
- the buyer's right or the seller's obligation to take or supply make-up quantities of LNG for any take-or-pay payments made by the buyer.

The prevailing market conditions – for example, whether it is a buyer's market or a seller's market – will determine the level of flexibility that a seller is willing to give the buyer in terms of its firm offtake obligations.[82] Because the seller typically has made a large investment in liquefaction facilities, the seller generally would prefer to limit the buyer's ability to use downward quantity flexibility to reduce its take-or-pay obligation for any year and require the buyer to take equivalent quantities at a later date as "make good."[83] In recent years, the increased growth in LNG trade, and in particular the increase in spot trade, as well as the increase in the number of import terminals and buyers, has injected more flexibility into the markets and facilitated the resolution of certain issues. For example, a buyer typically may want a period of two to four years to ramp up its ACQ obligations to the full level. Due to the growing spot and short-term market, it is now more feasible for sellers to accommodate such requests.[84]

Duration

Long-term SPAs generally are 20–25 years in length, which reflects the substantial investment required for LNG facilities. As the LNG industry has expanded and matured, there are more mid-term deals ranging from 3 to 15 years.[85] This reflects the fact that that many of the major LNG players are either supplying from a portfolio of supply sources and

---

82 Ibid. at p. 33.

83 Ibid.

84 Ibid.

85 See GIIGNL, The LNG Industry 2011, "Long-Term Contracts concluded 2011," at p. 6, indicating a number of contracts signed in 2011 have durations of 15 or less years with the maximum duration being 21 years (Australia & BG Portfolio and Chubu Electric).

diversified shipping fleets or buying for a portfolio of consumer destinations with diversified shipping fleets.[86]

Price

As previously discussed, unlike crude oil, natural gas and LNG are not priced on an international market basis, with most LNG being priced on a long-term basis of 20 years or more. The contractual and confidential nature of LNG pricing coupled with a lack of transparency of individual cargo prices means that a wide range of prices might exist even within the same country or region. For example, an LNG contract entered into many years ago may still be in effect under far different pricing structures that existed at the time.[87]

Historically, some Asian buyers have been able to introduce price caps or "S-curves" into their pricing mechanisms, which protect them against very high JCC prices and, in return, protect sellers against very low JCC prices. In addition, SPAs with Japanese and Korean buyers are generally viewed as incorporating the principle of "price review" or "price reopeners" for extraordinary circumstances.[88]

Price review or price "re-opener" clauses

A price review or price "re-opener" clause is found in many, but not all, long-term contracts. An example of the language typically used is as follows:

> If . . . economic circumstances in the [buyer's market] . . . have substantially changed as compared to that expected when entering into the contract for reasons beyond the parties' control . . . and the contract price . . . does not reflect the value of natural gas in the [buyer's market] . . . [then the parties may meet to discuss the pricing structure][89]

While price review clauses have been a feature of continental European long-term pipeline gas sale contracts for a long time, the reaction of most common-law lawyers to such clauses is "one of extreme discomfort" for the obvious reason that when invoked, the parties usually find that their interests were more divergent then expected![90] Nonetheless, price review

---

86 Susan H. Farmer and Harry W. Sullivan, *LNG Sale and Purchase Agreements*, at p. 36.

87 Ibid. at p. 40.

88 Ibid. at p. 42.

89 Ibid. at p. 43.

90 Ibid. at pp. 43–4.

or reopener clauses still remain a key clause of most long-term LNG SPAs and many, if not all, LNG suppliers and buyers enter into negotiations without fully grasping how difficult it is to negotiate such clauses, especially if they are to be enforceable if invoked.[91] As such, the following are key elements that must be addressed with regard to negotiating a price reopener clause:[92]

- The trigger event or conditions entitling a party to invoke the clause must be defined. Usually this is a change of circumstances beyond the control of the parties.
- The elements of the price mechanism which are subject to review must be defined and usually include:
    - the base price
    - indexation
    - floor price
    - ceiling price
    - inflection points of the S-curve formula.

If a requesting party has satisfied the trigger event or criterion, then there is the challenge of determining which benchmark should be applied to determine the revised price mechanism, and often the buyer's and seller's views of the relevant market differ significantly. Moreover, since the LNG market is not transparent, the parties may not always be able to access the necessary information and use data that are almost always confidential in price review discussions.[93]

If the parties cannot agree on a revised price mechanism, then they should consider referral of the matter to a third party or arbitration. However, many LNG contracts contain "meet and discuss" price review clauses that do not allow for such referral, leaving the parties without recourse unless some specific recourse is specified. For example, the parties could provide that if the parties are unable to agree on the revised price mechanism, then the seller has the right, upon written notice, to terminate the long-term LNG SPA.[94]

---

[91] Ibid. at p.44.
[92] Ibid. at p.45.
[93] Ibid.
[94] Ibid.

6.9.2 Other LNG Contracts and the Terminal Use Agreement (TUA)

In addition to negotiating the SPAs, lawyers specializing in LNG projects work with clients to ensure that the terms of the SPAs and/or other master sales agreements (MSAs) are coordinated with those of other major project agreements such as the engineering, procurement and construction (EPC) contracts, gas sale and purchase agreements, LNG shipping agreements and terminal use agreements (TUAs).[95]

Worth mentioning here is the TUA which contains terms and conditions similar to those found in tariff structures, such as LNG and gas quality specifications, LNG and gas measurement, and other technical specifications. In the US, the use of TUAs followed after several regulatory actions effectively gave a US terminal operator the right to develop a proprietary-access business model for LNG import terminals.[96] As such, the TUA also contains a key commercial term, which is the "fee for service" or use of the terminal.[97] In the US, this is generally a negotiated rate that is high enough to allow the terminal operator a financial rate of return above that which could be obtained in a regulated environment.[98]

6.9.3 The AIPN Model LNG Master Sale and Purchase Agreement (LNG MSPA)

In more recent years, new forms of master sales agreements (MSAs) have been utilized to address the commercial relationships developing under the growing LNG spot trade. MSAs typically consist of a main MSA that sets forth the "General Terms and Conditions" and a shorter "Confirmation

---

[95] Baker Botts LL.P., *LNG Projects, Deeper Understanding, Better Solutions*, 2009 (in-house publication). Baker Botts is a lawfirm with significant experience advising clients on all aspects of the LNG value chain, "including structuring LNG projects and investments in LNG projects and developing flexible contractual structures," www.bakerbotts.com.

[96] In 2002, in what is commonly referred to as the *Hackberry* decision, the US Federal Energy Regulatory Commission (FERC) "permitted, for the first time, the construction and operation of an LNG import terminal on a proprietary, individually negotiated basis, not subject to open access or service or rate regulation." Donna J. Bailey, *US LNG Import Terminals: the Perfect Storm*, LIQUEFIED NATURAL GAS: THE LAW AND BUSINESS OF LNG, 2nd edn (2012) (ed. Paul Griffin), at p. 164, citing *Hackberry LNG Terminal, LLC* 101 FERC ¶ 61,294 (2002), *order issuing certificates and granting reh'g* 104 FERC ¶ 61,269 (2003).

[97] Tusiani & Shearer, LNG: A NON-TECHNICAL GUIDE at p. 385.

[98] Ibid.

Memorandum" that constitutes a separate agreement between buyer and seller to purchase one or more cargoes of LNG. MSAs are an important part of the evolving LNG spot market and facilitate a number of transactions. For example, MSAs permit sales of LNG produced in excess of committed SPA sales for an LNG liquefaction project. Suppliers might also use MSAs to sell extra cargoes at higher prices to alternative destinations experiencing seasonal peaks in demand. MSAs might also be used by buyers to source additional cargoes to meet seasonal or temporary LNG demand.[99]

The Association of International Petroleum Negotiators (AIPN)[100] recently concluded a substantial industry-wide effort[101] to draft a Model LNG Master Sale and Purchase Agreement (LNG MSPA), which was first released in 2009 (LNG MSPA 2009) with the most recent updated version released in June 2012 (LNG MSPA 2012). The 2012 version is broader than the LNG MSPA 2009 and "addresses a variety of [LNG] trading arrangements."[102] According to the AIPN, the LNG MSPA 2012 is intended to be used for spot sales of LNG and will accommodate transportation by the Buyer ("FOB") or by the Seller ("DAP" or "DAT").[103]

As with other master agreements, Parties to the LNG Master MSA would execute the Base Contract (which includes a checklist of the MSA options and alternatives selected by the Parties, and which incorporates the provisions of the MSA by reference). Upon execution of the Base Contract, the MSA is effective (for example, provisions relating to confidentiality and certain other matters are intended to be binding from execution of the Base Contract), but the obligations to purchase and sell cargoes do not go into effect until such time, if ever, as the Parties or their Affiliates execute a Confirmation Memorandum setting forth the commercial terms applicable to a specific cargo. A form Confirmation

---

[99] Miles and Holmberg, *LNG Master Sales Agreements (MSAs)*.

[100] Association of International Petroleum Negotiators (AIPN), http://www.aipn.org.

[101] Co-chairs Steven Miles (partner, Baker Botts LLP) and Harry W. Sullivan, Jr. and their drafting committee held more than 15 meetings and 5 workshops in 7 countries on 5 continents over a several year period in order to develop and finalize the MSA. The drafting committee was composed of more than 150 industry representatives, including members from many of the major LNG sellers, buyers, transporters and traders worldwide.

[102] AIPN, Press Release, *2012 LNG Master Sale and Purchase Agreement Published*, July 6, 2012, https://www.aipn.org/PressReleases.aspx, quoting AIPN President William Lafferrandre, ConocoPhillips.

[103] AIPN, Model Contracts, LNG Master LNG 2012, http://www.aipn.org.

Memorandum is attached as Exhibit A to the MSA and Drafters' Notes are also provided as part of the Model LNG MSPA.[104]

Consistent with other AIPN model forms, the MSA is intended to reflect an overall balance between Buyer and Seller interests, and to be sufficiently flexible to be used for trades in any geographical region. The options and alternatives included in the MSA provide Buyers and Sellers with substantial flexibility to tailor the agreement to meet their needs.

The AIPN hopes that the Model LNG MSPA will help to enable the creation of a more efficient secondary market for LNG that will facilitate trading and arbitrage of LNG cargoes. According to one of the legal experts that assisted in drafting the Model LNG MSPA, this may happen since "the development of a unified form" among LNG buyers and sellers will "help promote the development of an LNG secondary market by reducing transactional cost and time involved in trading/hedging cargoes."[105]

Prior to the Model LNG MSPA, there was no industry standard in the LNG market, with more than 100 different forms being used by buyers and sellers of LNG. Moreover, many of the existing MSAs in the market were out of date and did not address key terms required by the current LNG industry, including destination flexibility, shipping optimization, offshore title transfer and counterparty risk mitigation. As such, the Model LNG MSPA should help the industry establish a uniform short-term and spot sales agreement, thereby reducing transaction time, cost and uncertainty.

6.9.4 Recent Contractual Developments – the World's First Bi-directional Contracts for US LNG Exports

The contracts entered into between Cheniere and four companies for Cheniere's Sabine Pass Liquefaction project (Figure 6.9) represent an unusual contract structure wherein the buyers agree to purchase an annual contract quantity of LNG and pay Sabine Pass Liquefaction a fixed sales charge of between $2.25/MBtu and $3.00/MBtu for the full contract quantity, regardless of whether the buyer purchases any cargoes of LNG. The fixed sales charge will be paid ratably on a monthly basis, and 15 percent of the fixed sales charge will be subject to annual adjustment for inflation.

---

[104] The LNG MSPA 2012 and other model contracts are available complimentary to AIPN members, and may be purchased by non-members for a nominal fee on AIPN's website, www.aipn.org.

[105] Baker Botts Press Release, "New Model LNG MSA Will Promote Development of LNG Secondary Market", Sept. 24, 2009, http://www.bakerbotts.com/new-model-lng-msa-will-promote-development-of-lng-secondary-market-steven-miles-09-24-2009/, quoting Baker Botts Partner Steven Miles.

Contracted Capacity at SPL
Sale and Purchase Agreements (SPAs)

Long-term, "take-or-pay" style commercial contracts equating to ~16 mtpa

	BG GROUP BG Gulf Coast LNG (1)	gasNatural fenosa Gas Natural Fenosa (1)	KOGAS Korea Gas Corporation (1)	GAIL GAIL (India) Limited (1)
Annual Contract Quantity (MMBtu)	286,500,000	182,500,000	182,500,000	182,500,000
Annual Fixed Fees (5)	~$723 MM	~$454 MM	~$548 MM	~$548 MM
Fixed Fees $/MMBtu (2)	$2.25 - $3.00	$2.49	$3.00	$3.00
Term (4)	20 years	20 years	20 years	20 years
Guarantor	BG Energy Holdings Ltd.	Gas Natural SDG S.A.	N/A	N/A
Guarantor Credit Rating (3)	A2/A	Baa2/BBB	A / A1	Baa2/NR/BBB-
Fee During Force Majeure	Up to 24 months	Up to 24 months	N/A	N/A
Contract Start Date	Train 1 + additional volumes with Trains 2,3,4	Train 2	Train 3	Train 4

(1) Conditions precedent must be satisfied by December 31, 2012 for BG Group and Gas Natural Fenosa and by June 30, 2013 for KOGAS and GAIL (India) Ltd. or either party can terminate. CPs include financing, regulatory approvals, positive final investment decision, issuance of notice to proceed and entering into common facilities agreements (other than KOGAS and GAIL (India) Ltd.).
(2) A portion of the fee is subject to inflation, approximately 15% for BG Group, 13.6% for Gas Natural Fenosa and 15% for KOGAS and GAIL (India) Ltd.
(3) Ratings may be changed, suspended or withdrawn at anytime and are not a recommendation to buy, hold or sell any security.
(4) SPAs have a 20 year term with the right to extend up to an additional 10 years. Gas Natural Fenosa has an extension right up to an additional 12 years in certain circumstances.
(5) BG will provide annual fixed fees of approximately $520 million for trains 1-2 and $203 million for trains 3-4.

Source: Cheniere

Figure 6.9 Cheniere's Sabine Pass Liquefaction sale and purchase agreements

If the buyer purchases LNG cargoes, then the buyer will also pay Sabine Pass Liquefaction a contract sales price for each MMBtu of LNG delivered under the SPA. The contract sales price will be equal to 115 percent of the final settlement price for the New York Mercantile Exchange Henry Hub natural gas futures contract for the month in which the relevant cargo is scheduled.[106]

The buyer will have the right to cancel all or any part of a scheduled cargo of LNG by a timely advance notice, in which case the buyer will continue to be obligated to pay the full monthly fixed sales charge but will forfeit its right to receive the cancelled quantity and will not be obligated to pay the contract sales price for the forfeited quantity.

Most unusual is the fact that buyers will receive bi-directional capacity rights (such as the right to use both import and export services). In addition, there is no destination clause so buyers are free to ship LNG to any location not prohibited by law, providing for true destination flexibility.

6.10 LNG IS GLOBALIZED BUT NOT YET COMMODITIZED

The trade in LNG seems likely to continue to globalize in the coming years and the IEA has recently recognized that "globalization is not only measured by the ratio between LNG trade and total demand, but also by the number of countries (or regions) involved."[107] As discussed throughout this book, the number of countries looking to both import and export LNG is increasing and new players are emerging. For example, over the past decade, the number of importing countries grew from 11 in 2001 to 25 in 2011, with 89 LNG regasification terminals (including ten floating facilities) in operation by the end of 2011, compared with 40 terminals in 2001.[108] The globalization of LNG trade is also evident when one compares trade movements over the past decade from 2001 (Figure 6.10) to 2011 (Figure 6.11).[109]

---

106 A summary of the key terms of Cheniere's export SPA with BG Group can be found in Chapter 12. Cheniere's contract with BG Group was submitted to the Securities and Exchange Commission (SEC) with Cheniere's SEC 8-K filing on October 26, 2011, and is available at www.chenierenergypartners.com.

107 IEA MTOGM 2011 at p. 185.

108 GIIGNL, The LNG Industry 2011 www.giignl.org.

109 While global natural gas trade increased by only 4% in 2011, LNG shipments grew by 10.1% with Qatar (+34.8%) accounting for virtually all (87.7%) of the increase. Pipeline shipments grew by just 1.3%. BP Statistical Review of World Energy 2012, Natural Gas Trade Movements, http://www.bp.com/extendedsectiongenericarticle.do?categoryId=9041232&contentId=7075237.

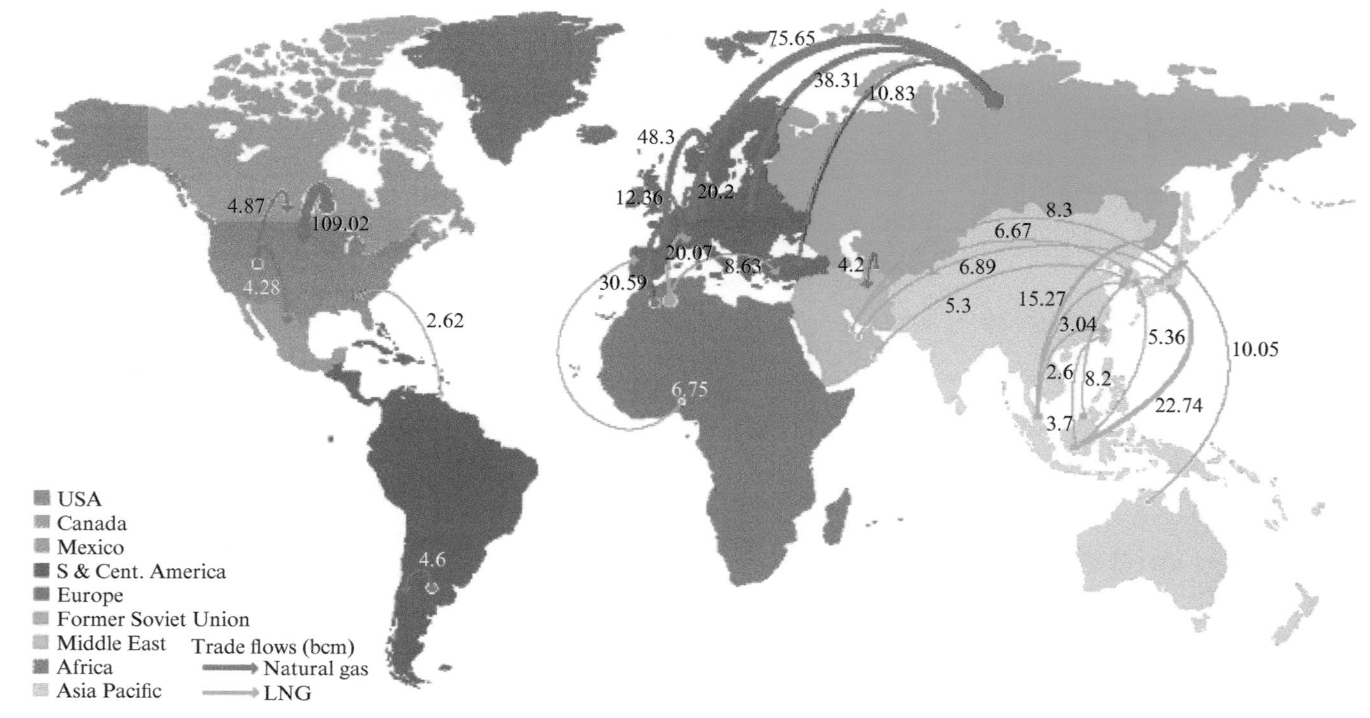

Source: *BP Statistical Review of World Energy 2002*

Figure 6.10 Major LNG trade movements, 2001

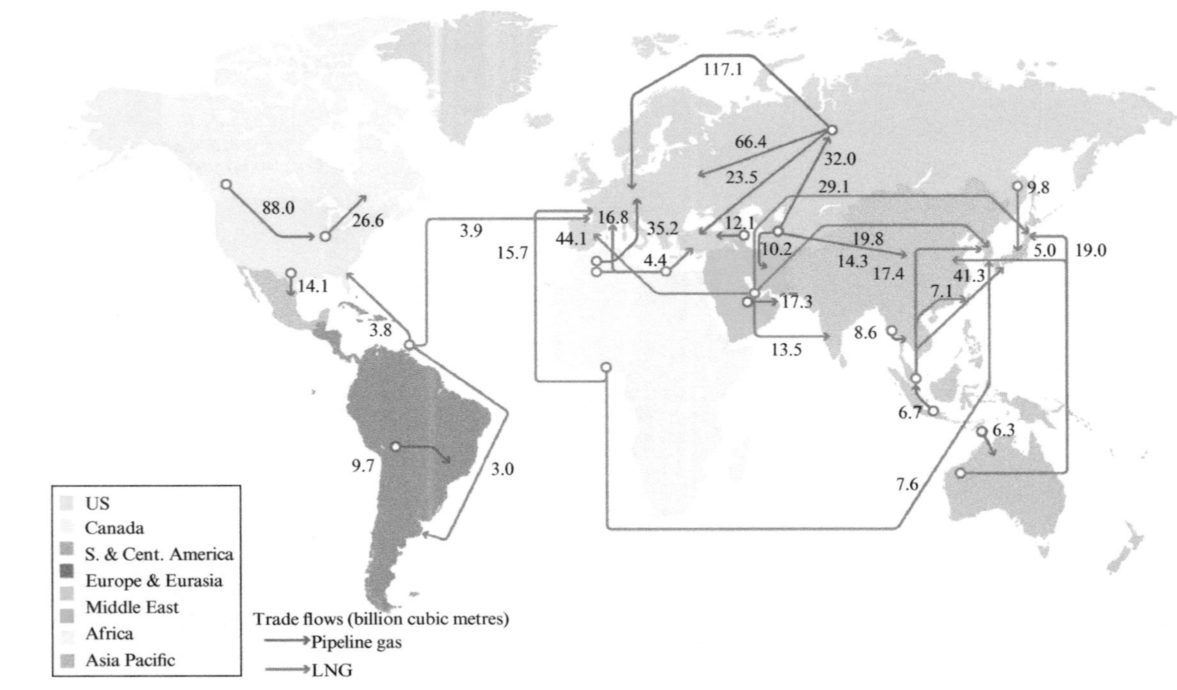

Source: *BP Statistical Review of World Energy 2012*

Figure 6.11 Major LNG trade movements, 2011

Moreover, there is growing recognition that LNG is the "glue" linking global gas markets and this trend is likely to continue into the future.[110] So while LNG has not yet become "commoditized" in the sense that it still does not trade on international markets based on a single pricing structure, there is no doubt that LNG trade has become globalized.

6.11 THE FUTURE EVOLUTION OF LNG MARKETS

Going forward, most experts expect the LNG markets to continue to evolve from the traditional market structure to a more dynamic, flexible structure that better reflects the LNG trade today (Figure 6.12).

Whether Asia, as the largest LNG market, will continue to follow its oil-linked pricing structure is always discussed at industry conferences. The general consensus seems to be that most people expect the Asian LNG market to develop more "flexible characteristics, including spot trading, a spot price index and more operational flexibility."[111] Nonetheless, at least for the foreseeable future, an oil-linked gas pricing mechanism is likely to remain dominant for long-term contract pricing in the Asia-Pacific market for several reasons.

First, LNG producers still require pricing certainty because of the high capital costs of developing LNG projects. Given the increase in capital costs over the past decade, and in particular the high costs for Australian projects, mechanisms for guaranteeing returns for LNG producers remain critical. While this does not necessarily require that gas prices be linked to oil prices, it supports a long-term contract price which locks in the existing mode of pricing for the duration of the contract. Most, if not all, of the recent Australian LNG projects are underpinned by long-term (15–25-year) contracts linked to oil.[112]

Second, in order for gas and oil prices to decouple in Asian markets, there needs to be an alternative mechanism on which to base prices. This mechanism needs to be transparent to ensure that no party (buyer or seller) is disadvantaged at price settlement. There is currently no suitable alternative mechanism in the Asian market.[113]

Third, key buyers, such as Japan and Korea, have limited import options and are currently constrained to LNG-based imports. As such,

---

110 IEA MTOGM 2010 at p. 158.

111 Pat Roberts, *CWC Dynamic Insights Report*, World LNG Series Asia Pacific Summit, September 19–21, 2011, Singapore.

112 BREE Gas Market Report 2012 at p. 42.

113 Ibid.

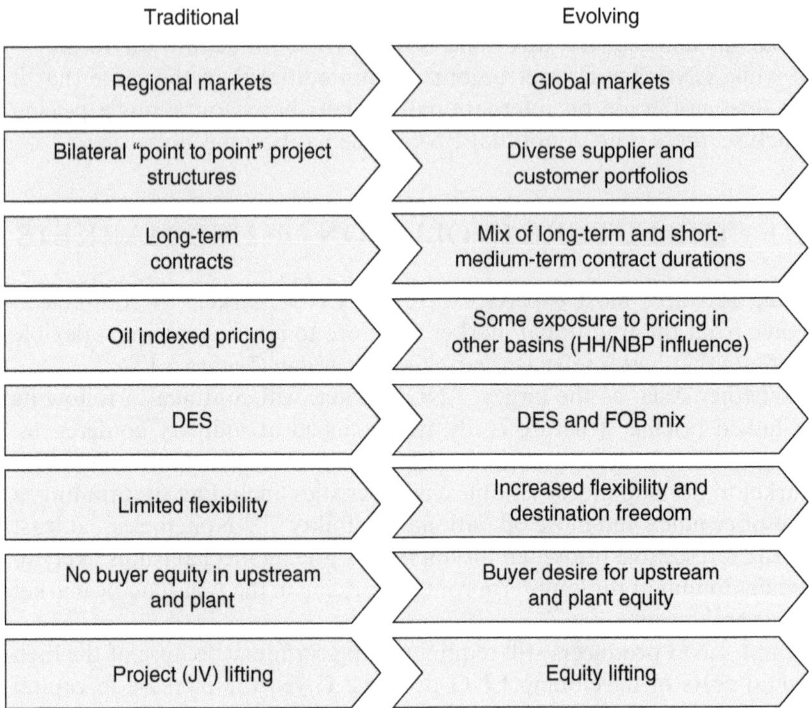

Source: Pat Roberts, *CWC Dynamic Insights Report*, World LNG Series Asia Pacific Summit, September 19–21, 2011, Singapore, citing Stephen Del Regno, Managing Director – Asia Region, Chevron Global Gas.

Figure 6.12 Evolution of LNG markets

security of supply is of paramount concern and there is a fear that any alternative pricing mechanism might jeopardize this.[114]

Given the dynamic nature of global gas markets and the significant price differentials that currently exist between the regions, there are a number of key questions going forward regarding how LNG trade and pricing will continue to evolve. These questions include:

- Will oil indexing continue to remain the dominant pricing structure or will other indexations emerge? Can oil indexation and hub pricing co-exist in Europe and the world?

114 Ibid.

- Sustained high prices in recent years have prompted many companies with additional capacity to secure short- and medium-term contracts at high rates. What implication does this have for pricing and the spot market? Will the spot market continue to expand and where will additional spot cargoes come from?
- What impact will North American LNG exports have on global LNG prices and will North American exports sustain the growing spot market?
- How will the addition of emerging markets impact LNG trade flows and prices? As more countries look to import LNG, there will be more buyers that can come into the markets when prices are low. Does this mean emerging markets will put a "floor" under LNG prices?

7. Safety and environmental sustainability of LNG

7.1 OVERVIEW

Regarding it as a clean-burning fuel, many policy leaders have suggested that LNG can play an important role as the world struggles to meet growing energy demand using more environmentally sustainable fuels. Others, however, have claimed that various safety issues and the environmental impact of LNG projects, including the life-cycle emissions of producing and shipping LNG, may nullify any clean-burning benefit LNG might otherwise provide. Over the same time period that the trade in LNG has grown, it has become increasingly clear that energy production and use will play key roles in moving toward an environmentally sustainable future. This chapter analyzes whether LNG can play a role in a sustainable energy future in the context of the above-mentioned issues.

Going forward, new projects in Australia and elsewhere will inform the discussion over the environmental sustainability of LNG and there is much room for research and debate on this topic. Some of the key issues that need to be analyzed will require a balancing of a number of factors including:

1. the emission benefits of natural gas
2. the economic benefits generated from LNG projects
3. security of supply and the role of natural gas
4. the environmental impact of LNG facilities
5. safety issues pertaining to LNG
6. methane and life-cycle emissions of LNG projects
7. technological breakthroughs that may make the production and transportation of LNG more efficient and sustainable.

Some of these factors have already been discussed in the chapters pertaining to the Golden Age of Gas and elsewhere throughout this book. For that reason, this chapter focuses primarily on the environmental and safety issues and the role of LNG in a carbon-constrained world.

7.2 ENVIRONMENTAL AND SAFETY ISSUES PERTAINING TO LNG

The construction of LNG facilities, whether liquefaction or regasification/import terminals, gives rise to numerous potential environmental impacts. While the potential impacts and necessary regulations vary depending on the project and the country, the International Finance Corporation (IFC) has issued guidelines that are illustrative of the many issues faced by all countries when assessing the environmental impact of proposed LNG facilities. The IFC's *Environmental, Health, and Safety Guidelines for Liquefied Natural Gas (LNG) Facilities*[1] are technical reference documents with general and industry-specific examples of Good International Industry Practice (GIIP). When one or more members of the World Bank Group are involved in a project, the IFC EHS LNG Guidelines are applied as required by their respective policies and standards.[2]

7.2.1 The Environmental Impact of LNG Facilities – IFC EHS LNG Guidelines

The IFC EHS LNG Guidelines contain the performance levels and measures that are generally considered to be achievable in new facilities by existing technology at reasonable costs. The applicability of the IFC EHS LNG Guidelines should be tailored to the hazards and risks established for each project on the basis of the results of an environmental assessment in which site-specific variables are taken into account. When host country regulations differ from the levels and measures presented in the IFC EHS LNG Guidelines, projects are expected to achieve whichever is more stringent.[3]

The IFC EHS LNG Guidelines provide that the following environmental issues associated with LNG facilities should be considered as part of a comprehensive assessment and management program:

- *Threats to aquatic and shoreline environments* Construction and maintenance dredging, disposal of dredge spoil, construction of piers, wharves, breakwaters and other structures, and erosion may

---

[1] International Finance Corporation (IFC), World Bank Group, *Environmental, Health, and Safety Guidelines for Liquefied Natural Gas (LNG) Facilities*, *http://www1.ifc.org/wps/wcm/connect/87e7a48048855295ac04fe6a6515bb18/Final%2B-%2BLNG.pdf?MOD=AJPERES&id=1323161924903*, hereinafter IFC EHS LNG Guidelines.

[2] Ibid.

[3] Ibid.

lead to short- and long-term impacts on aquatic and shoreline habitats. Additionally, the discharge of ballast water and sediment from ships during LNG terminal loading operations may result in the introduction of invasive aquatic species.

- *Hazardous materials management* Storage, transfer, and transport of LNG may result in leaks or accidental releases. LNG tanks and components should meet international standards for structural design integrity and operational performance to avoid failures and to prevent fires.
- *Wastewater* The Guidelines provide information on wastewater management, water conservation and re-use, along with wastewater and water quality monitoring programs.
- *Air emissions* Air emissions from LNG facilities related to combustion sources for power and heat generation in addition to the use of compressors, pumps, and reciprocating engines. Emissions due to flaring and venting may result from activities at both LNG liquefaction and regasification terminals. Principal gases from these sources include nitrogen oxides (NO_x), carbon monoxide (CO), carbon dioxide (CO_2), and in cases of sour gases, sulfur dioxide (SO_2).
- *Waste management* Waste materials should be segregated into non-hazardous and hazardous wastes and a waste management plan should be developed that contains a waste tracking mechanism from the originating location to the final waste reception location.
- *Noise* The main noise emission sources in LNG facilities include pumps, compressors, generators and drivers, compressor suction/discharge, recycle piping, air dryers, heaters, air coolers at liquefaction facilities, vaporizers used during regasification, and general loading/unloading operations of LNG carriers/vessels.
- *LNG transport* Common environmental issues related to vessels and shipping (such as hazardous materials management and wastewater) are covered in the IFC EHS Guidelines for Shipping. Emissions from tugs and LNG vessels, especially where the jetty is within close proximity to the coast, may also represent an important source affecting air quality.

7.2.2 Safety Issues Pertaining to LNG Facilities – Is LNG Safe?

While the LNG tanker industry can claim a record of relative safety since LNG shipping began in 1959, the safety record of onshore LNG terminals is a bit more mixed. In 1944, an accident at one of the United States' first LNG facilities in Ohio killed 128 people and led to public fears about the safety of LNG that still persist today. While technology has made LNG facilities much safer, a January 2004 accident at Algeria's Skikda LNG terminal that killed an estimated 27 workers and injured 74 others added to the ongoing controversy over LNG facility safety.[4]

Most of the debate and concerns pertaining to the safety of LNG originate from the United States where LNG is officially classified as a hazardous material. Since natural gas is combustible, an uncontrolled release of LNG poses a fire hazard or potentially an explosion if in confined spaces. While various experts have identified several potentially catastrophic events that could arise from an LNG release, the likelihood and severity of these events remains the subject of debate. Nonetheless, there appears to be consensus as to what the most serious physical and safety hazards are.

Physical and safety hazards of LNG

- *Pool fires* If LNG spills near an ignition source, the evaporating gas in a combustible gas–air concentration will burn above the LNG pool and the resulting "pool fire" will spread as the LNG pool expands away from its source and continues evaporating. Many experts believe that that pool fires, especially on water, pose the greatest LNG hazard.
- *Flammable vapor clouds* LNG that spills without immediately igniting could form a vapor cloud that may drift some distance from the spill site. If the cloud encounters an ignition source, those portions of the cloud with a combustible gas–air concentration will burn. An LNG vapor cloud fire would gradually burn its way back to the LNG spill and continue to burn as a pool fire.
- *Flameless explosions* If LNG spills on water, it could theoretically heat up and regasify almost instantly in a "flameless explosion" (also called a "rapid phase transition"). The effects of tanker-scale spills have not been studied extensively, but there is a general belief among experts that the hazards of a flameless explosion are not as great as those of a pool fire or vapor cloud.

Terrorism hazards

LNG tankers and land-based facilities could be vulnerable to terrorist attacks, including physical attacks on LNG tankers or tankers that are

---

4 Paul W. Parfomak, Cong. Research Serv., *Liquefied Natural Gas (LNG) in U.S. Energy Policy: Infrastructure and Market Issues*, Jan. 31, 2006, http://www.cnie.org/NLE/CRSreports/06feb/RL32386.pdf, citing Junnola, Jill et al., *Fatal Explosion Rocks Algeria's Skikda LNG Complex*. Oil Daily, Jan. 21, 2004, p. 6.

commandeered for use as weapons against coastal targets. LNG terminal facilities might also be physically attacked with explosives or by some other means.

The September 11, 2001 terrorist attacks in the United States focused more attention on the vulnerability of LNG infrastructure. A number of technical studies were commissioned after the attacks, which caused some controversy due to differing conclusions about the potential public hazard of LNG terminal accidents or terror attacks. Moreover, there was extensive media coverage and public debate due in part to the number of LNG import terminals that were being proposed at the time in the US.

In an effort to resolve the inconsistencies, the Department of Energy commissioned a comprehensive LNG hazard study from Sandia National Laboratories that was released in December 2004. Although the report concluded that "risks from accidental LNG spills . . . are small and manageable," it also concluded that "the consequences from an intentional [tanker] breach can be more severe than those from accidental breaches."[5] The Sandia study also determined that a worst-case, "credible" LNG tanker fire could emit harmful thermal radiation up to 2,118 meters (1.3 miles) away.[6] As a result of the mixed findings from Sandia, both proponents and opponents of new LNG terminals in the US cited the Sandia findings to support their positions and the controversy continued.

7.2.3 LNG Terminal Siting in the United States

In 2003, it was anticipated that much of the growth in demand for LNG would come from the United States, where LNG was expected to fill the supply gap caused by rising demand for natural gas coupled with falling indigenous natural gas reserves in the United States and Canada. While forecasts varied, many analysts expected LNG to account for 12 to 21 percent of total US gas supply by 2025, up from approximately 3 percent in 2005.[7]

Energy experts also suggested that the United States would need to rapidly invest in additional regasification terminals to accommodate the

---

[5] Sandia National Laboratories, *Guidance on Risk Analysis and Safety Implications of a Large Liquefied Natural Gas (LNG) Spill Over Water* (SAND2004-6258) (Sandia Report), Dec. 2004, Albuquerque, NM, at p. 14.

[6] Sandia Report, at p. 51.

[7] Paul W. Parfomak, Cong. Research Serv., *Liquefied Natural Gas (LNG) in U.S. Energy Policy: Infrastructure and Market Issues*, Jan. 31, 2006, http://www.cnie.org/NLE/CRSreports/06feb/RL32386.pdf.

expected increase in LNG imports[8] and numerous actions were undertaken by the US Congress, the Federal Energy Regulatory Commission (FERC), the Department of Energy, and other federal agencies to promote greater LNG supplies, including changing regulations, clarifying regulatory authorities, and streamlining the approval process for new LNG import terminals.[9]

For example, in December 2002, FERC exempted LNG import terminals from rate regulation and open access requirements.[10] This regulatory action, commonly called the "Hackberry decision," allowed import terminal owners to set market-based rates for terminal services, and allowed terminal developers to secure proprietary terminal access for corporate affiliates with investments in LNG supply. These regulatory changes greatly reduced investment uncertainty for potential LNG developers and assured access to their own terminals. In February 2004, FERC streamlined the LNG siting approval process through an agreement with the US Coast Guard and the Department of Transportation to coordinate review of LNG terminal safety and security. The agreement "stipulates that the agencies identify issues early and quickly resolve them."[11]

One of the key actions also undertaken to promote LNG development was the passage of the Energy Policy Act of 2005 (P.L. 109-58), which amended Section 3 of the Natural Gas Act of 1938, granting the Federal Energy Regulatory Commission explicit and "exclusive" authority to approve onshore LNG terminal siting applications (Sec. 311c) among other provisions. As a result, FERC is responsible for granting federal approval for the siting of new onshore LNG facilities and interstate gas pipelines, and also regulates prices for interstate gas transmission.[12]

FERC's approval process incorporates minimum safety standards for LNG established by the Department of Transportation (DOT). DOT's authority originally stemmed from the Natural Gas Pipeline Safety Act of 1968 (P.L. 90-481) and the Hazardous Liquids Pipeline Safety Act of 1979

---

[8] Daniel Yergin and Michael Stoppard, *The Next Prize*, Foreign Affairs, Vol. 82, No. 6 (2003) at pp. 109–11.

[9] Paul W. Parfomak, Cong. Research Serv., *Liquefied Natural Gas (LNG) in U.S. Energy Policy: Infrastructure and Market Issues*, Jan. 31, 2006, http://www.cnie.org/NLE/CRSreports/06feb/RL32386.pdf.

[10] Open access required terminal owners to offer services on a first come, first served basis, and could not discriminate against service requests to protect their own market activities.

[11] Federal Energy Regulatory Commission (FERC), Press release, R-04-3, Feb. 11, 2004.

[12] Natural Gas Act of 1938 (NGA), June 21, 1938, ch. 556, 52 Stat. 812 (codified as amended at 15 U.S.C. §§ 717 et seq).

(P.L. 96-129) but these acts were subsequently combined and recodified as the Pipeline Safety Act of 1994 (P.L. 102-508). The acts were further amended by the Pipeline Safety Improvement Act of 2002 (P.L. 107-355) and the Pipeline Safety Improvement Act of 2006 (P.L. 109-468). Under the resulting statutory scheme, DOT is charged with issuing minimum safety standards for the siting, design, construction, and operation of LNG facilities.[13]

The Energy Policy Act of 2005 (§ 311(c)) explicitly preserved states' authorities in LNG siting decisions under the Federal Water Pollution Control Act, the Coastal Zone Management Act of 1972, and other federal laws pursuant to which states could influence the siting of an LNG project by attaching various conditions or by denying certification.[14] For example, under the Federal Water Pollution Control Act, often referred to as the Clean Water Act (CWA), states have the authority to develop and enforce their own water quality standards.[15] Any federal permit applicant for a project that may discharge pollutants into navigable waters must provide the permitting agency with a certification from the state in which the discharge originates or will originate that the discharge is in compliance with the applicable provisions of the CWA, including the state's water quality standards.[16]

Despite the regulatory changes and encouragement by many US energy experts and politicians (see Chapter 3), the proposed construction of new LNG import terminals in the United States generated considerable public opposition in many of the communities where the terminals were proposed. Choosing acceptable sites for new LNG terminals was particularly challenging since many developers sought to build terminals near major consuming markets in California and the Northeast to avoid pipeline bottlenecks and to minimize transportation costs. However, those proposals struggled for approval due to community concerns about the safety and environmental impact of LNG facilities.[17]

For example, in California several proposed LNG terminals were

---

13 Paul W. Parfomak and Adam Vann, Cong. Research Serv., *Liquefied Natural Gas (LNG) Import Terminals: Siting, Safety, and Regulation* (RL 32205), Feb. 24, 2009.

14 Ibid.

15 33 U.S.C. § 1251(a),(b).

16 Ibid. at § 1341(a).

17 Paul W. Parfomak, Cong. Research Serv., *Liquefied Natural Gas (LNG) in U.S. Energy Policy: Infrastructure and Market Issues* (RL32386), Jan. 31, 2006, http://www.cnie.org/NLE/CRSreports/06feb/RL32386.pdf. As was noted at the time, LNG terminal opposition is not unlike that experienced by some other types of industrial and utility facilities. Ibid.

blocked, including a proposed terminal in Northern California's Humboldt Bay, Sound Energy Solution's proposed terminal in Long Beach, California, and BHP Billiton's proposed floating LNG facility off the coast of Malibu.[18]

Proposed LNG terminals on the East Coast of the United States did not fare much better, and in April 2008 the New York Secretary of State rejected a proposal by Broadwater Energy to construct a floating storage and regasification unit for imported LNG in Long Island Sound.[19] The proposed facility had been approved by FERC subject to more than 80 mitigation measures to enhance safety and security and minimize environmental impacts.[20] Terminals were blocked in other states as well and some developers abandoned proposals due to intense community opposition.[21]

In 2003, leading energy expert Daniel Yergin predicted that due to environmental concerns in the United States, import terminals would probably need to be built in neighboring countries such as Mexico and Canada to supply the United States. Mr Yergin's predictions were correct and, to date, the only West Coast import terminal to have been built is Sempra's LNG terminal, Energia Costa Azul, located in Baja California, Mexico (at the border of Southern California and Mexico).[22]

---

18 See Sierra Club, *Huge Victory Against Offshore LNG Terminal*, available at http://www.sierraclub.org/ca/coasts/victories/victory2007-04-19.asp, and Sierra Club, *Liquefied Natural Gas Threatens California's Coastal Communities*, available at http://www.sierraclub.org/ca/coasts/lng.

19 *New York Secretary of State Determines Broadwater's LNG Facility Not Consistent with the Long Island Sound Coastal Management Program*, New York Dept. of State Press Release, Apr. 10, 2008, available at http://www.dos.state.ny.us/pres/pr2008/41008.htm.

20 *FERC Approves Broadwater LNG Project Subject to Safety, Environmental Measures*, FERC News Release, March 20, 2008, available at http://www.ferc.gov/news/news-releases/2008/2008-1/03-20-08-C-1.asp. FERC's review of the project took 38 months and 25,000 staff hours and produced a final environmental impact statement (EIS) exceeding 2,200 pages. Ibid.

21 Due to local community opposition, LNG developers withdrew terminal projects proposed in California, Maine, North Carolina, Florida, and Mexico. Other terminal proposals in Rhode Island, New York, New Jersey and Canada faced stiff community opposition. In Alabama, a state assumed by many to be friendly to LNG development, community groups effectively blocked two onshore terminal proposals and called for LNG import terminals to be built only offshore. Paul W. Parfomak, Cong. Research Serv., *Liquefied Natural Gas (LNG) in U.S. Energy Policy: Infrastructure and Market Issues* (RL32386), Jan. 31, 2006, http://www.cnie.org/NLE/CRSreports/06feb/RL32386.pdf.

22 *Sempra Energy's New Baja California LNG Terminal Ready for Commercial Operations*, May 15, 2008, CNNMoney.com, available at http://money.cnn.com/news/newsfeeds/articles/marketwire/0397377.htm.

In short, no LNG project was successfully executed in the United States in the face of staunch state and local opposition. The importance of ensuring local stakeholder acceptance held true even though FERC was granted primacy over the siting of LNG terminals and approved a number of terminals. As it turned out, the US did not end up needing all of the proposed import terminals and most of the import terminals that were built are now seeking to try to add export capability! This amazing turn of events is discussed in detail in Chapters 10–12.

7.3 THE ROLE OF LNG IN A CARBON CONSTRAINED WORLD

Is LNG a contributor to a sustainable energy future? Initial analysis indicates that it may be. As the world grapples with issues related to climate change and carbon emissions, it has been widely recognized that natural gas is one of the cleanest-burning fuels and produces relatively low carbon dioxide emissions.[23] Nonetheless, there are two primary environmental considerations related to LNG emissions. The first is the methane emissions that come from all natural gas. The second consideration is the criticism that the life-cycle emissions – the entire supply chain emissions – diminish any clean-burning benefits LNG might provide.

7.3.1 Natural Gas Methane Emissions

Although natural gas is a relatively low-carbon, clean-burning fuel, the principal component of natural gas is methane (CH_4). Methane is a potent greenhouse gas (GHG) and has 23 times the radiative forcing impact of CO_2 on a weight basis over a 100-year period. Methane is the largest contributor to anthropogenic GHG emissions after carbon dioxide and accounts for about 16 percent of the total on a CO_2 equivalent basis. This factor makes the control of CH_4 emissions an important component of any GHG emissions mitigation strategy.[24]

---

[23] *LNG: An Essential Part of American's Clean Energy Mix*, Center for Liquefied Natural Gas, Media Advisory (April 24, 2008), available at http://www.lngfacts.org/Media-Center/042408_media_advisory.asp.

[24] IPIECA Workshop Report, *Natural Gas as a Climate Change Solution: Breaking Down the Barriers to Methane's Expanding Role* (2007), http://www.ipieca.org/publication/natural-gas-climate-change-solution-breaking-down-barriers-methane's-expanding-role, hereinafter IPIECA Workshop Report 2007. Co-sponsored by the Methane-to-Markets Partnership, US EPA and IPIECA,

Methane emissions occur in all sectors of the natural gas industry, from drilling and production, through processing and transmission, to distribution. Emissions primarily result from normal operations, routine maintenance, fugitive leaks and system upsets. As gas moves through the system, emissions occur through intentional venting and unintentional leaks. Venting can occur through equipment design or operational practices, such as the continuous bleed of gas from pneumatic devices (which control gas flows, levels, temperatures, and pressures in the equipment), or venting from well completions during production.

In addition to vented emissions, methane losses can occur from leaks (also referred to as fugitive emissions) in all parts of the infrastructure, from connections between pipes and vessels to valves and equipment.[25]

7.3.2 Life-Cycle Emissions of LNG

Although LNG burns cleanly, concerns have been raised that the environmental impact and emissions associated with LNG production may nullify the clean-burning benefits of LNG. To date, there is limited independent research that analyzes the environmental impact of the entire life-cycle emissions of LNG and most environmental impact statements (EIS) tend to focus on just one aspect of the LNG supply chain, such as the emissions associated with the liquefaction or regasification process. But at least one study, the Heede Cabrillo Study (discussed in detail below) has suggested that the *entire* supply chain emissions – or life-cycle emissions – from production through end-use of the delivered natural gas might be quite significant and should be considered in any environmental impact report.[26] As this book goes to print, more attention is being paid to life-cycle emissions and this appears to be the growing standard,

---

this workshop brought together experts from academia, business, governments, and international and non-governmental organizations to focus on the barriers to bringing methane to market, with attention to both increasing natural gas supply and decreasing fugitive methane emissions, on current strategies for breaching these barriers and on case studies that highlight successful implementation of these strategies in the oil and gas industry. The IEA has also recognized the importance of policy measures aimed at mitigating methane emissions. IEA (2009), *Energy Sector Methane Recovery and Use: The Importance of Policy*, OECD/IEA, Paris.

[25] Natural Gas STAR Program, available at http://www.epa.gov/gasstar/.

[26] Heede, Richard (2006), *LNG Supply Chain Greenhouse Gas Emissions for the Cabrillo Deepwater Port: Natural Gas from Australia to California*, commissioned by California Coastal Protection Network & Environmental Defense Center (Santa Barbara, CA), http://www.climatemitigation.com/publications/LNGrptMay06.pdf (hereinafter Heede Cabrillo Study).

although much more research and studies are needed with regard to LNG.

In the Heede study, an analysis was conducted of the life-cycle emissions resulting from BHP Billiton's proposed Cabrillo LNG terminal off the coast of southern California. In its permit application to the US Coast Guard and the State of California, BHP estimated greenhouse gas emissions only from the operation of its proposed Cabrillo Deepwater Port. The Heede study was commissioned to estimate the entire life-cycle emissions of the project from the production platform offshore Western Australia and across the Pacific Ocean to Southern California, including combustion by end-users in Southern California.[27]

The purpose of the Heede Cabrillo Study was not to attribute the entire supply chain emissions to BHP but, rather, to fully account for all the emissions attributable to the proposed project from start to finish – from production to combustion. The study ultimately found that the "supply chain emissions from production through end-use of the delivered natural gas [were] equal to 4.3 to 4.9 percent of California's total GHG emissions, and 5.3 to 5.9 percent of CO_2 emissions using EIA emissions data."[28]

The largest component of the supply chain emissions was the combustion of the natural gas delivered to the Southern Californian utility and its end-users. The emissions estimates for this segment ranged from 15.82 to 15.89 $MtCO_2$-eq plus 0.58 to 0.72 $MtCO_2$-eq of methane for an average total estimate of 16.50 $MtCO_2$-eq per year, or 72 percent of the total emissions.[29]

While not the largest component of emissions, the most relevant findings for LNG processing are the emissions estimates for the processing segment and the transportation segment. The emissions estimates for the processing segment range from 1.97 to 3.17 $MtCO_2$-eq for an average of 2.69 $MtCO_2$-eq per year, or 11.8 percent of the total. The emissions estimates for the transportation segment range from a low of 1.80 $MtCO_2$-eq to a high of 2.37 $MtCO_2$-eq for an average of 2.09 $MtCO_2$-eq per year, or 9.2 percent of the total.[30]

A major limitation of the Heede Study is that it is based on estimates assuming industry best practices or, in some cases, improvements over standard practice or industry benchmarks. The estimates were used since the facilities had not been designed or built and Heede did not have access

---

27 Heede Cabrillo Study.
28 Ibid.
29 Ibid.
30 Ibid.

to BHP engineering data other than limited information in the permit application. Nonetheless, the Heede study is instructive since the life-cycle analysis was used to support strong environmental opposition to BHP's proposed LNG facility – which was ultimately denied by the State of California.[31]

7.3.3 LNG versus Coal-fired Power Plants – Life-cycle Analysis of Emissions

In much of the world, coal is a plentiful resource and therefore is the dominant fuel source for electrical power production. Natural gas, and LNG as a supplement to domestic natural gas supplies, is increasingly playing a larger role in electrical power generation due to the perceived emissions benefits. At least two studies have accessed the GHG emissions from LNG versus coal-fired power plants and have reached different conclusions.

A study by researchers at Carnegie Mellon found that LNG imported from foreign countries to be used for electricity generation could have 35 percent higher life-cycle greenhouse gas emissions than coal used in advanced power plant technologies.[32] The Carnegie Mellon study "analyzed the effects of the additional air emissions from the LNG/SNG life-cycle on the overall emissions from electricity generation in the United States." The study found that with current electricity generation technologies, natural gas life-cycle GHG emissions are generally lower than coal life-cycle emissions, even when increased LNG imports are included. However, "the range of life-cycle GHG emissions of electricity generated with LNG is significantly closer to the range of emissions from coal than the life-cycle emissions of natural gas produced in North America." The study also found that upstream GHG emissions of NG/LNG/SNG[33] have a higher impact in the total life-cycle emissions than upstream coal emissions.[34]

---

[31] See Sierra Club, *Huge Victory Against Offshore LNG Terminal*, available at http://www.sierraclub.org/ca/coasts/victories/victory2007-04-19.asp, and Sierra Club, *Liquefied Natural Gas Threatens California's Coastal Communities*, available at http://www.sierraclub.org/ca/coasts/lng.

[32] Jaramillo, P., W.M. Griffin and H.S. Matthews (2007), *Comparative Life-cycle Air Emissions of Coal, Domestic Natural Gas, LNG, and SNG for Electricity Generation*, Environmental Science Technology, 41 (17), 6290–96, available at http://pubs.acs.org/cgi-bin/abstract.cgi/esthag/2007/41/i17/abs/es063031o.html (hereinafter Jaramillo Life-cycle Study 2007).

[33] NG = natural gas, LNG = liquefied natural gas, SNG = synthetic natural gas (produced via coal gasification–methanation).

[34] Jaramillo Life-cycle Study 2007.

The Carnegie Mellon Study also analyzed advanced technologies and suggested that as newer-generation technologies and carbon capture and storage (CCS) are installed, the overall life-cycle GHG emissions from electricity generated from coal, domestic natural gas, LNG or SNG could be similar. For SO_x, the study found that coal and SNG would have the largest life-cycle emissions. For NO_x, LNG would have the highest life-cycle emissions and would be the only fuel that could have higher emissions than the current average emission factor from electricity generation, even with advanced power design.[35]

In contrast to the Carnegie Mellon Study, a study commissioned by the Center for Liquefied Natural Gas (CLNG) found that existing US domestic coal power plants produce two-and-a-half times more emissions on a life-cycle basis than LNG does.[36] LNG emissions were even lower when compared with those of advanced ultra supercritical pulverized coal (SCPC) power plants and integrated gasification combined cycle (IGCC) coal-fired power plants (neither of which are currently commercially viable in the US). The production and combustion emissions were greater in all of the coal cases but the processing and transportation segment emissions were greater in the LNG cases.[37] (See Table 7.1.)

It has been recognized that comprehensive life-cycle analysis (LCA) of emissions is very complex and requires a variety of methodological and data assumptions. Thus, while there are published reports from both private and governmental agencies related to emissions, "the uncertainties, process changes and lack of comprehensive measurement in the value chain make it difficult to draw definitive conclusions related to emission contributions especially from the fuel value chain prior to the power plants."[38]

---

[35] Ibid.

[36] Pace, *Life Cycle Assessment of GHG Emissions from LNG and Coal Fired Generation Scenarios: Assumptions and Results*, Prepared for Center for Liquefied Natural Gas (CLNG), Feb. 3, 2009, http://www.lngfacts.org/resources/LCA_Assumptions_LNG_and_Coal_Feb09.pdf (hereinafter Pace CLNG Study).

[37] Pace CLNG Study.

[38] Working Document of the NPC North American Resource Development Study, made available September 15, 2011, Paper no. 4-2, *Life-cycle Emissions of Natural Gas and Coal in the Power Sector*, Prepared by the Life-Cycle Analysis Team of the Carbon and Other End-Use Emissions Subgroup, http://www.npc.org/Prudent_Development-Topic_Papers/4-2_Life-Cycle_Emissions_Coal_and_Gas_in_Power_Paper.pdf.

Table 7.1 Life-cycle emissions of coal power plants and LNG (lbs CO_2e/MWh)

	Production	Processing	Transportation	Combustion	Total
LNG	15	134	99	797	1,045
Coal IGCC	61	24	9	1,714	1,808
Advanced ultra SCPC	61	24	9	1,773	1,868
Existing coal technology	76	30	12	2,614	2,731

Source: Pace CLNG Study

7.4 GREENHOUSE GAS EMISSIONS AND NEW PROJECTS IN AUSTRALIA

Until recently, there have been very limited public data on emissions from LNG projects. New projects in Australia should provide some of the missing information. For example, Australia is planning a massive expansion in LNG liquefaction with several major projects underway including the massive Gorgon and Wheatstone projects. Going forward, assessing the environmental sustainability of LNG will mainly be driven by analysis of the data coming from the planned Australian LNG projects, primarily because Australia has a well-developed environmental and regulatory framework that is also transparent.

Since many of the Australian projects are still under construction, they are only just starting to shed some light on the potential emissions from LNG projects around the world. For example, the Wheatstone Draft EIS, which was released in July 2010, offered the following glimpse of what the emissions might be in that project as well as other LNG projects around the globe.[39] For the Wheatstone project, it is estimated that the annual greenhouse gas emissions from the project may increase Australia and

[39] Chevron Australia, DRAFT ENVIRONMENTAL IMPACT STATEMENT/ENVIRONMENTAL REVIEW AND MANAGEMENT PROGRAM FOR THE PROPOSED WHEATSTONE PROJECT: Volume I (Chapters 1 to 6) (hereinafter Wheatstone Draft EIS/EMP), July 2010, http://www.chevronaustralia.com/Libraries/Chevron_Documents/Wheatstone_Draft_EIS_ERMP_Volume_I_Chapters_1_to_6.pdf.sflb.ashx.

Western Australia's greenhouse gas emissions by 1.7 percent and 13.5 percent respectively.[40]

The Draft EIS argued that this increase should be considered in the context of the impact on *global* emissions and referenced a WorleyParsons study commissioned by Woodside Energy to compare exports of Australian LNG with exports of Australian black coal in terms of life-cycle emissions where the end-use was for electric power generation in China.[41] According to the WorleyParsons study, for each megawatt hour of electricity generated in China using LNG as a fuel (imported from Australia), between 440 and 600 kg of greenhouse gases were released to the atmosphere. For each megawatt hour of electricity generated using imports of Australian black coal, the range was between 720 and 1010 kg, or approximately 40 percent higher.[42] Thus, the Wheatstone Draft EIS suggested that exporting LNG to China was a better alternative in terms of emissions than exporting coal to China.[43]

The early Wheatstone Draft EIS also attempted to benchmark other LNG projects but acknowledged that very limited data were available.[44] The chart in Figure 7.1 shows the greenhouse gas emissions intensity associated with LNG processing for LNG projects currently in production as dark grey bars. The medium grey bars show the estimated LNG processing emissions intensity for the two Australian LNG projects that are currently under construction and the light grey bars show the estimated LNG processing emissions intensity for other Australian LNG projects that are currently undergoing environmental impact assessment. The estimated LNG processing emissions intensity of the Wheatstone project is shown in black. Where data are available, the white bar shows the emission intensity of the associated gas production operations. Projects where publicly available data on gas production emissions are not available are indicated with a circle.

---

40 Wheatstone Draft EIS/EMP.

41 WorleyParsons, *Woodside Energy Limited Greenhouse Gas Emissions Study of Australian LNG*, Originally prepared August 7, 2008, Modified for public release, March 2011, available at http://www.woodside.com.au/our-approach/climate-change/documents/worleyparsons%20(2008)%20greenhouse%20gas%20emissions%20study%20of%20australian%20lng.pdf (hereinafter WorleyParsons GHG LNG Study).

42 Ibid.

43 Wheatstone Draft EIS/EMP, at Section 4.2.6, p. 118.

44 Wheatstone Draft EIS/EMP.

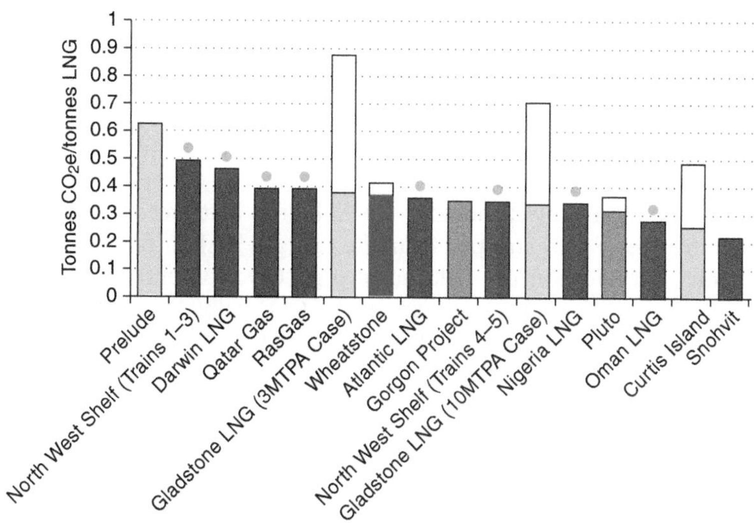

Source: Wheatstone Draft EIS/EMP

Figure 7.1 Greenhouse gas emissions from major LNG projects

7.4.1 Greenhouse Gas Emissions from Australian Projects[45]

According to a recent article that summarized the greenhouse gas emissions from several major projects in Australia, preliminary analysis shows that greenhouse gas emissions from Australia's booming LNG sector are "set to grow sharply."[46]

The article acknowledged that while LNG is a cleaner fuel than coal, extracting, processing, chilling and then shipping LNG releases large amounts of greenhouse gas emissions (the total life-cycle emissions), particularly carbon dioxide. The following are the estimated emissions from several major Australian projects either under construction or awaiting final investment decisions.[47]

---

45 The Australian projects, and in particular Chevron's Gorgon project, are discussed in more detail in Chapter 8.

46 David Fogarty, *Factbox: Projected CO_2 Emissions from Top Australia LNG Projects*, Reuters, May 10, 2011, http://www.reuters.com/article/2011/05/10/us-australia-lng-carbon-fb-idUSTRE7491FU20110510.

47 Ibid.

Gorgon[48]

- Initial capacity: 15 mtpa (likely to be expanded to 20 mtpa).
- Emissions: estimated CO_2-e emissions between 5.45 and 8.81 million tonnes per annum, depending on the level of CO_2 captured and stored.

The A$37 billion project includes a A$2 billion investment to pump 3.3 million tonnes of CO_2 stripped from the raw gas back into a deep natural reservoir. The Gorgon gas field has a high CO_2 content of between 13 and 14 percent, but capturing part of the CO_2 helps lower the project's emissions intensity to about 350 kilograms per tonne of LNG shipped.

Wheatstone

- Target total LNG capacity: 25 mtpa.
- Emissions: based on 25 mtpa production, the company estimates total greenhouse gas production of 10.3 million tonnes per year.

The gas fields tapped for the Wheatstone project have a much lower CO_2 content than Gorgon and the 2010 Draft EIS indicated that the lack of commercially viable reservoirs nearby and low CO_2 content made re-injection of the stripped CO_2 unfeasible. Estimated emissions intensity for the project is 370 kg of CO_2-e per tonne of LNG.

Pluto

- Initial capacity to come online in 2011: 4.3 mtpa. Second production train of 4.3 mtpa planned. Possibility of up to four trains of 4.3 mtpa.
- Emissions: about 1.4 million tonnes of CO_2-e for the first 4.3 mtpa.

The Pluto gas field has a CO_2 content of 2 percent and the company says the emissions intensity of the project is about 320 kg of greenhouse gas pollution per tonne of LNG produced. The company has signed a deal worth nearly A$100 million to create tree plantations to offset emissions from the project.

---

48 Additional information about greenhouse gas related issues from the Gorgon project can be found in Chapter 13 of the DRAFT ENVIRONMENTAL IMPACT STATEMENT/ENVIRONMENTAL REVIEW AND MANAGEMENT PROGRAMME FOR THE GORGON DEVELOPMENT, http://www.chevronaustralia.com/Libraries/Chevron_Documents/gorgon_ch13_LR.pdf.sflb.ashx.

Browse

- LNG capacity: between 11 and 50 mtpa.
- Emissions: between 7.1 million and 32 million tonnes of CO_2-e. Total CO_2 will depend on the design of the LNG plant and the eventual size of the operation.

Emissions from the project are estimated to be higher because of the high CO_2 content of the raw gas from the three main fields at 10 percent. The company says it is looking at re-injecting CO_2 back into underground rock formations.

Ichthys

- LNG capacity: 8.4 mtpa.
- Emissions: about 7 million tonnes of CO_2-e per year, or an estimated 278 million tonnes over the project's 40-year lifetime.

The Ichthys project will tap two main gas fields with 8 percent and 17 percent CO_2 content. About a third of the project's total carbon emissions will come from venting the stripped CO_2 into the atmosphere, although energy efficiency steps might reduce emissions on this project over time.

Prelude

- LNG capacity: 3.5 mtpa.
- Emissions: 2.3 million tonnes of CO_2-e.

This is among the more emissions-intensive LNG projects because of its 9 percent CO_2 content in the feed gas. Total greenhouse gas emissions are just over 600 kg per tonne of LNG produced.

Australia Pacific LNG Project (APLNG)

- LNG capacity: initially about 9 mtpa in two trains. Up to 18 mtpa in final development.
- Emissions: estimated at 640 kg CO_2-e per tonne of LNG shipped, or about 5.7 million tonnes for a 9 mtpa production capacity.

This is a coal seam gas to LNG project.

Curtis Island (QCLNG)

- LNG capacity: estimated at 11 mtpa by 2021 in three trains.
- Emissions: 4.5 million tonnes of CO_2-e.

This is a coal seam gas project.

Gladstone (GLNG)

- Capacity: 10 mtpa in three production trains.
- Emissions: project documents estimate between 5 and 7.2 million tonnes CO_2-e per annum.

This is a coal seam gas project.

7.4.2 Should Emissions be Considered on a Global or Local Basis?

In terms of countering claims that the new LNG projects planned for Australia will increase Australia's emissions, for its part the industry maintains that LNG can actually reduce overall emissions on a global basis with the contention that for every tonne of greenhouse gas emissions generated by LNG production in Australia, between 4.5 and 9 tonnes are avoided in Asia when this gas is substituted for coal in electricity generation.[49] In support of its position, the LNG industry primarily relies on the WorleyParsons study referenced in the Wheatstone Draft EIS and most recently modified for public release in March 2011.[50]

The study, entitled "Greenhouse Gas Emissions Study of Australian LNG", provides a comparison of Australian LNG and Australian black coal in terms of life-cycle greenhouse gas emissions, which includes the entire process from extraction and processing in Australia through to an end-use of combustion in China for power generation.[51]

---

49 Australian Petroleum Production & Exploration Association (APPEA), http://www.appea.com.au/policy/climate-change/how-gas-minimises-greenhouse-emissions.html, and see Chevron Australia, Gorgon Project, *Greenhouse Gas Management*, http://www.chevronaustralia.com/ourbusinesses/gorgon/environmentalresponsibility/greenhousegasmanagement.aspx, "Gorgon can help reduce net greenhouse gases worldwide by displacing other fossil fuels. Gorgon LNG sold globally will result in about 45 million ton[n]es less global greenhouse emissions per year, compared to the use of coal."

50 WorleyParsons GHG LNG Study.

51 Ibid.

Table 7.2 Life stages of fuel for GHG emissions assessment

North West Shelf Project LNG	Australian black coal
Extraction, processing & liquefaction in Australia	Mining & processing in Australia
Transport/shipping to China	Transport/shipping to China
Regasification in China	Combustion/power generation in China
Combustion/power generation in China	

Source: WorleyParsons GHG LNG Study

An assessment of the greenhouse gas emissions for the life-cycles of both LNG from the North West Shelf (NWS) Project and Australian black coal from the Hunter Valley was carried out to determine the total emissions attributable to each fuel source throughout its life-cycle, as indicated in Table 7.2.

The WorleyParsons study found that the displacement of coal with LNG for use for power generation in China results in substantial reductions globally in greenhouse gas emissions, albeit at the expense of some additional Australian greenhouse gas emissions. For the scenarios addressed in Table 7.3, between 5.5 and 9.5 tonnes of CO_2-e are saved globally by LNG replacement of coal for every tonne of CO_2-e released in Australia in the LNG process.

While the measurement of emissions on a global basis has some merit, it is far from clear that this is the consensus view. In addition, as discussed in Chapter 1, the IEA's Golden Age of Gas Report indicated that, absent additional actions, the world's increased use of natural gas would not result in the agreed-upon reduction of greenhouse gas emissions. More recently, at an energy conference in Australia, Christof Ruhl, chief economist at BP, told the conference that although natural gas is the fastest-growing fossil fuel, it is unlikely to reduce emissions enough to reduce climate change. Mr Ruhl also noted that while the replacement of coal by gas would have some effect, it would not be enough to stave off greenhouse gas emissions that can cause climate change. He indicated that BP's 2030 projections indicate CO_2 emissions will continue to increase despite advancements in energy efficiency. Moreover, "The problem of energy security won't subside – just shift," with India and China relying on imports for up to 80 percent of their oil consumption.[52]

---

[52] Rick Wilkinson, *WorleyParsons: LNG has Lower GHG-emission*

Table 7.3 Example scenarios of Australian–global emissions relationship for LNG

Scenario	Case	Technology	Global life-cycle GHG intensity t CO_2-e / MWh	Australian component of the GHG intensity t CO_2-e / MWh	Emissions ratio (global change in emissions: Australian emissions) t CO_2-e / t CO_2-e
New power generation plant in China	Base case	Super-critical	0.769	0.03	–
	LNG replacement	CCGT	0.440	0.06	−5.5: +1[1]
New power generation plant in China	Base case	Sub-critical	1.018	0.04	–
	LNG replacement	CCGT	0.440	0.06	−9.5: +1[2]

Notes:

1. This shows that through the utilization of LNG for a new power generation plant in China in place of coal, 5.5 tonnes of CO_2-e are saved globally, at the expense of every tonne of CO_2-e emitted in Australia.
2. This shows that through the replacement of a current coal-fired power generation plant in China with a LNG power plant, 9.5 tonnes of CO_2-e are saved globally, at the expense of every tonne of CO_2-e emitted in Australia.

Source: WorleyParsons GHG LNG Study

7.4.3 Australia's Proposed Carbon Tax

Policy makers throughout the world are considering a variety of legislative and regulatory options to mitigate greenhouse gas emissions. Assessing these options requires understanding their likely effectiveness, scale, and cost, as well as their implications for economic growth and quality of life.

For its part, the Australian government has acknowledged that the oil and gas industry is an important contributor to Australia's economy and employs about 12,000 people. LNG exports alone were valued at

Benefit, Oil&Gas Journal, April 21, 2011, http://www.ogj.com/articles/2011/04/worleyparsons--lng.html.

$7.8 billion in 2009–10 and are projected to increase to over $8.4bn in 2010–11.[53] While recognizing the economic benefits from the energy industry, the Australian government has also been seeking to impose a carbon tax on the industry in an effort to reduce emissions.

In July 2011, Australia's government unveiled its proposed plans for a carbon tax. Not surprisingly, the plans were not well received by the Australian LNG industry, which argues the tax will make the industry less internationally competitive.[54] The plan as proposed by Prime Minister Julia Gillard's government includes an initial carbon tax of A$23($24.74) per tonne from 2012, rising by 2.5 percent a year, moving to a market-based trading scheme in 2015. The plan could lead to the largest emissions trading scheme outside Europe as Australia tries to cut its emissions by 159 million tonnes by 2020, or by 5 percent based on 2000 levels.[55]

Because the Australian LNG producers will receive a supplementary allocation of emissions permits for 50 percent effective assistance on the tax, many industry analysts feel the impacts will be minor to moderate. Moreover, since the bulk of the costs for an LNG project are capital expenditures, the carbon tax may end up being a significant part of the operating cost but a relatively small portion of the overall project costs. Nonetheless, there is some concern that the tax will add yet one more additional cost to projects that are already considered to be some of the world's most costly and have gained a reputation for running over budget. However, although the carbon tax may erode the profit margins of some LNG projects, so far it appears that no major Australian project will become uneconomic as a result of the tax.[56]

The Australian Petroleum Production & Exploration Association (APPEA) has opposed the tax, claiming that the proposal does not do enough to protect the competitiveness of Australia's growing LNG industry. The industry also rejects the claim that it is "the big polluter"

---

[53] Global Methane Initiative, citing Australian Bureau of Statistics, "8155.0 – Australian Industry", 2008–9.

[54] Rebekah Kebede, *Q+A – Australia's Carbon Tax and the LNG Industry*, Reuters, July 11, 2011, http://uk.reuters.com/article/2011/07/11/carbon-tax-lng-idUKL3E7IB1CY20110711.

[55] Ibid.

[56] Ibid. As this book goes to print, potentially huge cost blowouts for some of the Australian projects are being announced with details yet to follow. Matt (Chambers), *Gorgon Project Profitable Despite Cost Blowout, Says Chevron*, The Australian, Dec. 1, 2012, http://www.theaustralian.com.au/business/mining-energy/gorgon-project-profitable-despite-cost-blowout-says-chevron/story-ebfrg9df-1226527781201.

when for every tonne of emissions produced in liquefying natural gas, up to 9.5 tonnes are saved from entering the atmosphere when LNG is substituted for coal in customer countries.[57] The industry also maintains that the carbon tax proposal would give Australia's competitors in LNG production – Qatar, Malaysia, and Indonesia – an edge over Australia, which is already suffering from a high cost environment.[58]

At this stage, it appears the proposed tax scheme will impact each project differently since carbon emissions vary for each project. For example, some projects in the Browse Basin off Western Australia have as much as 10 percent carbon dioxide by volume, which is extracted at the same time as the gas. Coal seam gas projects in Australia's eastern state of Queensland may produce more greenhouse gas due to the amount of energy required to extract the gas. In addition, some LNG trains are more energy and carbon efficient than others. Yet another difference is that some projects are limiting or offsetting carbon emissions. For example, Chevron plans to sequester some of the carbon it emits at its Gorgon projects, while Woodside's Pluto LNG will be offsetting its carbon emissions over 50 years by planting millions of trees at a cost of A$100 million.[59]

While the impact of Australia's proposed carbon tax will require closer analysis once it goes into effect on July 1, 2012,[60] at least one analyst has said that, for the moment, the carbon tax may be less of a worry than current labor shortages in Australia.[61]

---

[57] APPEA Media Release, *Carbon Pricing Mechanism to Add Cost to LNG Exports*, July 1, 2012, http://www.appea.com.au.

[58] Ibid.

[59] Ibid.

[60] Australia's carbon pricing mechanism commenced on July 1, 2012, and is expected to apply to about 300 of Australia's largest polluters. Starting in 2013, emitters will have to buy a permit from the government (a "carbon unit") for each tonne of pollution at the fixed price of $23 per tonne. For a summary of the key provisions, see The Weekend Australian Financial Review, June 30–July 1, 2012, pp. 32–3, www.afr.com.

[61] Myles Morgan, *Labour Shortage Tops Gas Industry Carbon Tax Fears*, ABC News Australia, Oct. 7, 2011, http://www.abc.net.au/news/2011-10-07/20111007-lng-conference-carbon-tax/3347750. Poten & Partners representative Stephen Thompson told a South East Asia Australia Offshore Conference in Darwin that "The shortage of labour and how to get enough people to do all of the projects Australia is currently embarked on is probably the most significant problem people have in mind." However, others at the conference expressed concerns that the carbon tax would hurt Australia's thriving LNG export business. Ibid.

7.5 ENHANCING ENVIRONMENTAL SUSTAINABILITY IN THE LNG INDUSTRY

While the Australian projects offer valuable insight into LNG emissions, much more analysis is needed of those projects, as well as other new projects around the world. Looking ahead, it is important that the LNG industry continues to utilize best practices to optimize energy efficiency and reduce greenhouse gas emissions. Although the ways in which the LNG industry can enhance environmental sustainability vary project by project,[62] in general some of the most significant advancements can be achieved with methane mitigation and improved efficiencies in processing and transportation.

7.5.1 Benefits of Methane (CH_4) Mitigation

Reducing methane emissions has many important energy, economic, environmental, and safety benefits. Methane is second only to carbon dioxide (CO_2) as a GHG resulting from human activities, and emissions of methane are expected to increase 23 percent to nearly 8 million tonnes (Mt) of carbon dioxide equivalent (CO_2-eq) by 2020, driven by growth in emerging economies, particularly in the natural gas and coal sectors.[63]

Methane is 23 times as effective at trapping heat in the atmosphere as CO_2 and also has a relatively short atmospheric lifetime of approximately 12 years. These two characteristics make methane emission reductions particularly effective in mitigating global warming in the next 25 years.[64] Since methane is the primary constituent of natural gas, the collection and utilization of methane provides a valuable, clean-burning and renewable energy source. Producing energy from recovered methane can also help to avoid the use of higher CO_2- and pollutant-emitting energy sources such as coal and oil.

Methane mitigation is also recognized as an important component under the Kyoto Protocol. The Kyoto Protocol's flexible mechanisms, the Clean Development Mechanism (CDM) and Joint Implementation (JI), provide carbon credits for fuel switching or CH_4 emission reduction projects. However, as of the end of 2005, only 2 percent of the credits from approved or proposed CDM projects were for reduction of CH_4 emissions

[62] See Chapter 8 for a discussion of the efficiency gains in the Chevron Gorgon project.

[63] Methane to Markets, *Global Methane Emissions and Mitigation Opportunities*, available at www.methanetomarkets.org/oil-gas/index.htm.

[64] Ibid.

from oil and gas operations. Credits from the flexible mechanisms can be used in the European Union Emissions Trading System (EU ETS) or sold in the carbon market. Under the EU ETS fuel switching from coal-fired electricity production to natural gas-fired electricity production may be promoted where the incremental rise in costs is met or exceeded by the value of carbon credits.[65]

7.5.2 Barriers to Methane Mitigation

There are several barriers to reducing CH_4 emissions. In some instances there is a lack of information about emission levels and the value of the fuel being lost. Traditional industry practices, such as using high-pressure natural gas in pneumatic devices, can also be a barrier to considering lower-emission alternatives. There may also be resistance from local communities to the building of the infrastructure (such as LNG terminals and pipelines) needed to transport recovered gas to market.[66]

Technical barriers to CH_4 emission reduction from oil and gas operation can arise from a lack of measurement techniques to characterize emissions and of familiarity with technical options and their benefits. Addressing technical barriers requires implementation of a phased approach to emission reduction. Initial steps require that all existing and potential sources of emissions are identified and inventoried. Following characterization of emission sources, the alternative technological options for reducing emissions can be evaluated, and project feasibility studies developed. Reviewing and evaluating experiences, as well as information sharing through partnerships, enables these barriers to be addressed.[67]

Additionally, regulatory barriers can block production of otherwise economically recoverable natural gas resources. For example, much of the potential natural gas production capacity in the US is currently inaccessible because of regulatory restrictions. Removing these restrictions requires public, political and industry support. To overcome the above barriers, the following partnerships and cooperative efforts have been formed.

Natural Gas STAR is a voluntary partnership between the US EPA and oil and gas industry partners with the goal of implementing cost-effective technologies and practices to reduce CH_4 emissions from oil and gas

---

[65] IPIECA Workshop Report, *Natural Gas as a Climate Change Solution: Breaking Down the Barriers to Methane's Expanding Role* (2007), http://www.ipieca.org/publication/natural-gas-climate-change-solution-breaking-down-barriers-methane's-expanding-role (hereinafter IPIECA Workshop Report (2007)).

[66] Ibid.

[67] Ibid.

operations. The Natural Gas STAR partners have identified approximately 100 emission reduction techniques, which have led to cumulative emission reductions of 460 Bcf.

Natural Gas STAR International works to promote technology transfer and provide assistance to international oil and gas companies to develop and implement cost-effective CH_4 emission reduction plans, and report on successes in controlling CH_4 emissions from their international natural gas operations.

The *Methane to Markets Partnership* is a cooperative agreement between countries designed to reduce CH_4 emissions from all anthropogenic sources, including the oil and gas industry. Private companies participate through a project network, which now includes more than 350 organizations. In 2005, the partner countries accounted for 56 per cent of global methane emissions from the oil and gas industry. In October 2010, the Methane to Markets Partnership became the Global Methane Initiative, and as of the time this book goes to print the website and outreach materials are in a transition phase.

The *Global Methane Initiative (GMI)* was launched on October 1, 2010, and replaced the Methane to Markets Partnership. The membership of the GMI includes 38 governments, the European Commission, the Asian Development Bank, and the Inter-American Development Bank, which launched the GMI to urge stronger international action to fight climate change while developing clean energy and stronger economies. GMI builds on the existing structure and success of the Methane to Markets Partnership to reduce emissions of methane, while enhancing and expanding these efforts and encouraging new resource commitments from country partners.[68]

The *Global Gas Flaring Reduction Partnership* is an agreement between countries, international oil companies, the World Bank, OPEC and the EU to support developing nation governments and the petroleum industry in their efforts to reduce gas flaring and venting. It has developed global standards for flaring and venting, regulatory best practices and a gas flaring data tool, and promoted carbon credit financing for flare reduction projects.[69]

The *Asia-Pacific Partnership on Clean Development and Climate* involves six countries (Australia, China, India, Japan, Korea and the US) and has the goal of promoting the development and deployment of clean energy technologies. One of its eight task forces is focused on cleaner fossil

---

68 The Global Methane Initiative, http://www.globalmethane.org/.

69 World Bank, *Global Gas Flaring Reduction Partnership*, http://web.worldbank.org.

fuel technology, including greater use of LNG. Experience drawn from projects and initiatives highlights the importance of information in overcoming the technical barriers to CH_4 emission reduction, and the need for an improved regulatory environment and stakeholder support for projects to increase natural gas supply.

7.5.3 Methane Mitigation Opportunities in the Gas and LNG Industry

A wide variety of technologies are available to reduce CH_4 emissions from natural gas production and use. The Natural Gas STAR programme has identified almost 100 emission reduction techniques. Many of these technologies are relatively simple and cost effective, such as vapour recovery for storage tanks and the use of electric motors rather than gas-fired engines to power wellhead equipment.[70]

Methane emission reduction opportunities generally fall into one of three categories: (1) technologies or equipment upgrades that reduce or eliminate equipment venting or fugitive emissions; (2) improvements in management practices and operational procedures; or (3) enhanced management practices that take advantage of improved technology. In all cases, reducing methane emissions makes additional gas available for sale and use.[71]

Methane losses from LNG facilities are estimated at 4 Bcf and can be broken down as follows:[72]

Emissions from LNG liquefaction equipment

- compressor seals
- CO_2 removal systems
- dehydration systems
- tank overpressure
- ship loading displacement vapors

Emissions from LNG storage tanks

- tank overpressure venting
- leaks from pressure relief valves
- vapor recovery compressors

---

70 IPIECA Workshop Report (2007).

71 Methane to Markets, *Global Methane Emissions and Mitigation Opportunities*, available at www.methanetomarkets.org/oil-gas/index.htm.

72 ConocoPhillips and EPA's Natural Gas STAR Program, Kenai, AK, May 25, 2006, *Liquefied Natural Gas Emissions Reduction Opportunities*, available at http://www.epa.gov/gasstar/documents/workshops/kenai-2006/lng-opportunities.pdf (hereinafter ConocoPhillips Nat Gas STAR Program).

Emissions from LNG marine terminals

- fugitives
- venting, if boil-off vapor cannot be consumed as fuel

Emissions from LNG tankers

- flanges and fitting leaks during cool-down
- leaking vapor recovery systems, not operating
- leaking cargo tank relief valves
- cargo tank venting during delays

Emissions from LNG send-out

- vaporizer fuel system leaks
- pressure relief valves.

Methane savings in the LNG sector can come from the use of centrifugal compressor seals and by implementing a directed inspection and maintenance (DI&M) program. Other LNG emission prevention opportunities include improved connect/disconnect practices, improved tank pressure management, improved vapor recovery system maintenance and availability, and strict enforcement of ship venting rules.[73]

7.5.4 Efficiencies in LNG Liquefaction Processes

The liquefaction process transforms natural gas into LNG by cooling it to approximately −161 °C, after which it is stored until it can be shipped on board LNG tankers to its final destination, where it is regasified at the import terminal. There are a variety of proprietary processes marketed for large-scale LNG liquefaction plants, which can be broken down into the following broad categories:[74]

- pure-refrigerant cascade process
- propane-precooled mixed-refrigerant processes
- propane-precooled mixed-refrigerant, with back-end nitrogen expander cycle
- other mixed-refrigerant processes
- nitrogen expander-based processes.

While a detailed technical discussion of the various methods is beyond the scope of this book, the propane-precooled mixed-refrigerant (C3-MR)

---

[73] ConocoPhillips Nat Gas STAR Program.

[74] Tusiani, Michael D. and Shearer, Gordon (2007), LNG: A Nontechnical Guide, PennWell Books.

process is worth noting since it is considered the "workhorse" of the LNG industry with over 80 percent of the world's completed trains utilizing a variation of this process due to its proven technology in a variety of process and environmental settings and high efficiency.

In general, the C3-MR system uses a multi-component refrigerant to condense and evaporate natural gas in one cycle over a wider range of temperatures. The mixed refrigerant used is Air Products & Chemicals, Inc. (APCI)'s proprietary Multi-Component Refrigerant (MCR). High efficiencies are achieved by adding the propane pre-cooling stage for both feed gas and the mixed-refrigerant loop, allowing the MCR vaporization temperature curve to closely match the natural gas liquefaction curve.[75]

Since the LNG business is extremely capital intensive, economies of scale are critical, especially when it comes to the size of LNG trains. Since about 2001, LNG trains have more than doubled in size from 2 million tons per year to 5 million tons per year. The drive toward increased liquefaction economies of scale led APCI to develop a variant of their C3-MR process that increases the liquefaction train capacity from 5 to almost 8 MMt/y. The new AP-X process adds a third cycle of nitrogen expander (N2) refrigeration to the back end of the C3-MR process's propane (C3) mixed-refrigerant (MCR) cycles. This (N2) cycle takes the LNG subcooling duty off the MCR cycle, increasing the natural gas capacity and reducing the refrigeration loads on the first two cycles.[76]

To gain a competitive advantage in the LNG market, ExxonMobil and RasGas began the development years ago of so-called "mega" trains using the APCI AP-X process that will be able to process 7.8 MMt/y. These larger, more economical trains use state-of-the-art turbines, compressors and heat exchangers uniquely combined to result in greater efficiencies.[77]

In 2009, the first of the Qatari mega trains (RasGas Train 6) began production with other trains following in more recent years. It remains to be seen whether the production from these mega trains will result in significant efficiencies and further studies and research are needed to determine whether the mega trains offer any net environmental benefits.[78]

---

[75] Ibid.

[76] Ibid.

[77] ExxonMobil Corp. News, *A Sea Change for LNG Carriers*, available at http://www.exxonmobil.com/ corporate news_features_20070901_lngcarriers.aspx.

[78] As this book was going to print, additional information was becoming available about methane emissions from the natural gas industry, including LNG. For example, building on the prior workshops, the IPIECA held another workshop in 2012 that focused on recent estimates of life-cycle GHG emissions

7.5.5 Efficiencies in LNG Transportation

For more than 30 years, the size of LNG ships remained virtually unchanged with capacity of about 140,000 cubic meters. In 2001, joint venture partners Qatar Petroleum and ExxonMobil wanted to expand beyond the primarily Asian market for Qatar's LNG and thus needed to develop a better way to deliver Qatar's LNG to more distant ports. Since shipping accounts for about one-third of the cost of LNG, a new class of carrier that was more efficient needed to be designed and built.

Over the course of several years, the design team settled on two similar ship platforms called the Q-Flex (Q for Qatar and Flex for the flexibility to access most LNG ports) and the slightly larger Q-Max (Max for the largest ship that can use the Qatar LNG terminal). The new large LNG ship technologies include a number of industry breakthroughs and significant enhancements, including increased ship size, onboard reliquefaction units, slow-speed diesel engines, twin propellers and rudders, and the latest in hull and antifouling protection and improved fire-protection systems.

The Q-Flex carries 50 percent more LNG than the average carrier operating today while the Q-Max transports 80 percent more. The Q-Max carriers are longer than three football fields, tower twenty stories tall from keel to masthead and are equipped with the largest membrane containment tanks ever built. With a total capacity of up to 266,000 cubic meters, each ship carries enough natural gas to meet the energy needs of 70,000 US homes for one year.

The innovative Q-Max ships carry up to 80 percent more cargo, yet require approximately 40 percent less energy per unit of cargo than conventional LNG carriers due to economies of scale and efficiency of the engines. The end result of these new-generation ships is a 20–30 percent reduction in transportation cost with improved efficiency and emission reductions. Improved efficiency and emission reduction are key as the shipping industry is certain to encounter more stringent guidelines from the International Maritime Organization (IMO) going forward with recent measures to reduce GHG emissions from international shipping.[79] (See Chapter 12.)

---

from natural gas production, transport and use, in relation to competing systems such as gas and coal fired power generation. There was also special consideration of the implications for gas from unconventional sources and LNG systems. IPIECA Workshop Report, *The Expanding Role of Natural Gas: Comparing Life-Cycle Greenhouse Gas Emissions* (2012), http://www.ipieca.org/event/20111202/expanding-role-natural-gas-comparing-life-cycle-greenhouse-gas-emissions.

[79] IMO MEPC, 59th Session, July 13–17, 2009, *IMO Environment*

In addition to increasing the size of the ship, a major initiative was undertaken to design, test and implement the on-board reliquefaction plant, which reliquefies natural gas that is vaporized during transit, re-injecting it as liquid into the cargo tanks rather than using is as vaporized gas to power the tanker itself – allowing for delivery of nearly 100 percent of the cargo. This is particularly beneficial for the long-haul voyages from Qatar to Europe and the Americas.

The on-board reliquefaction facilities created an opportunity to shift from steam boilers and turbines used for propulsion by conventional LNG ships to highly efficient slow-speed diesel engines. The Q-Max ships are equipped with two diesel engines driving twin propellers and rudders. This leads to more energy efficient, reliable and maneuverable ships, reducing fuel consumption by up to one-third.[80]

While the large tankers offer economies of scale and greater efficiencies, a limitation is that these tankers will require appropriate accommodation at the loading and unloading facilities. Unless existing LNG importers reconfigure and reinforce their terminals' berthing facilities, the new-generation LNG tankers will be tied to newly constructed import terminals designed specifically to accommodate the larger ships.[81]

7.5.6 Efficiencies in LNG Import Terminals

The LNG receiving or import terminal is the final link in the LNG chain and the point of connection to the consumers. Whereas the liquefaction plants serve as enormous refrigerators to cool natural gas into a liquid, the import terminals "regasify" or warm the gas back up so that it can be sent through the gas pipeline system. All baseload onshore LNG import terminals basically feature the following components: tanker berthing and unloading facilities, storage tanks, a regasification system, facilities to handle vapor and boil-off gas, high-pressure LNG pumps, a metering and pressure regulation station, gas delivery infrastructure, gas odorization, calorific value control and LNG truck loading facilities.

While there are many aspects of efficiencies to be gained at import

---

Meeting Issues Technical and Operational Measures to Address GHG Emissions from Ships, available at http://www.imo.org/About/ mainframe.asp?topic_id=1773&doc_id=11579.

[80] ExxonMobil Corp. News, *A Sea Change for LNG Carriers*, available at http://www.exxonmobil.com/ corporate news_features_20070901_lngcarriers.aspx.

[81] Tusiani, Michael D. and Shearer, Gordon (2007), LNG: A NONTECHNICAL GUIDE, PennWell Books.

terminals, the design of more efficient ambient air vaporizers, which heat the LNG into its gaseous state, has received the most focus. There are essentially two designs of ambient air vaporizers:

- *direct ambient air vaporizers* This design transfers heat from the ambient air directly into the LNG through the heat transfer surface of a heat exchanger.
- *indirect ambient air vaporizers* In this design, heat from ambient air is transferred to an intermediate fluid which in turn transfers heat to LNG through a separate heat exchanger.

Although the final determination of the vaporization design is site specific, recent analysis suggests that direct ambient air vaporizers are a more efficient solution as they add some, if not all, of the heat required to vaporize LNG under the range of operating conditions. Also, since the amount of supplemental heat that will need to be added (by natural-gas-fired heaters) will be lower in systems that use direct ambient air vaporizers, there is a corresponding reduction in emissions to the environment.[82]

---

[82] Oregon LNG Import Terminal, *Vaporizer Alternatives Study*, prepared by CH-IV International, available at http://www.oregonpipelinecompany.com.

8. Global LNG mega projects and players – Qatar and Australia

8.1 OVERVIEW

In recent years, two countries have emerged as the dominant players in LNG markets – Qatar and Australia. Although Qatar[1] is one of the world's smallest countries, it has been the world's largest LNG producer since 2006. Qatar's dominance in the LNG markets was confirmed with the December 2010 announcement that Qatar had reached its long-stated goal of a combined LNG production capacity of 77 million tonnes per annum. While Qatar's LNG production capacity would be difficult, if not impossible, for most countries to beat, Australia is giving it a try with over $200 billion of LNG projects under construction or planned. Whether Australia can catch Qatar remains to be seen but it will be an interesting race to follow.

1 Qatar is a small Middle Eastern peninsula bordering the Persian Gulf and Saudi Arabia. According to the US CIA World Factbook, Qatar has been ruled by the Al Thani family since the mid-1800s and has transformed itself from a poor British protectorate noted mainly for pearling into an independent state with significant oil and natural gas revenues. During the late 1980s and early 1990s, the Qatari economy was crippled by a continuous siphoning-off of petroleum revenues by the Amir, who had ruled the country since 1972. His son, the current Amir Hamad bin Khalifa Al Thani, overthrew him in a bloodless coup in 1995. Qatar has not experienced the level of unrest or violence seen in Tunisia, Egypt, Libya and other Middle East and North Africa (MENA) countries in 2010–11, due in part to the fact that its energy sector has fueled the Qatari economy and resulted in Qatar having the highest per capita income in the world. Doha, Qatar, is the home of the Al Jazeera news network and also the Gas Exporting Countries Forum (GECF), which is discussed in Chapter 12. US CIA World Factbook, *Qatar*, https://www.cia.gov/library/publications/the-world-factbook/geos/qa.html.

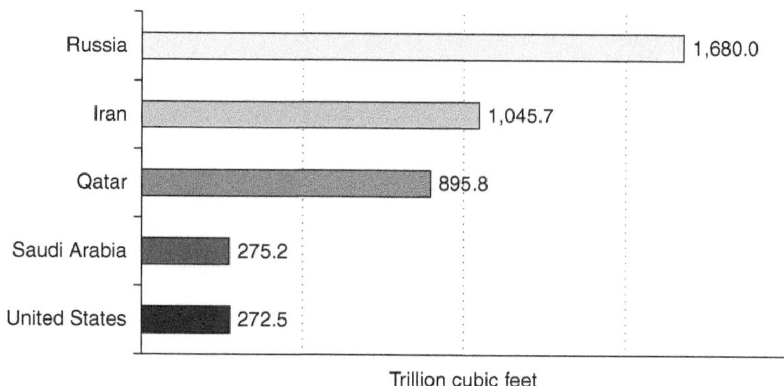

Source: US EIA

Figure 8.1 World natural gas reserves by country, Jan. 1 2011

8.2 QATAR – THE WORLD'S LEADING LNG PRODUCER

Qatar, with proven natural gas reserves of approximately 896 Tcf, currently holds the third-largest gas reserves in the world behind Russia and Iran[2] (Figure 8.1). The majority of Qatar's natural gas is located in the massive offshore North Field, which occupies an area roughly equivalent to Qatar itself. The North Field is part of the world's largest non-associated natural gas field and is a geological extension of Iran's South Pars field, which holds an additional 450 Tcf of recoverable natural gas reserves (Figure 8.2). In recent years, Qatar's energy-fueled GDP growth and small population have catapulted the country toward the top of the global per capita GDP rankings with a 2009 estimate of $121,700.[3]

8.2.1 Qatar's LNG Industry

Although Qatar has been exporting LNG only since 1997, government emphasis on the LNG sector – in terms of both making investments and attracting foreign investors – resulted in the rapid development of

[2] EIA Country Analysis Briefs, *Qatar*, http://205.254.135.24/countries/cab.cfm?fips=QA, last updated Jan. 2011. Qatar holds almost 14% of the world's total natural gas reserves.

[3] Christopher M. Blanchard, *Qatar: Background and U.S. Relations*, Congressional Research Service, RL31718, May 5, 2010.

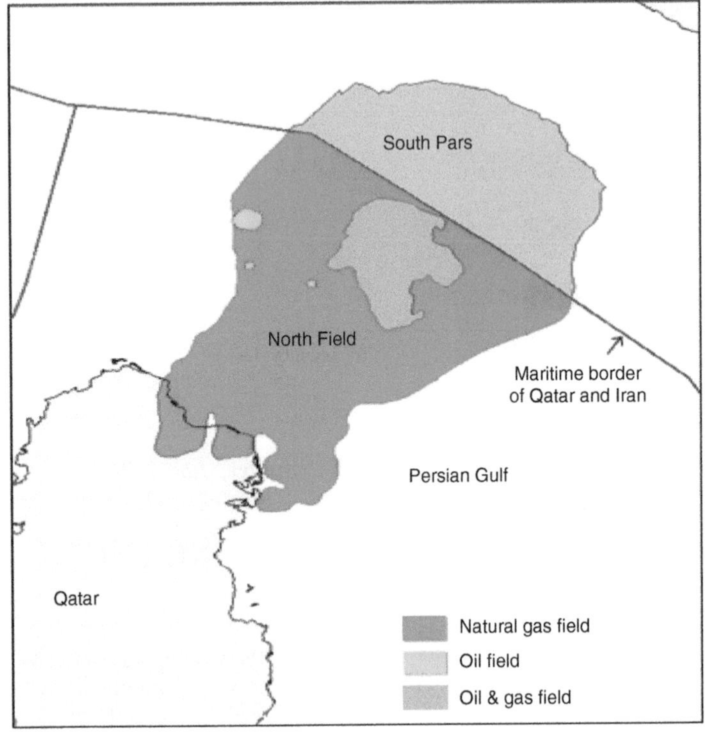

Source: US EIA

Figure 8.2 Qatar's North Field

Qatar's LNG capacity, thus enabling Qatar to become the world's largest LNG producer from 2006.[4] Qatar's position as the world's largest LNG exporter was solidified in December 2010 when Qatar announced that it had reached its long-stated goal of combined LNG production capacity of 77 million tonnes per annum (mtpa).[5] In February 2011, Qatargas and Shell announced that the first cargo of LNG from Qatargas 4 Project

4 IEA WEO-2009 at p.490.

5 *Qatar Celebrates Achieving 77 Mta LNG Production Capacity*, LNG World News, Dec. 15, 2010, http://www.lngworldnews.com. The 77 Mta Celebration Event was the highest-profile energy event in Qatar in 2010 and celebrated the vision set out by the Emir of Qatar for Qatar to reach LNG production capacity of 77 million tonnes of LNG per annum by the end of 2010, a milestone achieved in a mere 14 years.

(Train 7) was loaded and en route to the Hazira receiving terminal in India, thereby completing Qatar's vision of becoming the largest LNG producer in the world.[6]

Qatar's energy sector is dominated by state-owned Qatar Petroleum (QP), which has played an important role in developing Qatar's LNG industry. While QP has maintained a majority share in most of its LNG projects, there is extensive foreign company involvement by international oil companies with the technology and expertise in integrated mega projects, including ExxonMobil, Shell, and Total.[7]

Qatar's LNG projects are shared, along with international partners, by two sister companies, Qatargas Operating Company Limited (Qatargas) and Ras Laffan Company Limited (RasGas), both of which are owned by QP. Qatargas and RasGas handle all upstream to downstream natural gas transportation, while the Qatar Gas Transport Company – Nakilat (which means "carriers" in Arabic) – is responsible for shipping Qatari LNG to customers around the world.[8]

Qatargas operates four major LNG ventures (Qatargas I–IV) and RasGas operates three major LNG ventures (RasGas I–III). RasGas is 70 percent owned by QP and 30 percent owned by ExxonMobil, while the Qatargas consortium includes QP, Total, ExxonMobil, Mitsui, Marubeni, ConocoPhillips, and Shell. Each venture has an individual ownership structure, though QP owns at least 65 percent of all the above ventures.[9]

Along with numerous other technological advancements, the world's largest trains, capable of producing 7.8 MMt/y (10.6 bcm/y) of LNG were constructed for the Qatargas and RasGas projects. Qatargas also pioneered the development of two new classes of LNG tankers – the Q-Flex and the Q-Max. The Q-Flex and Q-Max were designed by a team of engineers that made a quantum leap in the capacities of LNG carriers, with each ship having a cargo capacity of between 210,000 and 266,000 cubic meters. A total of 32 of these ships (19 Q-Flex and 13 Q-Max) have been delivered for Qatargas.[10]

RasGas and Qatargas currently have 14 LNG trains online, with a total

---

6 Qataragas Press Release, *Qatargas, The World's Largest LNG Producer, Sends First Train 7 Cargo to India*, Feb. 20, 2011, http://www.qatargas.com/PressReleases.aspx?id=248014&tmp=88&folderID=164.

7 EIA Country Analysis Briefs, *Qatar*, http://www.eia.gov/countries/cab.cfm?fips=QA.

8 EIA Country Analysis Briefs, *Qatar*, http://www.eia.gov/countries/cab.cfm?fips=QA.

9 Ibid.

10 Qatargas Operating Company Ltd.,http://www.qatargas.com.

Table 8.1 Qatargas and RasGas LNG projects

Project	Start-up date (initial target)	Capacity MMt/y (bcm/year)	Partners
Qatargas (QG) projects			
QG 1: Trains 1–3	1997–98	3 × 3.2 (12.9)	QP, XOM, Total, Marubeni, Mitsui
QG 2: Train 4	Q2 2009 (2008 Q1)	7.8 (10.6)	QP, XOM
QG 2: Train 5	Q3 2009 (2008)	7.8 (10.6)	QP, XOM, Total
QG 3: Train 6	2010 (2009)	7.8 (10.6)	QP, ConocoPhillips, Mitsui
QG 4: Train 7	2011 (2010)	7.8 (10.6)	QP, Shell
RasGas projects			
RasGas: Trains 1-2	1999	2 × 3.2 (9.0)	QP, XOM, Kogas, Itochu, LNG Japan
RasGas: Train 3	2004	4.7 (6.4)	QP, XOM
RasGas: Train 4	2005	4.7 (6.4)	QP, XOM
RasGas: Train 5	2007	4.7 (6.4)	QP, XOM
RasGas: Train 6	Q3 2009 (2008)	7.8 (10.6)	QP, XOM
RasGas: Train 7	2010 (2009)	7.8 (10.6)	QP, XOM
Total LNG		77 (104.8)	

Sources: IEA, EIA, Qatargas, RasGas

LNG liquefaction capacity of 3,750 Bcf/y (77 MMt/y). Five of these trains were added in 2009 and 2010. RasGas III, Train 7, with a liquefaction capacity of 380 Bcf/y (7.8 MMt/y) of LNG began operations in February 2010. Qatargas III, Train 6, came online in November 2010 with the same liquefaction capacity. In March 2011, the inauguration of Qatargas IV, Train 7, completed Qatar's monumental cycle of LNG infrastructure expansion.[11] (See Table 8.1.)

---

[11] EIA Country Analysis Briefs, *Qatar*, http://www.eia.gov/countries/cab.cfm?fips=QA.

8.3 OVERVIEW OF QATARGAS LNG PROJECTS[12]

The LNG industry in Qatar was pioneered by state-owned Qatargas and today Qatargas is the largest LNG producing company in the world, with an annual LNG production capacity of 42 million tonnes per annum (MTA). Qatargas has seven LNG trains, four of which are the largest trains in the world (known as "mega trains") each with a production capacity of 7.8 MMt/y (10.9 bcm/y). Since the establishment of Qatargas 1, Qatargas has completed an expansion program that has included the development of Qatargas 2, Qatargas 3 and Qatargas 4 projects; the Laffan Refinery; a dedicated fleet of ships; a regasification terminal; liaison Offices in Japan and China; and the move to corporate headquarters in central Doha, Qatar.

Qatargas operates all its existing trains and facilities, including Offshore, on behalf of its shareholders in all of its assets – Qatargas 1, Qatargas 2, Qatargas 3 and Qatargas 4. The North Field Bravo offshore complex is the heart of the Qatargas offshore operation. Commissioned in 1996, the main facilities in this complex include living quarters, two production facility platforms, three wellhead platforms (two of which are connected by bridges to the production facilities) and one remote platform located about 5 km away.

Qatargas's onshore operations occupy a site within Ras Laffan Industrial City. The original plant consisted of only three trains to process the natural gas from offshore into LNG, but has since expanded to accommodate an additional four trains. In 2009, Qatargas Trains 4 and 5, each with a capacity of 7.8 million tonnes per annum (MTA), started operating, bringing the combined production capacity of Qatargas to 26 MTA. In late 2010 Qatargas Train 6 started producing LNG, followed by Qatargas Train 7 in early 2011. Qatargas Trains 6 and 7 each have a capacity of 7.8 MTA. With the start of production of LNG from Qatargas Train 7, Qatargas's overall production capacity is 42 million tonnes of LNG per annum.

8.3.1 Qatargas 1

The principal market for current LNG production from Qatargas 1 is Japan with Chubu Electric Power Co., Inc. currently being the biggest

[12] Qatargas Corporate Brochure, http://www.qatargas.com/uploadedFiles/QatarGas/Media_Center/QG%20Corporate%20Brochure%20Sep%202011%20-%20EN.pdf.

Table 8.2 Japan's receiving terminals for Qatargas imports

No.	Terminal	Buyer
1	Kawagoe	Chubu Electric Power Co. Inc.
2	Yokkaichi	Chubu Electric Power Co. Inc.
3	Chita	Chubu Electric Power Co. Inc., Toho Gas Co., Ltd.
4	Niigata	Tohoku Electric Power Co., Inc.
5	Futtsu	The Tokyo Electric Power Co., Inc.
6	Higashi-Ohgishima	The Tokyo Electric Power Co., Inc.
7	Sakai	The Kansai Electric Power Co., Inc.
8	Himeji	The Kansai Electric Power Co., Inc.Osaka Gas Co. Ltd.
9	Yanai	The Chugoku Electric Power Co., Inc.
10	Mizushima	The Chugoku Electric Power Co., Inc.
11	Sodegaura	Tokyo Gas Co., Ltd.
12	Ohgishima	Tokyo Gas Co., Ltd.
13	Negishi	Tokyo Gas Co., Ltd.
14	Senboku	Osaka Gas Co. Ltd.

Source: Qatargas

customer in terms of contracted volumes. Japan receives LNG from Qatargas at 14 terminals (Table 8.2).

Qatargas 1 has a fleet of ten purpose-built vessels, each with a capacity of about 135,000 cubic meters (4.8 million cubic feet) currently in operation for the transportation of LNG from Qatargas to its Japanese buyers. Each vessel contains five Moss-Rosenberg design spherical LNG tanks with a maximum cruising speed of about 20 knots, which translates into a return voyage time between Qatar and Japan of about one month.

8.3.2 Qatargas 2 – the World's First Integrated LNG Project[13]

The Qatargas 2 (QG2) project is the world's first fully integrated value chain LNG venture. It includes two world-class LNG mega trains, each with a capacity of 7.8 million tonnes per annum (MTA) of LNG and 0.85 MTA of liquefied petroleum gas (LPG), condensate production of 90,000 bpd, a fleet of 14 Q-Flex and Q-Max ships and Europe's largest LNG receiving terminal – South Hook.

---

[13] Summaries of Qatargas 2–3 are adapted from various Qatargas publications available at http://www.qatargas.com.

In April 2009, the QG2 project was inaugurated in a special ceremony in Doha by His Highness the Emir Sheikh Hamad Bin Khalifa Al Thani, in the presence of His Royal Highness the Duke of York, Prince Andrew. The receiving terminal, South Hook, was inaugurated in the UK five weeks later in May 2009 by His Highness the Emir of the State of Qatar in the presence of Her Majesty the Queen.

QG2 includes 30 offshore wells and three platforms in Qatar's North Field. The offshore platforms are unmanned and produce 2.9 billion cubic feet of gas per day. Total production is piped to shore via two wet-gas pipelines. The LNG is processed using Air Product's proprietary APX process technology. As part of the total expansion of Ras Laffan capacity, Qatargas 2 also led to the construction of facilities for expanded LNG storage and loading, including five 145,000 cubic meter tanks and three LNG berths, a 12,000 tonne/day common sulfur system serving all Ras Laffan ventures and an export pipeline and mooring buoy for loading condensate ships some 55 km offshore.

Once the gas is processed and turned into LNG, it is loaded and shipped in a specially designed fleet of ships to markets in the United Kingdom, the United States, Asia and continental Europe. Upon arrival in the United Kingdom, the LNG is off-loaded into a purpose-built LNG receiving terminal, South Hook, at Milford Haven, Wales. The terminal is the largest LNG receiving terminal in Europe and is linked to the UK's national pipeline grid serving approximately 20 percent of the current UK natural gas demand.

8.3.3 Qatargas 3

The Qatargas 3 (QG3) project involved the construction of a new LNG mega train (Train 6) with a capacity of 7.8 million tonnes per annum (MTA). Production from Train 6 started in November 2010. The LNG produced by QG3 is transported to market on a fleet of ten ships, each with a capacity of approximately 210,000–266,000 cubic meters. The upstream platforms and infrastructure consist of three unmanned platforms, 33 wells and two subsea pipelines, all of which are shared with the Qatargas 4 project. Qatargas 3 produces 1.4 billion standard cubic feet of gas per day, delivering LNG and substantial volumes of condensate and LPG.

Qatargas 3 utilizes the same Air Products proprietary APX process technology as Qatargas 2, which helps to achieve economies of scale and integration. The Qatargas 3 and Qatargas 4 projects were developed and executed by a joint asset development team to capture synergies between the two projects.

8.3.4 Qatargas 4 – Completing the Vision

Qatargas 4 (QG4) started producing LNG in January 2011 and completed Qatargas's planned LNG expansion projects. The QG4 project involved the construction of a new LNG mega train (Train 7), similar to Qatargas 2 and Qatargas 3 with a production capacity of 7.8 MTA. The upstream platforms and infrastructure consist of three unmanned platforms (each containing 11 wells) and two subsea pipelines, which are shared with Qatargas 3.

Qatargas 4 will produce 1.4 billion standard cubic feet of gas per day, delivering LNG and substantial volumes of condensate and LPG, as well as high purity grade sulfur. QG4 utilizes the same Air Product's proprietary APX process technology as QG2 and QG3, helping to achieve economies of scale and integration not previously possible in the LNG industry. The QG3 and QG4 projects were developed and executed by a joint asset development team to capture synergies between the two projects. The LNG from QG4 will be transported to global markets via a fleet of eight Q-Flex or Q-Max ships (approximately 210,000–266,000 cubic meters capacity each) that have been constructed in Korean shipyards. Qatargas's seventh train will predominantly supply North America, the Middle East and Asia.

8.3.5 Qatargas's Role in LNG Markets

The significance and role of Qatargas in world LNG markets was highlighted after Japan lost most of its nuclear power after the March 2011 tsunami and Fukushima disaster. Qatargas was able to step in quickly to support its long-term foundation customers in Japan and committed to supply additional LNG volumes equivalent to over 60 conventional LNG cargoes to Japan. According to Qatargas, 60 conventional LNG cargoes represent approximately 4 million tonnes of LNG, which will generate approximately 28 terrawatt-hours (TWh) of electricity, or enough to meet the average electricity consumption of five million Japanese households for a year.[14] The ability to swiftly secure the additional LNG supplies for Japan demonstrates Qatargas's ability to respond to sudden changes in the market and also highlights the important role Qatargas plays as the world's largest LNG supplier.

---

[14] The Pioneer, The Magazine of Qatargas Operating Company, May 2011, Issue 132, http://www.qatargas.com/uploadedFiles/QatarGas/Media_Center/Pioneer%20132%20May%202011%20English.pdf.

8.4 RASGAS LNG PROJECTS[15]

RasGas Company Limited (RasGas) was established in 2001 as a Qatari joint stock company by Qatar Petroleum (70 percent) and ExxonMobil (30 percent). It acts as the operating company for and on behalf of various project owners including the following: Ras Laffan Liquefied Natural Gas Company Limited, which was established in 1993 to produce LNG and related products from its two trains – Trains 1 and 2; Ras Laffan Liquefied Natural Gas Company Limited (II), which was established in 2001 and owns Trains 3, 4, and 5; and Ras Laffan Liquefied Natural Gas Company Limited (3), which was established in 2005 and owns Trains 6 and 7.[16]

8.4.1 RasGas Trains 1 and 2

Trains 1 and 2 were RasGas's first onshore LNG trains, capable of producing a combined 6.6 Mta of LNG. In August 1999, the first LNG produced by Train 1 was exported to KOGAS (RasGas's first customer), marking the end of a development process that had begun back in 1971 when natural gas was first discovered off the northeast coast of Qatar.

The construction of Trains 1 and 2 set many new benchmarks for the LNG industry. Although the trains themselves are onshore, there is a substantial offshore infrastructure supplying the raw natural gas that they process. Three wellhead platforms supply gas and condensate from 15 offshore wells with a production capacity of 1.2 billion standard cubic feet per day (Bscfd). The gas and condensate extracted from the North Field are dehydrated offshore, then transported onshore together through a 32-inch diameter, 92 km export pipeline. The offshore facilities also include two 16-inch intrafield pipelines, a riser, utilities, living-quarter platforms and a flare structure.

8.4.2 RasGas Trains 3, 4 and 5

In February 2004, RasGas Train 3 began exporting LNG to India. Train 3 was built to fulfill the major part of the agreement with India's Petronet LNG to supply 5 Mta of LNG for a period of 25 years. The following year, in August 2005, RasGas commissioned Train 4 to fulfill its European sales commitments. The inauguration of RasGas Train 5

---

15 Summaries of RasGas's LNG projects are adapted from RasGas, http://www.rasgas.com.

16 RasGas, *About RasGas*, http://www.rasgas.com/L_2.cfm?L2_ID=1.

in March 2007 took Qatar's LNG production capacity to 20.7 Mta and firmly placed Qatar at the top of the list of global LNG producers. The LNG from Train 5 is largely exported to RasGas's growing portfolio of customers in Europe.

Trains 3, 4 and 5 each have a capacity of 4.7 Mta, representing a per train increase of 40 percent over Trains 1 and 2. Together the three trains form part of the first RasGas expansion project, known as RGX, and individually each marks a number of major technological and commercial milestones. Trains 3, 4 and 5 represent one of the largest and most innovative LNG projects ever completed – and set the pace for the development of RasGas's two mega trains: Trains 6 and 7.

8.4.3 RasGas Train 6

Train 6 was RasGas's first in a new generation of LNG mega trains. Each of these new trains is capable of producing 7.8 Mta of LNG. RasGas Train 6 produced its first LNG in July 2009. The following month its first cargo, a 212,000m^3 shipment of LNG, was loaded aboard the RasGas Q-Flex tanker *Al Utouriya*. The new train was formally inaugurated in October 2009 and the start-up of Train 6 boosted RasGas's LNG production capacity to 28.5 Mta.

Train 6 forms part of the RasGas Expansion Phase 2 Project (RGX2), a multi-billion dollar expansion project that brought the company's LNG production capacity to approximately 37 Mta in 2010, with the completion of seven operational trains. The new mega trains build on the success of existing RasGas expansion projects in terms of technology, design, project specifications, existing infrastructure and location. Train 6 is fed by an expansion to RasGas's offshore facilities that also serves Train 7, including two offshore 12-well platforms and two 100 km 38-inch export pipelines to the shore. Each pipeline has an eventual transmission capacity of up to 2.1 Bscfd. Helium recovered from the raw input stream will supply a second Qatar helium refining plant.

8.4.4 RasGas Train 7

Train 7 came on-stream in February 2010, and together Trains 6 and 7 have greatly extended RasGas's global reach. Like Train 6, RasGas Train 7 is one of the new generation of LNG mega trains operated by RasGas, each with a production capacity of 7.8 Mta, and is part of the RasGas Expansion Phase 2 Project (RGX2), which brought the company's LNG production capacity up to approximately 37 Mta. Building the new mega trains was an enormous accomplishment that required:

- more than 1,200 major pieces of equipment
- a 500,000 square meter site – roughly equal to 100 football fields
- 44,000 tonnes of structural steel
- up to 28,000 site workers at peak periods.

As well as generating 15.6 Mta of LNG, the two trains together produce 110,000 barrels per day of condensate and 1.5 Mta of liquefied petroleum gas (LPG). A significant amount of the LNG produced by Train 7 was expected to be shipped to the United States aboard RasGas's own chartered LNG carriers, which deliver to the new Golden Pass regasification terminal in Texas.

8.5 RAS LAFFAN INDUSTRIAL CITY[17]

In the mid-1990s, Qatar invested millions in port facilities and infrastructure to create the Ras Laffan Industrial City (RLIC) to support the exploration, storage and export of its vast natural gas resources. Covering an area of more than 295 square km, RLIC lies 80 km northeast of Doha and is strategically close to the North Field, between the Far East and Europe on the international maritime shipping route. Over 100,000 work at RLIC, which is an energy industry hub, home to international companies such as RasGas, Qatargas, ExxonMobil, Shell, Total and Dolphin Energy.

In 2006 the management team completed a master plan that anticipates the city's next 20 years of growth. Implementation is under way, with extensive development work at Ras Laffan Port to handle the anticipated increases in volume of LNG and associated liquid products. Qatar's achievement of its long-term goal of a total LNG production capacity of 77 MTPA was made possible by the facilities at RLIC, including seven RasGas LNG trains, seven Qatargas LNG trains, storage facilities and Ras Laffan Port, the largest LNG-exporting facility in the world.

In 2010 RLIC became the home of the Ras Laffan Emergency and Safety College, under an agreement between Qatar Petroleum and Texas A&M Engineering Extension Service. The new college, the first of its kind in the Middle East, will include training facilities for dealing with industrial and hazardous materials and for municipal, rescue and emergency medical services.

---

[17] Summaries of RasGas's LNG projects are adapted from RasGas, http://www.rasgas.com.

8.6 PROSPECTS FOR FUTURE LNG PROJECTS AND OTHER QATARI GAS PROJECTS

The prospects for further growth in Qatari LNG production beyond 2012 are clouded by the uncertainty created by a moratorium on new export projects, which was imposed in 2005, while the effect of existing projects on North Field reservoirs was studied. Most recently, Qatari government officials have indicated that they do not anticipate building any more LNG facilities in the near term, and that any additional capacity increases will be the result of improvements in the existing facilities.[18] It is worth noting that, in addition to its LNG projects, Qatar is the supplier for the Dolphin Project and also commissioned the world's biggest GTL plant, Oryx, in 2006.

8.6.1 Dolphin Gas Project[19]

The Dolphin Gas Project is a unique strategic energy initiative that involves production and processing of natural gas from Qatar's offshore North Field and transportation of the processed gas by subsea pipeline to the United Arab Emirates (UAE) and Oman (Figure 8.3).[20]

Dolphin Energy, in which Oxy has a 24.5 percent interest, is a development company established in Abu Dhabi to implement the Dolphin Gas Project. Dolphin Energy processes gas produced from wells offshore Qatar at the onshore plant in Ras Laffan, Qatar, the biggest initial build gas plant ever built. This massive plant processes approximately 2.5 billion cubic feet of natural gas and liquids per day. The gas is transported through a 48-inch-diameter, 230-mile-long subsea pipeline – the longest large-diameter pipeline ever built for subsea use – to markets in the UAE and Oman.[21]

The Dolphin Gas Project is the first cross-border natural gas pipeline in the Gulf Arab region and the pipeline currently exports 2 Bcf/d from

---

[18] Although Qatar's most recent train additions were originally intended with US markets as the primary target, low US gas prices due to the shale gas boom have significantly reduced US demand for imported LNG. For example, in the first ten months of 2010, the United States had imported only 33 Bcf of LNG from Qatar. The reduced demand from the US has caused Qatar to pursue LNG contracts with other countries – particularly China and India. EIA Country Analysis Briefs, *Qatar*, http://www.eia.gov/countries/cab.cfm?fips=QA.

[19] Dolphin Energy, http://www.dolphinenergy.com/.

[20] Ibid.

[21] Oxy, *Dolphin Gas Project*, www.oxy.com.

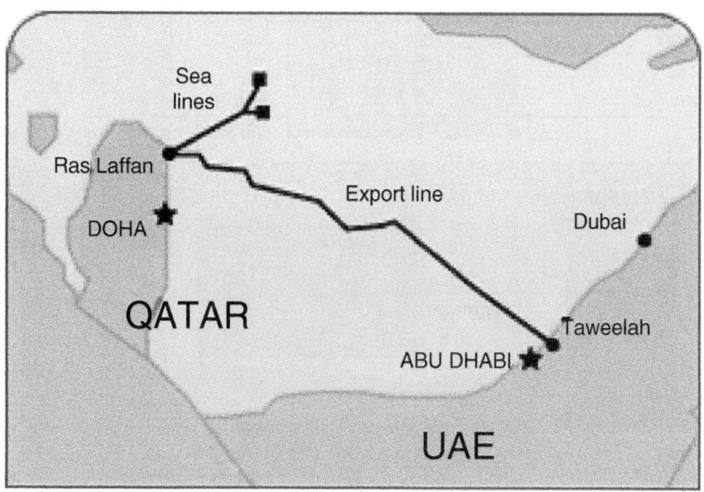

Source: www.oxy.com, Dolphin Gas Project

Figure 8.3 Dolphin Gas Project map

Qatar, though it has a design capacity of 3.2 Bcf/d. Dolphin Energy has been trying to secure additional Qatari gas to meet the rapidly growing demand for gas in the UAE, but increased supplies from Qatar are uncertain.[22]

8.6.2 Gas-to-Liquids (GTL) Projects

Gas-to-liquids (GTL) technology uses a refining process to turn natural gas into liquid fuels such as diesel, naphtha, and LPG (Figure 8.4).

Qatar is one of only three countries – South Africa, Malaysia, and Qatar – to have operational GTL facilities. Qatar's Oryx GTL plant (QP 51 percent, Sasol-Chevron GTL 49 percent) came online in 2007 but, due to initial problems, was not fully operational until early 2009. At full capacity, the Oryx project uses about 330 MMcf/d of natural gas feedstock from the Al Khaleej field to produce 30,000 bbl/d of GTL.[23]

The Pearl GTL project (QP 51 percent, Shell 49 percent) is expected to

---

22 EIA Country Analysis Briefs, *Qatar*, http://www.eia.gov/countries/cab.cfm?fips=QA.

23 EIA Country Analysis Briefs, *Qatar*, http://www.eia.gov/countries/cab.cfm?fips=QA; Oryx GTL, www.oryxgtl.com.qa.

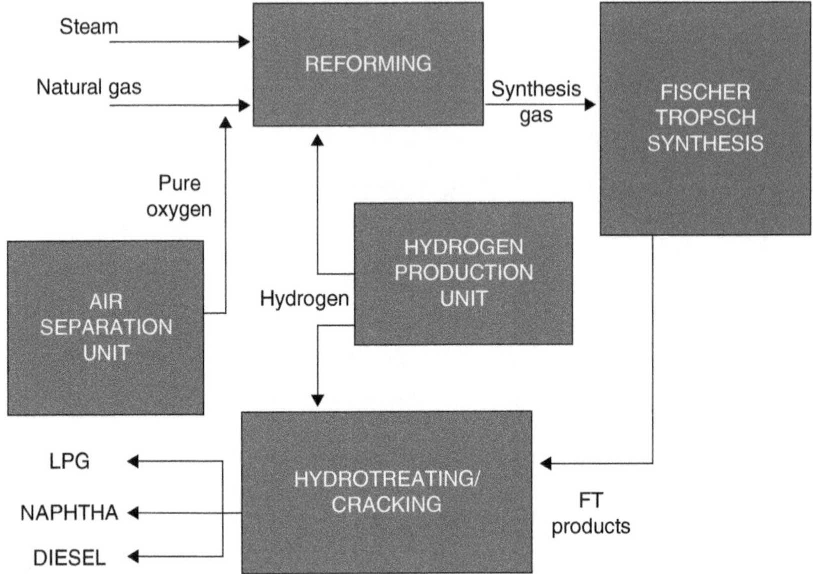

Source: Oryx GTL, www.oryxgtl.com.qa

Figure 8.4 Gas-to-liquids (GTL) process

use 1.6 Bcf/d of natural gas feedstock to produce 140,000 bbl/d of GTL products as well as 120,000 bbl/d of associated condensate and LPG.[24] The Pearl GTL project is being developed in two phases. The first phase started up in early 2011 and exported the first commercial shipment of gas oil in June 2011. Phase 2 of the plant started up in early November 2011 by bringing in sour gas from offshore wells. The whole plant is currently ramping up towards full production by mid-2012.[25] In addition to being the largest GTL plant in the world, the Pearl project will also be the first integrated GTL operation, meaning it will have upstream natural gas production integrated with the onshore conversion plant.[26]

---

24 EIA Country Analysis Briefs, *Qatar*, http://www.eia.gov/countries/cab.cfm?fips=QA.

25 Shell, *Pearl GTL – An Overview*, http://www.shell.com/home/content/aboutshell/our_strategy/major_projects_2/pearl/overview/.

26 EIA Country Analysis Briefs, *Qatar*, http://www.eia.gov/countries/cab.cfm?fips=QA.

8.6.3 Barzan Gas Project

Rather than plan additional LNG export, GTL, or pipeline projects, in recent years the Qatari government has been more concerned about ensuring adequate natural gas supplies to the domestic market, including a burgeoning petrochemical sector. In 2007, gas blocks at the North Field assigned to a canceled GTL project (a joint venture between Qatar Petroleum and ExxonMobil) were reassigned to the domestic Barzan Gas Project.[27] In January 2011, Qatar Petroleum and Exxon Mobil entered into a formal agreement to develop the Barzan gas plant to produce gas from Qatar's North Field for utilization in the domestic gas related and petrochemical industries.[28]

While not directly related to Qatar's LNG industry, the Barzan Gas Project is a significant project designed to fuel the future growth of Qatar. Over the next decade a range of factors will drive growth in the demand for energy in Qatar. For example, a new airport and seaport will be built, and major initiatives will take place in the transport, health and education sectors. In addition, a host of new facilities will need to be built in readiness for the FIFA World Cup in 2022.

When its two trains enter production, Barzan will supply gas to generate the power needed to drive the next phase of Qatar's physical and economic development. For that reason, the Barzan project is "much more than an important business partnership: it is a giant step forward, in line with Qatar's National Vision 2030, in Qatar's overall industrial and human development."[29]

In December 2011, Qatar Petroleum (QP) announced the successful closing on the financing of the US$10.4 billion Barzan Gas Project. The project will be financed with up to 30 percent equity and the remaining 70 percent by a combination of banks and export credit agencies in the form of a syndicated loan expected to total US$7.2 billion. ExxonMobil, QP's partner in the project, also provided a pro rata portion of the senior debt.

The financing of the Barzan Gas Project was the world's largest project financing to close in 2011 and one of the largest oil and gas project financings ever concluded. It comprises a commercial bank facility of US$3.34

---

27 *Energising Qatar*, RasGas Magazine, http://www.rasgas.com/rg/files/articles/RG34_En_Energising%20Qatar_LR.pdf.

28 Qatar Petroleum, *Barzan Gas Project*, http://www.qp.com.qa/en/homepage/qpactivities/epsa_dpsa/20-3767916404.aspx.

29 *Energising Qatar*, RasGas Magazine, http://www.rasgas.com/rg/files/articles/RG34_En_Energising%20Qatar_LR.pdf.

billion, a US$850 million Islamic facility and US$2.55 billion of export credit agency (ECA) financing.

The ECA financing was arranged from Japan (US$1.2 billion facility, with US$600 million provided by the Japan Bank for International Cooperation and US$600 million funded by banks and insured by Nippon Export and Investment Insurance), Korea (US$1 billion, of which US$700 million was provided by K-Exim and US$300 million was bank-funded against a K-Exim guarantee), and Italy (US$355 million funded by banks and guaranteed by SACE). The US$1 billion K-Exim facility is the largest project financing that K-Exim has ever concluded.

RasGas, the project manager and operator, will develop and operate the Barzan Gas Project on behalf of its shareholders, Qatar Petroleum and ExxonMobil, which have a 93 percent and 7 percent stake in the project, respectively.

When finished, the Barzan Gas Project will consist of onshore and offshore gas-processing facilities with the initial gas production line, Train 1, expected to be completed in 2014 and Train 2 scheduled for completion in 2015. Barzan will eventually produce 1.4 billion cubic feet of sales gas a day. The project is designed to accommodate a maximum of six trains.[30]

8.7 AUSTRALIA – THE NEXT QATAR?

In 2010, Australia was the fourth-largest LNG producer, behind Qatar, Malaysia and Indonesia.[31] In the coming years, however, Australia is poised to assume the number two spot and could even overtake Qatar as the world's largest LNG exporter. Australia is by far the most rapidly growing LNG producer in the world, with 8 of the 14 liquefaction plants under construction around the world located in Australia.[32] Stated another

---

[30] Qatar Petroleum, *QP Completes Funding for the Barzan Gas Project*, Dec. 22, 2011, http://www.qp.com.qa/en/homepage/mediacentre/news/11-12-22/Qatar_Petroleum_completes_funding_for_the_Barzan_Gas_Project.aspx.

[31] IGU World LNG Report 2010; see also EIA Country Analysis Briefs, *Australia*.

[32] Deloitte Access Economics, *Advancing Australia: Harnessing our Comparative Energy Advantage* (June 2012), available at APPEA, http://www.appea.com.au. "Commissioned by the Australian Petroleum Production & Exploration Association (APPEA) and produced by Deloitte Access Economics, the report – Advancing Australia: Harnessing our comparative energy advantage – analyses the economy-wide effect of the industry's expansion and outlines the policy framework needed to secure long-term benefits for all Australians." APPEA Media Release, *Multi-speed Economy Analysis Reveals Multiple Fast Lanes*, June

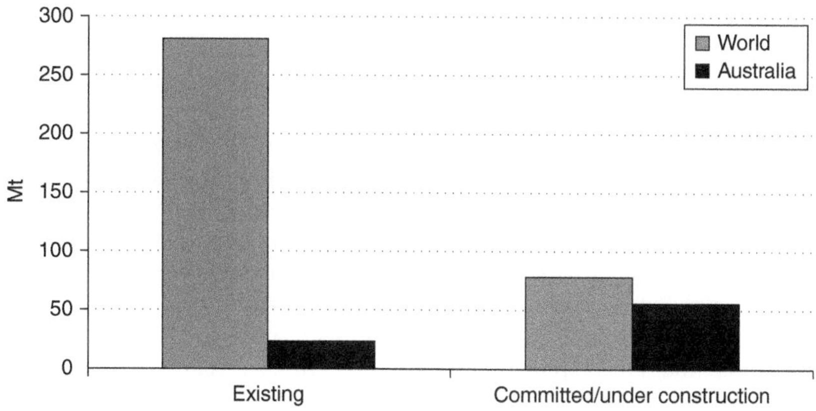

Source: BREE Gas Market Report 2012

Figure 8.5 Australian and world LNG export capacity

way, about two-thirds of the total world LNG liquefaction capacity currently under construction is located in Australia[33] (Figure 8.5).

With the massive amount of LNG capacity under construction, Australia is projected to become the world's second-largest exporter of LNG, behind Qatar, as early as 2015–16.[34] According to the Australian Petroleum Production & Exploration Association (APPEA),[35] the Australian LNG industry is targeting production of at least 60 million tonnes per annum

---

25, 2012, http://www.appea.com.au/images/stories/media/120625_dae%20report.pdf.

33 Australian Government, Bureau of Resources and Energy Economics (BREE), Gas Market Report July 2012, www.bree.gov.au, (hereinafter BREE Gas Market Report 2012). "This is the first of what is planned to be an annual report on the current state and projected developments in international and domestic gas markets." "The Bureau of Resources and Energy Economics (BREE) is a professionally independent, economic and statistical research unit within the Australian Government's Resources, Energy and Tourism (RET) portfolio. The Bureau was formed on 1 July 2011 and its creation reflects the importance placed on resources and energy by the Australian Government and the value of these sectors to the Australian economy." BREE, www.bree.gov.au.

34 BREE Gas Market Report 2012.

35 The Australian Petroleum Production & Exploration Association Ltd (APPEA) is the peak national body representing Australia's upstream oil and gas exploration and production industry. APPEA, http://www.appea.com.au/about/appea.html.

(mtpa) by 2020, up from 2009 production of 16.71 mtpa.[36] The IEA forecasts that "by 2016, Australia's LNG production capacity should reach 75.8 bcm/y (55.7 mtpa) of LNG, making it the world's second-largest LNG exporter after Qatar."[37] Moreover, "there are a number of other LNG projects under consideration which, if brought into operation, could increase Australia's LNG capacity to over 100 Mt and could make Australia the world's largest LNG exporter by the end of the decade."[38]

Australia's potential as a major LNG exporter is underpinned by several significant factors. First, Australia's petroleum, natural gas and coal reserves are so significant relative to its domestic demand that it is one of only three member states of the Organisation for Economic Co-operation and Development (OECD) that are net energy exporters.[39] According to the *Oil & Gas Journal*, as of January 2011, Australia's proven natural gas reserves are 110 Tcf, making Australia the twelfth-largest holder of conventional natural gas reserves in the world. Australia also has significant unconventional natural gas reserves such as shale gas and coal bed methane.[40]

Second, Australia's stable political environment has made it successful in attracting foreign investment to develop capital-intensive LNG projects.[41] Australia's management of oil exploration and production is divided between the states' and the Federal (Commonwealth) governments. Australia's states manage the applications for onshore exploration and production projects, while the Commonwealth shares jurisdiction over Australia's offshore projects with the adjacent state or territory. The Department of Resources, Energy and Tourism (RET) and the Ministerial Council on Energy (MCE) function as regulatory bodies over Australia's oil and gas sector. In place of a national oil company, the Australian government supports privately held Australian companies, of which the largest are Woodside Petroleum and Santos. ExxonMobil is the largest foreign oil producer; other international oil companies

---

36 APPEA, http://www.appea.com.au/industry/lng.html.

37 IEA Medium-Term Oil and Gas Markets (MTOGM) 2011 at p.246. The IEA's forecast does not include the Prelude FLNG, which is expected to start operations in 2017.

38 BREE Gas Market Report 2012.

39 Martin Ferguson, Australian Minister for Resources and Energy and Minister for Tourism, *Australia's Energy and Resources Future*, June 23, 2010, available at http://minister.ret.gov.au/MediaCentre/Speeches/Pages/Australis%27sEnergyandResourcesFuture.aspx. Canada and Norway are the other two OECD members that are net energy exporters.

40 EIA Country Analysis Briefs, *Australia*.

41 IEA MOTGM 2011.

include Shell, Chevron, ConocoPhillips, Japex, Total, BHP Billiton, and Apache.[42]

Lastly, Australia is geographically close to the growing Asia-Pacific LNG market and many of Australia's LNG projects have already been committed to buyers in this region.[43] Australia's major LNG markets are Japan, China and South Korea, while India is an important emerging market. The Australian LNG industry has been promoting LNG to the buyers in the Asia-Pacific region as a more environmentally friendly alternative to coal for power generation. According to the APPEA, for every tonne of greenhouse gas emissions generated by the production, liquefaction, and transport of Australian LNG, up to 9 tonnes of emissions are avoided in customer countries when LNG is substituted for coal in electricity generation.[44]

8.8 CURRENT AUSTRALIAN LNG PROJECTS

At the end of 2011, Australia had two producing LNG developments and several other confirmed LNG projects, with more than a dozen other LNG projects either under construction or at various stages of planning. The country's first LNG project – the North West Shelf Venture (NWSV) – began shipping LNG in 1989. The project has since grown to include five production trains and now produces up to 16.3 mtpa of LNG. The country's second LNG development – Darwin LNG – began production in 2006. This one-train project produces up to 3.5 mtpa.

In addition, six large LNG projects are currently under construction. Four are in northern Western Australia (Pluto, Gorgon, Wheatstone, and Prelude) and two in Queensland (Queensland Curtis and Gladstone LNG). (See Figure 8.6.) The Australia Pacific LNG project is currently planned for Queensland but is still in the approval process, with FID not yet taken at the time of this writing.

As well as the existing and committed projects, Australia currently has around $200 billion worth of LNG projects on the drawing board,[45] leaving no doubt that the Australian economy is in the relatively early stages of the largest oil and gas investment boom in history.[46] The existing

---

42 EIA Country Analysis Briefs, *Australia*, http://205.254.135.24/EMEU/cabs/Australia/pdf.pdf, last updated Oct. 2011.

43 IEA MTOGM 2011.

44 APPEA, http://www.appea.com.au/industry/lng.html.

45 APPEA, http://www.appea.com.au/industry/lng.html.

46 Deloitte Access Economics, *Advancing Australia: Harnessing our*

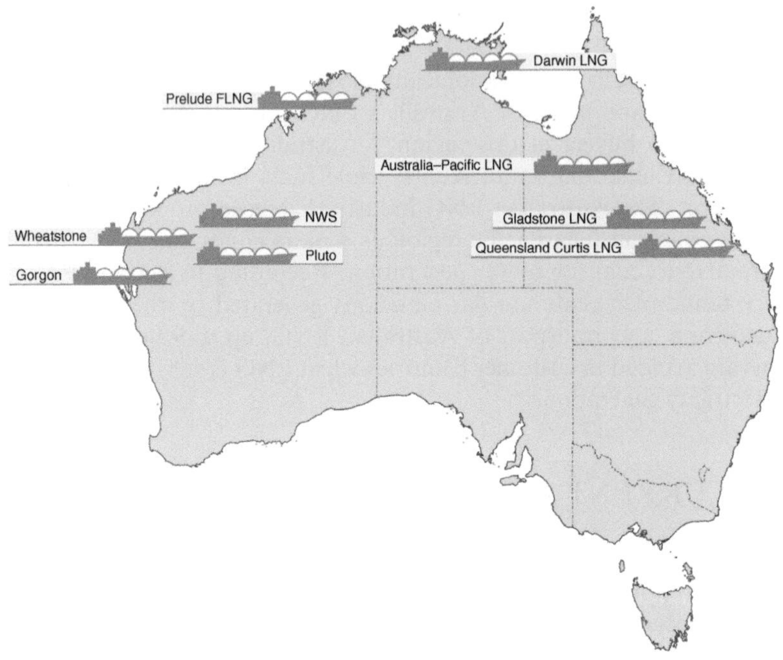

Source: APPEA

Figure 8.6 Australia's existing and committed LNG plants

and planned LNG projects involve a continual flow of LNG tankers to overseas customers and in aggregate trade terms the flow of LNG (and related liquids) from Australia represents an important contribution to Australia's export income. An example of the economic value of a typical LNG tanker is set forth in Figure 8.7.[47]

8.9 OVERVIEW OF AUSTRALIAN CONVENTIONAL LNG PROJECTS

Four new conventional LNG production projects are under construction or are in advanced planning in Australia that will use conventional gas

---

Comparative Energy Advantage, June 2012, available at APPEA, http://www.appea.com.au.

[47] Ibid.

- A standard LNG tanker holds about 140,000 cubic metres of LNG. At present production levels, around one tanker leaves Australia's shores every day destined for export markets – this will increase substantially as additional LNG terminals come online over the next five years.
- Each tanker cargo contains an export value of LNG equal to approximately $33 million at current prices, with a value added economic contribution of around $31 million.
- Each cargo provides a total tax contribution of about $8.7 million (based on industry averages), comprising $4.9 million in company tax and $3.8 million in associated production taxes such as royalties and resource rent payments.

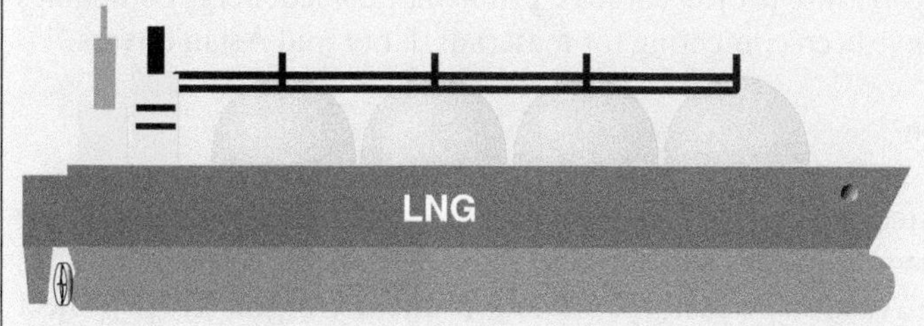

Source: Deloitte Access Economics

Figure 8.7 Economic breakdown for standard LNG tanker cargo

from offshore the northwest coast – Pluto, Gorgon, Wheatstone, and Prelude.

8.9.1 Pluto LNG[48]

The Pluto LNG project was approved for development in July 2007 and will process gas from the Pluto and Xena gas fields, located about 190 km northwest of Karratha in Western Australia, into LNG and condensate. The Pluto and Xena gas fields are estimated to contain 4.8 trillion cubic feet (Tcf) of dry gas reserves and an additional 0.25 Tcf of contingent resources.

Pluto LNG is expected to come online in 2012 with a single 4.3 mtpa LNG processing train but there are tentative plans to add further trains.[49]

---

[48] Woodside, Pluto LNG Project, available at http://www.woodside.com.au/Our-Business/Pluto/Pages/default.aspx.

[49] According to reports, Woodside Petroleum Ltd.'s A$15 billion Pluto

Woodside Energy owns 90 percent of the venture, which is supported by 15-year sales contracts with Kansai Electric and Tokyo Gas, each with a 5 percent equity ownership. The Pluto project includes an offshore platform connecting 5 subsea wells and a 112-mile pipeline to an onshore LNG facility on the Burrup Peninsula.[50]

Plans for a second train for Pluto have recently been placed on hold as additional gas supplies are sought. In July 2011, Woodside Petroleum announced that it had slowed progress on an expansion of its flagship Pluto gas-export project as it searches for more gas to support a larger Pluto project. While Woodside is betting billions of dollars on projects like Pluto that could transform the company into a major gas exporter in the Asia-Pacific region, significant delays to expanding the A$14.9 billion (US$15.8 billion) Pluto project could be problematic since energy companies in Australia have been competing for materials, labor and Asian buyers.[51]

8.9.2 Gorgon[52]

At an estimated cost of $37 billion, the Gorgon project is one of the world's largest natural gas projects and the largest single resource natural gas project in Australia's history.[53] The greenfield Gorgon LNG project is based on three liquefaction trains, each capable of producing 15 million tonnes per year. Chevron took the final investment decision on Gorgon LNG in September 2009[54] and groundbreaking for the LNG plant took place in December 2009.[55]

---

facilities began production of LNG in May 2012 with the lead-up to the first sales cargo, although no date for the shipment has been announced. Rick Wilkinson, *Woodside Starts Pluto LNG Production, Explores for more Reserves*, Oil&Gas Journal, May 7, 2012, http://www.ogj.com/articles/2012/05/woodside-starts-pluto-lng-production-explores-for-more-reserves.html.

50 EIA Country Analysis Briefs, *Australia*, http://205.254.135.7/countries/country-data.cfm?fips=AS.

51 Ross Kelly, *Woodside Delays Orders for Pluto LNG Expansion*, Wall Street Journal Online, July 19, 2011, http://online.wsj.com/article/SB10001424052702303795304576454992463750856.html.

52 Chevron Australia, *Gorgon*, http://www.chevronaustralia.com/ourbusinesses/gorgon.aspx.

53 Chevron Australia, *Business Portfolio*, http://www.chevron.com/countries/australia/businessportfolio/.

54 Chevron Press Release, *Chevron Makes Final Investment Decision to Construct Gorgon Gas Project*, Sept. 13, 2009, http://www.chevron.com/news/press/release/?id=2009-09-13.

55 Offshore Technology, *Gorgon*, http://www.offshore-technology.com/projects/gorgon/.

The Gorgon project is currently under construction with an expected completion date of 2014.[56] The massive Gorgon project is operated by Chevron and is a joint venture of the Australian subsidiaries of Chevron (approximately 47 percent), ExxonMobil (25 percent) and Shell (25 percent), together with Osaka Gas (1.25 percent), Tokyo Gas (1 percent) and Chubu Electric Power Company (0.417 percent). In the beginning of 2011, Chevron signed long-term sales agreements with Nippon and Kyushu corporations for sales of Gorgon LNG.[57]

The Gorgon project will develop the Gorgon and Jansz/Io gas fields, located within the Greater Gorgon area, about 130 km off the northwest coast of Western Australia. The huge Gorgon gas field is believed to contain 40 Tcf of natural gas and is currently Australia's largest known natural gas resource. The Gorgon project includes development of the gas fields, with connection by subsea pipelines to Barrow Island, where gas processing facilities will have production capacity of 700 Bcf per year to supply gas to Western Australia. Plans also include the construction of a 15 million tonne per annum (MTPA) LNG plant on Barrow Island. Gorgon LNG will be offloaded via a 4 km-long loading jetty for transport to international markets. (See Figure 8.8.)

The complexity and scale of Gorgon makes it one of the most technically challenging LNG projects of the 21st century.[58] For example, one of the challenges for Gorgon is what to do with the carbon dioxide produced from the project. The raw gas from Gorgon contains 12–15 percent carbon dioxide (CO_2). In order to avoid freezing in the LNG process, the carbon dioxide contained in the natural gas produced from the field will be removed prior to liquefaction. Instead of being vented into the

---

[56] Chevron Australia, *Gorgon*, http://www.chevronaustralia.com/ourbusinesses/gorgon.aspx.

[57] EIA Country Analysis Briefs, *Australia*.

[58] For example, General Electric (GE) is supplying the project with more than $1 billion in heavy duty mining and LNG equipment. In July 2012, GE announced it had won another $600 million contract to provide Chevron with "mission critical services" at Gorgon for the next 22 years. Among the equipment GE is supplying are five 2,300 ton power generation modules that will produce a combined 650 megawatts of electricity for the Barrow Island terminal and heavy-duty gas turbines that will drive Gorgon's refrigeration units that turn the natural gas into LNG. GE subsea technology such as wellheads, pipeline termination systems and other equipment will help move gas through an underwater pipeline network that measures hundreds of miles long. GE will also power Chevron's unique CO_2 sequestration system that will pipe carbon dioxide into empty gas wells. GE Reports, *Gorgontuan: New $600 Million LNG Services Deal with Chevron in Australia Brings GE's Total to $1.8 Billion*, July 20, 2012, http://www.gereports.com/gorgontuan/.

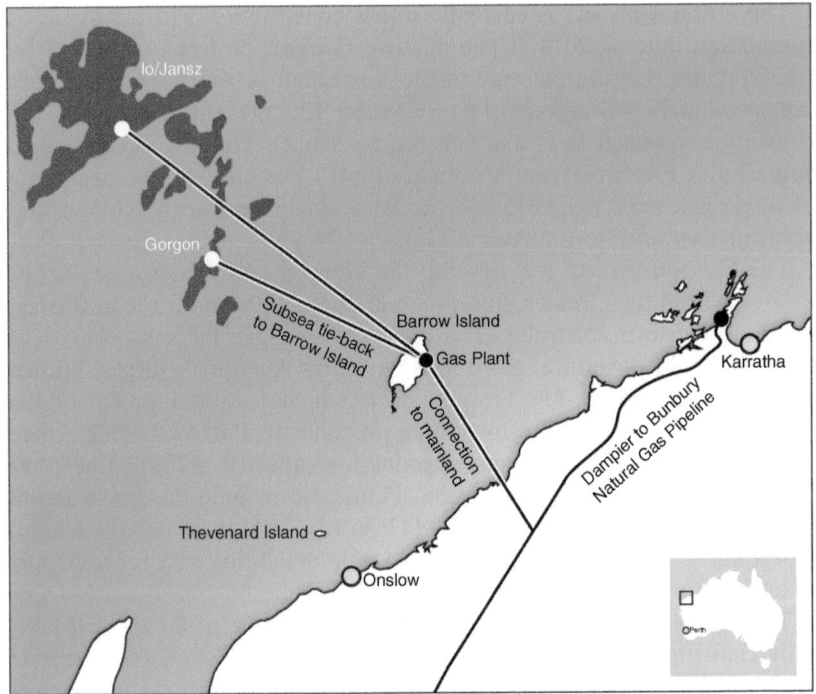

Source: Chevron Australia

Figure 8.8 Overview of the Gorgon project

atmosphere, the carbon dioxide will be injected into the Dupuy saline reservoir below Barrow Island[59] (Figure 8.9).

The CO_2 injection location is on the central eastern coast of Barrow Island near the gas processing plant. The CO_2 injection wells will be directionally drilled from surface locations to reduce the surface area impact and monitoring wells will provide a sample point within the injection area.

Once the CO_2 is injected into the subsurface, it will continue to move through the host reservoir, driven by the injection pressure and natural buoyancy until it becomes trapped. According to Chevron, there are four mechanisms that can trap injected CO_2 within the host reservoir: solution trapping;

[59] Offshore Technology, Gorgon, http://www.offshore-technology.com/projects/gorgon/.

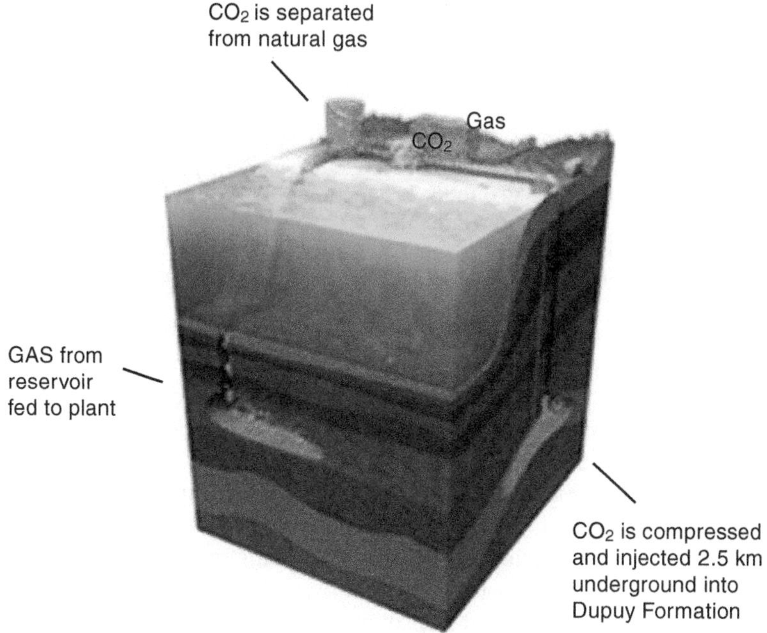

Source: Chevron Australia, Gorgon, "Carbon Dioxide Injection," http://www.chevronaustralia.com/ourbusinesses/gorgon/environmentalresponsibility/carbondioxideinjection.aspx.

Figure 8.9 Gorgon CO_2 injection

residual gas trapping; mineralogical trapping; and large-scale geometric trapping. A range of monitoring activities are planned, including routine observation and surveillance to detect CO_2 migrating to the surface.[60]

Chevron plans to inject and permanently store about 2 trillion cubic feet of CO_2 more than 8,200 feet beneath Barrow Island, which is four times more CO_2 than any previous project, making Gorgon the largest commercial scale CO_2 injection project in the world.[61] If this is successful, Chevron and Australia could become global leaders in underground carbon dioxide injection technology.[62]

---

60 Chevron Australia, Gorgon, *Carbon Dioxide Injection*, http://www.chevronaustralia.com/ourbusinesses/gorgon/environmentalresponsibility/carbondioxideinjection.aspx.

61 Chevron, Next Magazine, Issue 4, November 2010, http://www.chevron.com/documents/pdf/nextissue4.pdf.

62 Chevron Press Release, *Chevron Makes Final Investment Decision to*

According to Chevron's benchmarking analysis, the proposed LNG facility on Barrow Island has the potential to be among the most greenhouse gas efficient of its kind in the world.[63]

> The greenhouse efficiency of the LNG component of the reference case for the Gorgon Development is 0.353 tonnes of CO_2e per tonne LNG. This efficiency includes all emissions related to the production of the natural gas from the offshore fields, the energy required to inject reservoir CO_2 and the volume of reservoir CO_2 vented. As such, it represents the greenhouse efficiency of the overall LNG component of the proposed Development, not just the manufacture of LNG.[64]

Based on this data, the Gorgon project emissions will be comparable to those of the North West Shelf Train 4 and proposed Train 5 expansion, with only Oman LNG and Snohvit having appreciably better LNG greenhouse gas efficiency (Figure 8.10).

While the CO_2 injection and storage are significant components in reducing emissions from Gorgon, other efforts were also made to reduce emissions from the original 1998 concept, including replacing the offshore gas processing platform with an all subsea development, changes in LNG process technology, and improving waste heat recovery on the gas turbines resulting in a significant reduction in the use of supplementary boilers and heaters[65] (Figure 8.11).

Aside from emissions, the Gorgon project remains challenging due to numerous environmental constraints, including the fact that the selected project site of Barrow Island was declared a Class A nature reserve in 1910 and is home to many rare fauna and flora.[66] As a result, the Gorgon project underwent a rigorous and thorough environmental assessment that culminated in some of the most stringent conditions imposed on a major project anywhere in the world.[67] Nonetheless, some environmental

---

Construct Gorgon Gas Project, Sept. 13, 2009, http://www.chevron.com/news/press/release/?id=2009-09-13.

63 Chevron Australia, Draft Environmental Impact Statement/Environmental Review and Management Programme for the Gorgon Development, pp. 616–17, http://www.chevronaustralia.com/Libraries/Chevron_Documents/gorgon_ch13_LR.pdf.sflb.ashx.

64 Ibid.

65 Chevron Australia, Gorgon, Greenhouse Gas Management, *Minimising Emissions*, http://www.chevronaustralia.com/ourbusinesses/gorgon/environmentalresponsibility/greenhousegasmanagement.aspx.

66 Chevron, Next Magazine, Issue 4, November 2010, http://www.chevron.com/documents/pdf/nextissue4.pdf.

67 Chevron Australia, Gorgon, *Environmental Approvals*, http://www.chev

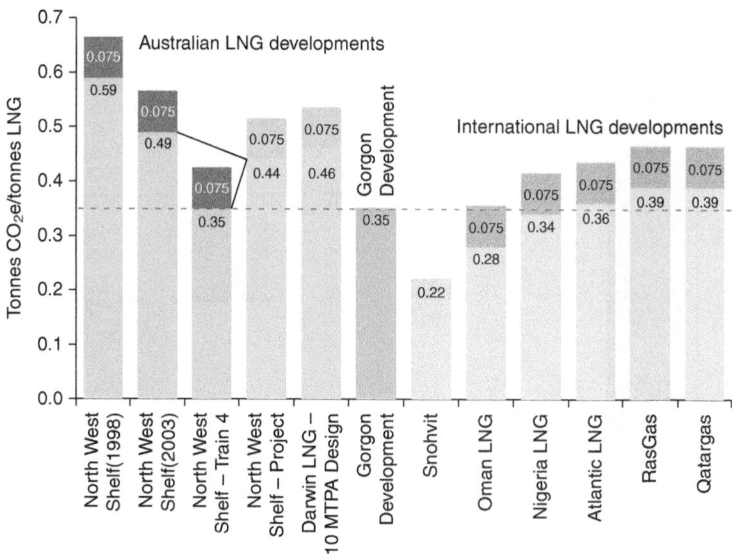

Source: Chevron Australia, *Draft Environmental Impact Statement/Environmental Review and Management Programme for the Gorgon Development*, p. 617, http://www.chevronaustralia.com/Libraries/Chevron_Documents/gorgon_ch13_LR.pdf.sflb.ashx.

Figure 8.10 Benchmarked greenhouse gas efficiency

groups have challenged the project from its inception citing numerous environmental concerns and, in particular, concerns about the project's impact on the Australian flatback sea turtle, which nests on the beaches near the project.[68] While some are opposed to the project, significant opposition has failed to materialize, in part due to Gorgon's remote Western Australian location and the general acceptance of lucrative energy and mining projects.[69]

---

ronaustralia.com/ourbusinesses/gorgon/environmentalresponsibility/environmentalapprovals.aspx.

[68] Sea Turtle Restoration Project, *Threats to Endangered Sea Turtles from Oil and Gas Exposed in Chevron Alternative Annual Report and Ads: International Activists Call for Halt to Natural Gas Facilities on Turtle Beaches*, http://www.seaturtles.org/article.php?id=1622. See also The Wilderness Society of Western Australia, www.wilderness.org.au and Save The Kimberley, www.savethekimberley.com.

[69] As one reporter has noted, "It's frankly a challenge to rally opposition to a project on an island almost no one is allowed to set foot on in a territory four times the size of Texas but home to only 2 million people. Mining for iron, copper, gold

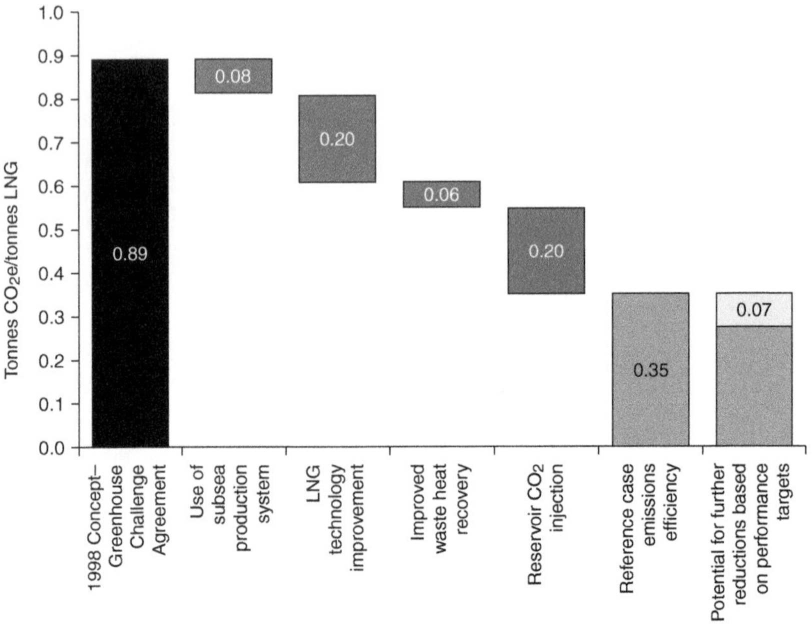

Source: Chevron Australia

Figure 8.11 Gorgon greenhouse gas emissions efficiency improvements

8.9.3 Wheatstone[70]

In conjunction with the Gorgon project, the Chevron-led Wheatstone project in northwestern Australia reinforces Chevron's and Australia's position as leading LNG suppliers in the Asia-Pacific region. Chevron is the operator of the Wheatstone project and holds an approximate 73 percent interest in the project along with Apache (23 percent), Kuwait Petroleum (KUFPEC) (7 percent) and Shell (6.4 percent). KUFPEC

---

and coal is the keystone of the Western Australia economy. Flush with cash from mining and energy projects (a unionized welder can bring home $500,000 a year), the citizenry is relatively sanguine about environmental damage." Christopher Helman, *Drilling into Eden*, Forbes Magazine, August 22, 2011, http://www.forbes.com/forbes/2011/0822/features-chevron-barrow-island-natural-gas-drilling-eden.html.

[70] Chevron, Wheatstone, http://www.chevronaustralia.com/ourbusinesses/wheatstone.aspx.

and Apache joined the project as gas suppliers from their nearby Julimar and Brunello fields, which will extend the life of the project. Wheatstone is supported by long-term LNG sales contracts with TEPCO and KOGAS.[71]

Gas will be processed at an onshore facility located at Ashburton North, 12 km west of Onslow in Western Australia's Pilbara region. The foundation project will include two LNG trains with a combined capacity of 8.9 million tonnes per year and a domestic gas plant with a targeted completion date of 2016.[72] While some concerns were raised in 2011 about potential cost increases and labor shortages in Australia that could negatively impact planned LNG projects, in September 2011 Chevron took a final investment decision (FID) on the A$29 billion Wheatstone project.[73] Chevron indicated that construction on Wheatstone was to begin immediately with an estimated start-up date of 2016.[74]

According to analysts at Wood Mackenzie,[75] Australia could have 73 million tonnes a year of capacity once all the projects that are producing or have been sanctioned, including Wheatstone, are taken into account. This compares with 77 million tonnes a year of capacity in Qatar. However, if the Japanese Ichthys project is sanctioned by the end of 2011 as expected, Australia could achieve 81.4 millon tonnes a year of capacity by 2017–18, surpassing Qatar.[76]

With projects like Wheatstone, the Australian economy is likely to remain strong if recent years are any indication. For example, in the first half of 2011, higher prices and greater volumes pushed the value of Australia's LNG exports up 34 percent to A$10.5 billion. The Australian government's Bureau of Resources and Energy Economics has forecast a further 11 percent increase to A$11.6 billion in 2011–12.[77]

---

[71] EIA Country Analysis Briefs, *Australia.*

[72] EIA Country Analysis Briefs, *Australia.*

[73] Chevron, Wheatstone, Press Release dated Sept. 26, 2011, *Chevron Gives Wheatstone Project Green Light*, http://www.chevronaustralia.com/media/media-statements/wheatstone.aspx?NewsItem=85dc19d5-a958-47f9-a168-670fef9ac5f1.

[74] Ibid.

[75] "Wood Mackenzie is the most comprehensive source of knowledge about the world's energy and metals industries." Wood Mackenzie website, http://www.woodmacresearch.com/cgi-bin/wmprod/portal/energy/aboutEnergy.jsp.

[76] Peter Smith and Sylvia Pfeifer, *Chevron Approves A$29bn Australia LNG Project*, Financial Times, Sept. 26, 2011, http://www.ft.com/intl/cms/s/0/bf12c3aa-e813-11e0-9fc7-00144feab49a.html#axzz1cm4TEeil.

[77] Ibid.

8.9.4 Prelude – the World's First FLNG Project

Australia also seems likely to become the first country in the world to host a floating LNG liquefaction project. Shell Development Australia will use a specially designed ship – the largest floating structure ever built – with LNG production facilities to develop its Prelude gas field off the north coast of Western Australia. This project is expected to begin producing 3.5 mtpa from 2016 or 2017.[78] For additional discussion about FLNG and the Prelude project, see Chapter 12.

8.10 THE NEXT WAVE OF AUSTRALIAN LNG PROJECTS

There are several other LNG projects in the planning stage in Australia with some aimed at taking FID in late 2011 or later[79] (Table 8.3).

Some projects have secured buyers and are in advanced stages of obtaining permits. For example, Woodside's Browse LNG is in talks with Asian customers and aims to finalize contract terms in 2011.[80] The Australian state environmental approval for Browse was granted in December 2010 and is now proceeding to the Australian Federal government approval process. Shell and PetroChina are jointly developing the CS CSG project after Arrow Energy's board of directors and shareholders approved a takeover bid in 2010.[81]

The Bonaparte project is a 2.7 bcm floating LNG project that GDF Suez expects to take FID on in 2014. The pre-FEED contract was awarded in 2011 and this project could reinforce GDF Suez's position in the Asia-Pacific region after signing short-term contracts with Kogas and CNOOC.[82] Woodside's Greater Sunrise project seems to be facing challenges as East Timor continues to insist that the project be built on East Timor's national territory, although the project sponsors are looking at a floating option due to project economics.[83]

The Icthys project, located off the northwest coast in the Browse Basin, is now expected to begin construction in early 2012. The project is led by Japan's INPEX (74 percent) and Total (26 percent). A 528-mile undersea

---

[78] IEA MTOGM 2011.
[79] Ibid.
[80] Ibid.
[81] Ibid.
[82] Ibid.
[83] Ibid.

Table 8.3 Emerging Australian LNG projects

Project	Partners (share)	Expected FID	Capacity (bcm)	LNG buyers	Sales volume (bcm)	Start-up	CAPEX ($b)*
Wheatstone	Chevron 80%, Apache 13%, KUFPEC 7%	2011	11.7	Tokyo Electric, KOGAS, Kyushu Electric	5.6 2.7 1.1	2016	29.0
Australia Pacific LNG	ConocoPhillips 50%, Origin 50%	2011	6.1 plus	Sinopec	5.8	2015+	19.6
Pluto Train2	Woodside 100%	2011	6.5	None yet	N/A	2016	15.3
Ichthys	Inpex 76%, Total 24%	2011	10.9	None yet	N/A	2016+	30.8
Browse	Woodside 50%, Chevron 16.7%, BP 16.7%BHP and Shell 8.3% each	2012	16.3	Osaka Gas, CPC	1.6 4.1	2016+	
CS CSG	Shell 50%, PetroChina 50%	2012	10.8 plus	Petro China, Shell	5.4 5.4	2016+	
Greater Sunrise	Woodside 33.4%, Shell 26.6%, ConocoPhillips 30%, OG 20%	2012	6.8	None yet	N/A	2016+	
Bonaparte	GDF Suez 60%, Santos 40%	2014	2.7	None yet	N/A	2018	

Sources: IEA MOTGM 2011; Deloitte Access Economics Study*

pipeline will connect the gas fields to a new export LNG terminal to be built near Darwin. The project is expected to come on-stream in 2016, with an expected production of 380 Bcf of LNG and 19 million barrels of LPG per year, as well as 100,000 bbl/d of condensate. The Australian government gave environmental approval in June 2011 and the final investment decision is expected in the fourth quarter of 2011.[84]

8.11 AUSTRALIAN LNG PROJECTS BASED ON CSG[85]

Australia's energy industry is making multibillion-dollar investments pioneering the development of coal seam gas (CSG) into LNG. CSG is a form of natural gas (methane) found in coal seams, rather than in the sandstone reservoirs that hold conventional natural gas. The amount of gas held in coal seams may be several times greater than the current reserves of natural gas. Much of the attention is focused on Queensland since it is one of the most abundant regions for coal seam gas in the Pacific Basin.[86] In Australia, approximately 20 percent of natural gas currently being used comes from coal seam gas.[87]

There are currently four LNG projects based on CSG planned for development in Queensland. Two of these schemes – Gladstone LNG and Queensland Curtis – have already received all necessary federal and state approvals and the project proponents have made positive final investment decisions.

The Gladstone project,[88] located onshore Queensland, will be the world's first major CSG to LNG operation. The project is a joint venture between Santos (30 percent), Petronas (27.5 percent), Total (27.5 percent), and KOGAS (15 percent) and, so far, a long-term sales agreement with KOGAS has been signed.[89] The project will initially produce 7.8 million tonnes per annum (mtpa) of LNG, with a maximum potential production of 10 mtpa. Construction is expected to start in 2011, with the first cargoes scheduled to be exported from 2015.[90]

The Queensland Curtis LNG (QCLNG) project is being developed

---

84 EIA Country Analysis Briefs, *AustraliaI.*
85 APPEA, http://www.appea.com.au/industry/csg/introduction.html.
86 QCLNG Project, http://www.qgc.com.au/01_cms/details.asp?ID=5.
87 Gladstone LNG, http://www.glng.com.au/Content.aspx?p=57.
88 Gladstone LNG, http://www.glng.com.au.
89 IEA MOTGM 2011.
90 Gladstone LNG, http://www.glng.com.au/Content.aspx?p=55.

by QGC, a BG Group business and a leading Australian coal seam gas explorer and producer.[91] Queensland Curtis LNG is a priority project for QGC and involves expanding exploration and development in southern and central Queensland and transporting gas via a 540 km buried pipeline to Curtis Island near Gladstone, where it will be liquefied. In 2009, QGC produced approximately 20 percent of Queensland's natural gas. QGC has agreements in place to supply China, Chile and Singapore.[92]

The project's first stage will consist of two LNG trains, at the Curtis Island plant. These trains, which have a design life of at least 20 years, will produce a combined 8.5 million tonnes of LNG a year. The QCLNG site can accommodate an expansion to 12 million tonnes of LNG a year. Construction is scheduled to begin in 2010, with the first LNG delivery expected in 2014. The proposed LNG plant will occupy less than 1 percent of Curtis Island and will be separated from public areas by an environmental protection zone.[93]

Two other large CSG–LNG projects, Australia Pacific LNG and Arrow LNG, are also being planned for Queensland, although these have not yet finished working their way through the federal and state approvals processes and FIDs have not yet been made on these projects. However, the project proponents are actively developing gas fields and pipelines and are marketing their gas to potential buyers.

The four CSG–LNG projects will pipe the gas from the CSG fields of inland Queensland to the port city of Gladstone. Plants at Gladstone will then liquefy the natural gas into LNG. The Queensland Government expects these projects will create around 18,000 jobs, increase State Domestic Product by 1 percent, and generate around $1 billion per annum in state revenue.

As an end-use product, CSG–LNG is the same as conventional natural gas and is used to generate electricity and to power natural gas appliances such as heaters and stoves. CSG already comprises about 90 percent of Queensland's gas production and the industry is forecast to become a substantial contributor to New South Wales (NSW) energy production. CSG companies are undertaking appraisal and early-stage development work in several traditional coal-mining regions of NSW, as well as in the Gunnedah and Clarence-Moreton basins in northern NSW.

---

91 QCLNG Project, http://www.qgc.com.au/01_cms/details.asp?ID=5.
92 Ibid.
93 Ibid.

8.11.1 CSG and Economic and Regional Development[94]

The Australian government believes that CSG can play an important role in regional development by bringing infrastructure and investment to parts of rural and regional Queensland and NSW, and by providing new jobs and strengthening and diversifying regional economies. The CSG industry has already become a major source of employment and income in parts of inland south Queensland. Several Queensland and NSW CSG companies have built or are planning to build significant gas-fired power stations and associated infrastructure near regional centers. Over the next two decades, the CSG industry is likely to spend as much as $6 billion on NSW gas field developments and a further $3.9 billion of operating spending, according to a report commissioned by APPEA.[95]

According to APPEA, the Commonwealth Scientific and Industrial Research Organisation (CSIRO)[96] has estimated eastern Australia's CSG resources to be more than 250 trillion cubic feet; this is enough to power a city of one million people for 5,000 years and more than enough to feed Australia's domestic and export CSG projects for many decades to come.[97] Nonetheless, the CSG industry has recently come under greater scrutiny as cost blow-outs have been announced amid doubts about the CSG industry's ability to supply enough gas to the three LNG export projects based on CSG.[98]

8.12 ENVIRONMENTAL ISSUES RELATED TO CSG

Although the Australian government is optimistic about the benefits of CSG, numerous environmental concerns have been raised about CSG

---

94 APPEA, http://www.appea.com.au/industry/csg/introduction.html.

95 APPEA, *Economic Significance of Coal Seam Gas in New South Wales.*

96 The Commonwealth Scientific and Industrial Research Organisation (CSIRO) is Australia's national science agency and one of the largest and most diverse research agencies in the world. CSIRO, http://www.csiro.au.

97 APPEA, http://www.appea.com.au/csg/about-csg/what-is-csg.html.

98 Matt Chambers, *CSG Industry in Crisis as Gas Plays Face $20bn Blowout*, The Australian, July 7, 2012, http://www.theaustralian.com.au/business/mining-energy/csg-industry-in-crisis-as-gas-plays-face-20bn-blowout/story-e6frg9df-1226419419888, noting that "QUEENSLAND'S burgeoning $60 billion coal-seam gas export industry is facing growing doubts about its ability to supply enough gas to three giant export plants due to start in Gladstone in 2014, and to contain development costs."

and opposition to Australia's CSG industry appears to be growing with calls for new legislation to regulate the industry. A recent poll released by Australia's Green party, which is strongly opposed to the industry, showed that 70 percent of respondents want coal seam gas mining banned in urban areas, while 68 percent want a moratorium on CSG in Australia until the effects on the environment are more fully known.[99]

The concerns raised about CSG are quite similar to the concerns raised about the production of shale gas (see Chapter 10). First, coal seam gas production involves hydraulic fracturing, which requires large volumes of water, sand and chemicals to be injected into the well at high pressure to free the gas trapped in the coal seams. Opponents say the process could cause the water table in key farming areas to drop and may also cause water contamination. Concerns have also been raised that the extensive dredging required to build the CSG projects could harm sites including Australia's Great Barrier Reef.[100]

Landowner rights are another point of contention. Under Australian law, landowners have surface land rights, but the government owns the mineral rights. Coal seam gas companies have said they prefer to make voluntary agreements with landowners rather than force their way onto farmland to gain access to resources, and companies typically negotiate land access agreements with landowners.[101] While the terms of these agreements are usually confidential, Queensland AgForce, a group that represents agricultural interests, has suggested that compensation varies widely, ranging from A$500 (U$522.77) to A$5,000 a year per well.[102]

In terms of restrictions, both Queensland and New South Wales have banned the chemical combination known as BTEX, which has been used for hydraulic fracturing in the United States. These states have also ruled out using evaporation ponds to dispose of the extremely saline water produced by coal seam gas production. Other restrictions vary from state to state.[103] In Queensland, State Premier Anna Bligh recently banned mining and coal seam gas exploration within 2 km of urban areas with populations over 1,000. Farmers have pushed to get the ban extended to

---

99 Rebekah Kebede, *Q+A – Regulating Australia's Coal Seam Gas Sector*, Reuters, Aug. 26, 2011, http://www.reuters.com/article/2011/08/26/australia-coalseam-idUSL4E7JO1ZF20110826.

100 Ibid.

101 Ibid.

102 Ibid.

103 Ibid.

farmland, but experts say Bligh is unlikely to do so, given the huge amount of investment and jobs involved.[104]

New South Wales, where the coal seam gas industry is still relatively new, has extended an existing moratorium on hydraulic fracturing through the end of 2011.[105] The Australian Green party is currently leading the charge to impose more stringent regulations on CSG development and recently introduced legislation that would allow farmers to refuse to have coal seam gas drilling on their land.[106] Analysts say the recent tightening of regulations in Queensland is unlikely to affect projects significantly, but the growing public opposition to the coal seam gas sector could be a factor down the road and might lead to project delays and higher costs.[107]

For its part, the Australian CSG industry contends that the CSG industry has been around for decades and it has been a significant source of gas production in Queensland for more than ten years.[108] Moreover, industry and government have a detailed knowledge of the hydraulic fracturing process and of the chemicals being used. Water and sand comprise more than 99 percent of the volume of fracking fluid and companies must identify the chemicals being used in any fracking operation and detail any likely interactions with the water and rock formations in the area being fracked.

In terms of landowners' issues, the industry notes that by law companies must notify landholders before entering their land. They are generally required to give at least ten business days' notice before entering land for any form of activity. For any activities entailing some form of disturbance of the land, companies must have a negotiated conduct and compensation agreement with the landholder.

The industry also maintains that wells are constructed in a way that ensures there can be no migration of gas to neighboring bores and aquifers. A recent wellhead safety report in Queensland surveyed more than 2,900 wells and found only 35 leaks of concern. These leaks posed no threat to human health but the companies were required to rectify them.

In terms of arguments that the CSG production activity could cause the Great Artesian Basin to dry up, threatening farming activity, the industry maintains that given the huge volume of water held in the Great Artesian

---

104 Ibid.

105 Ibid.

106 Ibid.

107 Rebekah Kebede, *Insight: Australia's Coal-seam Gas Industry Feels Political Heat*, Reuters, Nov 16, 2011, http://www.reuters.com/article/2011/11/16/us-australia-gas-repeat-idUSTRE7AF1QJ20111116.

108 APPEA, http://www.appea.com.au/industry/csg/introduction.html.

Basin, the actual volume predicted to be extracted by CSG production is a tiny fraction of this total. If there are any localized impacts on pressure in some waterbores, CSG producers are required to make good any loss of access to water, ensuring a continued reasonable supply of water to landholders.

9. New players and projects – Russia, Peru, Yemen, and Papua New Guinea

9.1 OVERVIEW

The projects in Qatar demonstrate how vision and technology can transform a country into the world's largest LNG exporter. Similarly, the huge waves of LNG projects underway in Australia demonstrate vision and technologies that will propel the LNG industry to even greater achievements. At the same time, however, the Australian projects are challenged by many environmental and emissions concerns and, for that reason, these projects will no doubt be studied well into the 21st century.

While Qatar and Australia are global leaders in LNG, a number of other countries have recently stepped onto the world LNG stage for the first time, including Russia, Peru, Yemen, and Papua New Guinea. These new projects present numerous opportunities and challenges that differ significantly from those presented in Qatar and Australia. For example, in Russia, the challenges of working through an opaque bureaucracy that requires many layers of approvals can delay or even derail projects. Russia also demonstrates the difficulties of working in extremely harsh Arctic conditions. The Peru, Yemen, and Papua New Guinea LNG projects highlight the opportunities and challenges associated with bringing natural gas to the global market while balancing economic growth, environmental protection and social development demands.

9.2 RUSSIA – THE WORLD'S LARGEST NATURAL GAS RESERVES

Russia is a key player in the global energy markets and is a major exporter of oil and natural gas. Over the past decade, most of Russia's economic growth has been driven by its energy exports. Natural gas exports remain a major driver of Russia's gas industry since most of the industry's revenues come from exports, whereas natural gas for the domestic market is sold at regulated prices.[1] According to the IEA's most recent *World Energy Outlook* (WEO-2011), which featured a special section on Russia, the energy policy choices made by Russia in the coming years will have major implications on global energy security and environmental sustainability.[2]

The state-run Gazprom dominates Russia's upstream energy sector and Gazprom controls more than 65 percent of Russia's proven natural gas reserves, with additional reserves being controlled by Gazprom in joint ventures with other companies. In addition, almost 90 percent of Russia's total natural gas output is produced by Gazprom. While independent producers have gained some footing in recent years, with producers such as Novatek and LUKoil contributing increasing volumes to Russia's production, the upstream sector remains fairly limited to independent producers and other companies.[3] Gazprom's position is further cemented by its legal monopoly on Russian gas exports (discussed below).

9.2.1 Russia's Natural Gas Reserves

According to the *Oil & Gas Journal*, Russia holds the world's largest natural gas reserves, with 1,680 trillion cubic feet (Tcf) or approximately 48 trillion cubic meters (tcm)[4] (Figure 9.1). Russia's reserves account for about a quarter of the world's total proven reserves, with the majority of these reserves located in Western Siberia, with the Yamburg, Urengoy, and Medvezh'ye fields alone accounting for about 45 percent of Russia's total reserves.[5]

---

1 Interfax, Corporate News Agency, *Russia's Gas Industry in 2010–2011*, July 2011. Interfax Group (www.interfax.com) is a leading provider of information on Russia, China and the emerging markets of Central Asia. In September 2011, Interfax launched a new natural gas service, Interfax Global Energy Services (www.interfaxenergy.com), and pdf publication, *Natural Gas Daily*, which provides global coverage of every phase of the gas chain. Interfax's information on Russia's natural gas industry was particularly helpful for this book.

2 IEA WEO-2011, Chapter 7, *Russian Domestic Energy Prospects*.

3 EIA Country Analysis Brief, *Russia* (Nov. 2010), available at http://www.eia.gov/EMEU/cabs/Russia/.

4 Ibid. Most recently, the IEA has indicated that proven reserves of natural gas in Russia are generally quoted to be about 45 tcm with varying estimates including: 45 tcm, BP, 2011; 48 tcm in O&GJ; 46 tcm in Cedigaz, 2011; and 48 tcm in Government of Russia, 2009. IEA WEO-2011. A useful website for metric conversions is www.metric-conversions.org.

5 EIA Country Analysis Brief, *Russia* (Nov. 2010), available at http://www.eia.gov/EMEU/cabs/Russia/.

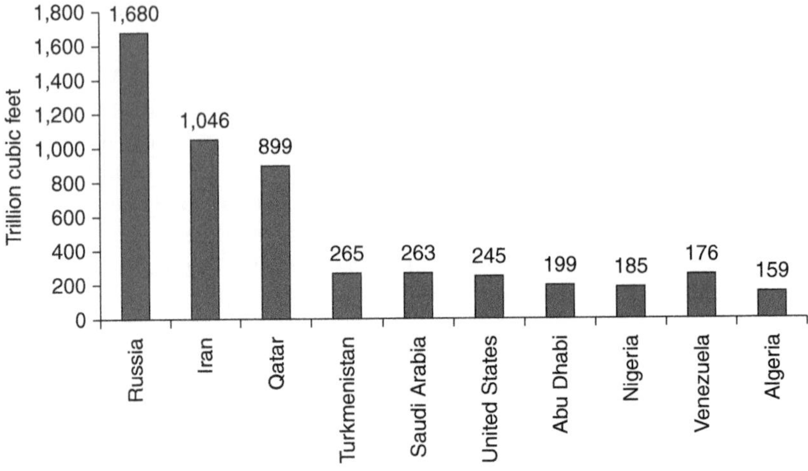

Source: US EIA

Figure 9.1 Top proven natural gas reserve holders, 2010

9.2.2 Exploration and Production

Russia has been the world's long-standing leader in terms of natural gas production. However, in 2009, and largely due to the increase in shale gas production, the United States overtook Russia and became the world's largest natural gas producer with production of 21 Tcf versus Russia's production of 19.3 Tcf, although Russia was still the world's largest exporter of natural gas (7.3 Tcf).[6] According to the *BP Statistical Review of World Energy 2012*, the United States remains the world's top natural gas producer for the third consecutive year. (See Chapter 11, Table 11.1.)

9.2.3 Russian Natural Gas Flaring

Natural gas associated with oil production is often flared. According to the US National Oceanic and Atmospheric Administration, Russia flared an estimated 1,432 Bcf of natural gas in 2008, the highest of any country in the world (Figure 9.2). The Russian government has taken steps to reduce natural gas flaring and set a target of 95 percent utilization of associated gas by 2012.[7]

---

6 Ibid.
7 Ibid.

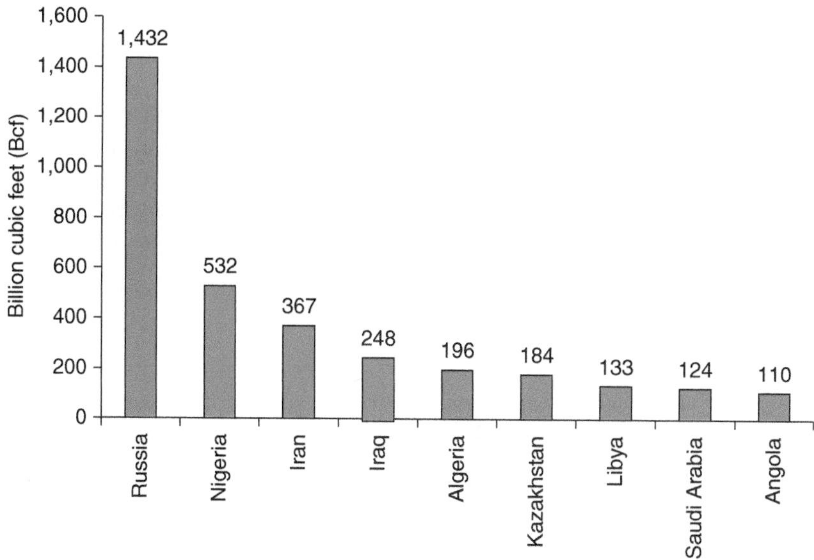

Source: US EIA

Figure 9.2 Volumes of gas flared, 2008

9.2.4 Natural Gas Exports and Export Disputes

Russia supplies Europe with most of its natural gas and, in particular, supplies significant amounts of natural gas to the Commonwealth of Independent States (CIS).[8] Germany is the biggest export market with a 33.4 percent share.[9] In addition, Gazprom, through its subsidiary Gazexport, exports natural gas to Turkey, Japan, and other Asian countries.[10] According to published data, in 2009 Russia exported more than 6.5 Tcf of natural gas, which includes 4.5 Tcf to Eastern and Western Europe and 2.2 Tcf to CIS countries.[11] The share of Russia's natural gas exports as a percentage of Russia's gas output was the highest in 1999, amounting to 33.5 percent.[12] While this percentage has dropped in recent

---

8 Ibid.

9 Interfax, Corporate News Agency, *Russia's Gas Industry in 2010–2011*, July 2011.

10 EIA Country Analysis Brief, *Russia* (Nov. 2010), available at http://www.eia.gov/EMEU/cabs/Russia/.

11 Ibid.

12 Interfax, Corporate News Agency, *Russia's Gas Industry in 2010–2011*, July 2011.

years (2010 – 26.8 percent), Russia/Gaszprom still remains the largest natural gas supplier in the European market with a market share of 23 percent.[13]

Most of Europe's natural gas imports from Russia transit through Ukraine and Belarus via pipeline. In recent years, pricing disputes between Russia and these natural gas hubs have resulted in natural gas being cut off to much of Europe. As a result, most European countries are seeking alternate sources of natural gas and alternate pipeline routes to ensure security of natural gas supplies in the future.[14]

9.2.5 Russia's Natural Gas Pipelines

As in the upstream operations, Gazprom also dominates Russia's natural gas pipeline system. There are currently nine major pipelines in Russia, seven of which are export pipelines. The Yamal-Europe I, Northern Lights, Soyuz, and Bratrstvo pipelines all carry Russian gas to Eastern and Western European markets via Ukraine and/or Belarus. These four pipelines have a combined capacity of 4 Tcf. Three other pipelines, Blue Stream, North Caucasus, and Mozdok-Gazi-Magomed, connect Russia's production areas to consumers in Turkey and the former Soviet Union (FSU) republics in the east.[15] As noted below, there are also a number of pipeline expansion projects underway in Russia.[16]

Yamal-Europe II

A proposed expansion of the current 1 Tcf Yamal-Europe I, which carries natural gas from Russia to Poland and Germany via Belarus, would add another 1 Tcf of capacity. Gazprom and Poland currently disagree on the exact route of the second branch as it travels through Poland, with Gazprom wanting a route via southeastern Poland to Slovakia and on to Central Europe, while Poland wants the branch to travel through its own country and then on to Germany.[17]

---

13 Ibid. Other major natural gas suppliers to Europe include Norway (19%), Algeria (10%), and Qatar (6%). Ibid.

14 EIA Country Analysis Brief, *Russia* (Nov. 2010), available at http://www.eia.gov/EMEU/cabs/Russia/.

15 Ibid.

16 Ibid.

17 Ibid.

South Stream

The first component of the South Stream project plans to send natural gas from the same starting point as the Blue Stream pipeline at Beregovaya for 560 miles under the Black Sea, achieving a maximum water depth of over 6,500 feet. The second, onshore component will cross Bulgaria with two alternatives: one directed towards the northwest, crossing Serbia and Hungary and linking with existing gas pipelines from Russia; and the other directed to the southwest through Greece and Albania, linking directly to the Italian network. As a result of the Russia–Ukraine disputes, the pipeline will be constructed through Turkey's waters, avoiding Ukraine's territory altogether. Gazprom expects the pipeline to be completed by 2015.[18]

Nord Stream pipeline

A northern pipeline extending over 2,000 miles from Russia to Germany via the Baltic Sea was initially approved in 2005. Once completed, the pipeline will be the longest subsea pipeline, with a capacity to transport 1.9 Tcf of natural gas.[19] Although the project has suffered delays, the first phase of the Nord Stream gas pipeline was launched in November 2011, with some calling into question the project economics for the pipeline as well as Europe's need for the new pipeline.[20] One consultancy report predicts that gas price growth will stop in 2015 as a "gas glut" hits Europe, thereby obviating the need for the Nord Stream pipeline.[21]

By far, Germany appears to be the primary beneficiary of the Nord Stream pipeline as an eventual 55 bcm/y of extra gas (Nord Stream's total capacity once fully operational) will flow directly to Germany, with no transit states and to the benefit of local German companies Wintershall and E.ON Ruhrgas, which are shareholders of Nord Stream. This should certainly help German energy security, especially since Germany has replaced planned nuclear power with gas-fired power.[22]

---

[18] Ibid.

[19] Ibid.

[20] Josh Posaner, *What Does Nord Stream Mean for Europe?* Interfax Global Energy Services, Natural Gas Review, Dec. 1, 2011, available at http://interfaxenergy.com/natural-gas-news-analysis/russia-and-the-caspian/nord-streams-role-in-europe-in-question/.

[21] Ibid., citing a report unveiled in Germany by consultancy firm AT Kearney, called the *European Gas Market Study*.

[22] Josh Posaner, *What Does Nord Stream Mean for Europe?* Interfax Global Energy Services, Natural Gas Review, Dec. 1, 2011, available at http://interfaxenergy.com/natural-gas-news-analysis/russia-and-the-caspian/nord-streams-role-in-europe-in-question/.

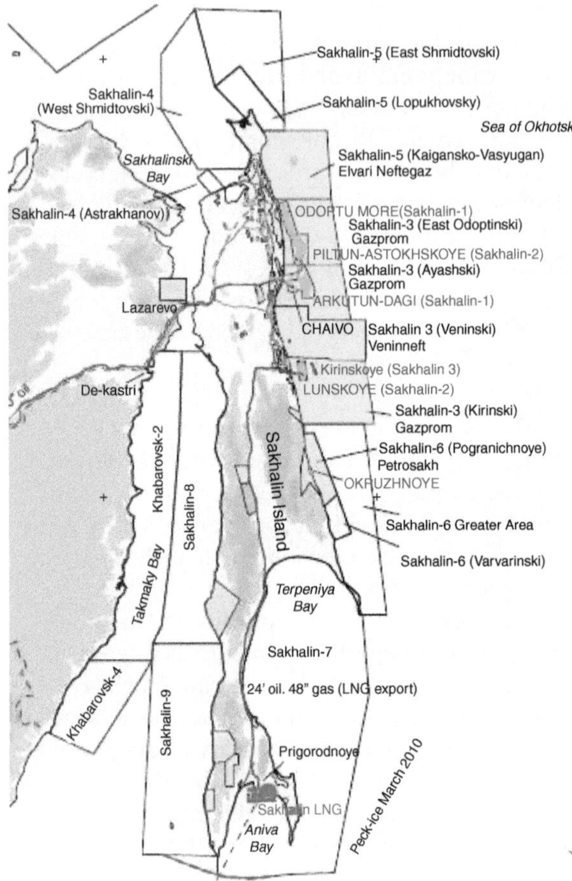

Source: US EIA/Wood Mackenzie

Figure 9.3 Sakhalin Island

9.2.6 Sakhalin Island, Sakhalin LNG and Sakhalin I–VI

Sakhalin Island is a former penal colony located on the Far East side of Russia and is a major oil and gas producing area. It is also the location of Russia's first LNG terminal – Sakhalin LNG.[23] (See Figure 9.3.)

---

[23] EIA, Country Analysis Brief, *Sakhalin Island* (last updated June 2011), available at http://www.eia.gov/emeu/cabs/Sakhalin/pdf.pdf.

According to energy experts Wood Mackenzie, technically and commercially recoverable oil reserves around Sakhalin Island are estimated at almost 5 billion barrels. Natural gas reserves are estimated to be approximately 34 trillion cubic feet. Both Russian exploration companies and international consortia are involved in the development of the various projects on Sakhalin Island. Although there are extensive export plans via LNG and pipelines to the mainland, little progress has been made beyond the first two developments on Sakhalin – Sakhalin I and Sakhalin II.[24]

Sakhalin I[25]

The Sakhalin I project is led and operated by Exxon Neftegaz, a subsidiary of ExxonMobil, which holds a 30 percent interest. Other participants include: Russian oil company Rosneft acting via its affiliates RN-Astra (8.5 percent) and Sakhalinmorneftegas-Shelf (11.5 percent); Japanese consortium SODECO (30 percent), and Indian state-owned oil company ONGC Videsh Ltd. (20 percent).[26]

Due to its harsh Arctic climate, Sakhalin-1 is one of the most challenging projects in the world. Although they were discovered several decades ago, the Sakhalin-1 fields were not developed until new technologies became available as a result of the partnerships between Russian and international companies.[27] The first phase of the project development is already underway and Exxon Neftegaz is extracting resources from both the Chayvo and Odoptu fields. Commercial production of oil and natural gas in the Odoptu field began in September 2010, offsetting the declining output from the project's Chayvo field. Subsequent phases of the project will include the development of the Arkutun-Dagi field as well as Phase II of the Chayvo development. The Arkutun-Dagi field development will help maintain production as the Chayvo field production naturally declines. According to Exxon Neftegaz, oil production is expected to begin in 2014. Phase II of the Chayvo field will allow for expanded natural gas production for both domestic consumption and exports.[28]

So far, the mode of natural gas exports from Sakhalin 1 has not been decided, but the project is supplying natural gas to the local area.

---

[24] Ibid.

[25] Exxon Neftegas Limited, Sakhalin-1, http://www.sakhalin-1.com/Sakhalin/Russia-English/Upstream/default.aspx.

[26] Ibid.

[27] Ibid.

[28] EIA, Country Analysis Brief, *Sakhalin Island* (last updated June 2011), available at http://www.eia.gov/emeu/cabs/Sakhalin/pdf.pdf.

ExxonMobil would like to send the natural gas to the south via pipeline to China, but other shareholders and Gazprom would rather market the natural gas as LNG via Sakhalin II, which would require an expansion of the facilities there.[29]

Sakhalin II (Sakhalin LNG)[30]

The Sakhalin II project is being developed under a production-sharing agreement (PSA) that includes Gazprom (50 percent), Shell (27.5 percent), Mitsui (12.5 percent), and Mitsubishi (10 percent), all of which participate in the Sakhalin Energy consortium. The project includes the development of two offshore fields, the Piltun-Astokhskoye (mostly oil) field and the Lunskoye (mostly gas) field, as well as construction of onshore pipelines to carry oil and gas to the south of the island, an onshore processing facility, LNG liquefaction plants, and offloading terminals for crude oil and LNG. Sakhalin II's estimated commercial and technical reserves are 14 Tcf of natural gas and 1 billion barrels of liquids.[31]

Sakhalin II is one of the largest integrated projects in the world and was undertaken amid some of the world's harshest conditions in an area prone to earthquakes.[32] Perhaps not so surprisingly, the projects faced major delays and cost overruns.[33] The cost overruns, along with other challenges encountered throughout the project, ultimately resulted in Shell losing control of the majority interest in the project in 2007. Shell originally held a 55 percent stake in the project until Gazprom announced in late 2007 that it was cutting Shell's stake in Sakhalin II to 27.5 percent and reducing the interest of the Japanese partners as well. While details are limited, Gazprom's actions appeared to be due to Russia asserting control over its natural resources as well as the cost of the project under Shell, which ballooned from $10 billion in 1997 to $20 billion in 2005. The cost overruns, plus some other environmental

---

29 Ibid.

30 Sakhalin Energy, http://www.sakhalinenergy.com/en/.

31 EIA, Country Analysis Brief, *Sakhalin Island* (last updated June 2011), available at http://www.eia.gov/emeu/cabs/Sakhalin/pdf.pdf.

32 Shell News Release, *Sakhalin II Gears up for Full Production*, May 20, 2009, available at http://www.shell.com/home/content/media/news_and_media_releases/archive/2009/sakhalin_20052009.html.

33 EIA, Country Analysis Brief, *Sakhalin Island* (last updated June 2011), available at http://www.eia.gov/emeu/cabs/Sakhalin/pdf.pdf. For an interesting article describing the challenging conditions at Sakhalin, including the long journey just to get there, see *Sakhalin Island: Journey to Extreme Oil*, Bloomberg Businessweek, May 15, 2006, available at http://www.businessweek.com/magazine/content/06_20/b3984008.htm.

concerns and mishaps, apparently caused anger and resentment towards Shell's leadership.[34]

In terms of environmental issues, Sakhalin II faced intense opposition from an environmental group called Sakhalin Environment Watch (SEW). SEW was highly critical of Sakhalin Energy's pipeline route, oil-spill preparedness and construction techniques. Moreover, there were concerns about the huge project's impact on the salmon spawning habitat and the overall marine ecosystem.[35]

Despite the challenges and delays, Sakhalin II eventually came online in March 2009, two years later than its targeted date of 2007.[36] In April 2009, the first Sakhalin II LNG cargo was delivered to Japan, marking the very first Russian export of LNG. Virtually all of the gas from Sakhalin II has been sold under long-term contracts to customers in the Asia-Pacific region, including Japan and South Korea. In 2010, Sakhalin LNG announced it had reached full production capacity of 9.6 MMT/y, which is about 5 percent of the world LNG output.[37]

Sakhalin III

The Kirinskoye and Veninskoye blocks in the Sakhalin III project are estimated to contain 1.5 billion barrels of oil and 2.7 Tcf of natural gas in commercial and technical reserves. The Veninskoye block, with oil reserves of 1.4 billion barrels, will be developed by a Rosneft-Sinopec consortium, with four wells drilled so far. Once exploration has been completed, production could start from the Veninskoye field in 2017. Gazprom began exploration drilling at the Kirinskoye offshore field in July 2009, and it is the first Sakhalin project to be carried out completely by a Russian firm. The company plans to deliver natural gas produced at the field to the Sakhalin-Khabarivsk-Vladivostok gas system by 2014.

---

[34] Abraham Lustgarten, *Shell Shakedown*, CNNMoney.com, Feb. 1, 2007, available at http://money.cnn.com/magazines/fortune/fortune_archive/2007/02/05/8399125/index3.htm.

[35] Ibid.

[36] Interestingly, Sakhalin received a commissioning cargo in October 2007 from the tiny Kenai LNG facility in Kenai, Alaska, when Marathon Oil's *Arctic Sun* berthed at Sakhalin Island. Since Alaskan LNG is mainly methane, it was ideally suited for the initial commissioning and testing of the Sakhalin LNG plant. Sakhalin Energy, *Alaska to Aniva . . . Second LNG Shipment Arrives!* (Oct. 2007), available at http://www.sakhalinenergy.ru/en/default.asp?p=channel&c=1&n=230.

[37] Sakhalin Energy, *In 2010 Sakhalin Energy Produced 100 LNG* (August 2010), http://www.sakhalinenergy.ru/en/default.asp?p=channel&c=1&n=378.

Gazprom also holds licenses to the Vostochno-Odoptu and Ayashski blocks.[38]

Sakhalin IV–VI

Areas of Sakhalin Island lying to the north and southeast of Sakhalin I and II are at various preliminary stages of development and have thus far shown mixed results. The Rosneft-BP efforts in the East and West Shmidtovski fields have been disappointing. However, exploration in the Kaigansko-Vasyugan continues with promising results.[39]

Table 9.1 summarizes Sakhalin I–VI.

9.2.7 Yamal LNG

Two gas mega projects are currently underway in Russia, the Yamal LNG project and the Shtokman project.[40] The Yamal LNG project was launched in late 2008 in response to output declines in the Yamburg, Urengoy, and Medvezh'ye fields.[41] The Yamal project is being developed by Novatek, Russia's largest independent natural gas producer and the second-largest natural gas producer in Russia after Gazprom.[42]

In 2010, Novatek accounted for approximately 6 percent of the natural gas produced in Russia, providing approximately 10 percent of total natural gas deliveries through the UGSS (United Gas Supply System) to the domestic market. Novatek is currently required to sell 100 percent of its natural gas to customers in the Russian Federation and its main customers are power generation companies, industrial users, regional gas distributors and wholesale gas traders.[43]

As an independent natural gas producer, Novatek is not subject to the government's regulation of natural gas prices; however, the regulated price as set by the Federal Tariff Service (FTS) significantly influences the market conditions in Novatek's regions of delivery as well as the price in its natural gas contracts with end-customers and wholesale traders. Novatek transports its natural gas through its own pipelines into the UGSS, which it then uses to deliver its gas to end-customers in accordance with the relevant

---

38 EIA, Country Analysis Brief, *Sakhalin Island* (last updated June 2011), available at http://www.eia.gov/emeu/cabs/Sakhalin/pdf.pdf.

39 Ibid.

40 IEA WEO-2011.

41 Ibid.

42 Novatek, *About Us*, available at http://www.novatek.ru/en/about/general/.

43 Ibid.

Table 9.1 Sakhalin project summary

	Sakhalin I	Sakhalin II	Sakhalin III	Sakhalin IV	Sakhalin V	Sakhalin VI
Primary field/block names	Odoptu (Onshore), Chayvo (Onshore and offshore), Arkutun-Dagi	Piltun-Astokskoye, Lunskoye	Kirinkii, Veninskoye, Vostochno-Odoptu, Aiyashkii	Pogranichny Block Okruzhnoye, West Schmidt	Kaigansko-Vasyukansk East Schimdt	Pogranichny
Oil reserve estimate	2.1 billion bbl (Source: Wood Mackenzie)	1.0 billion bbl (Source: Wood Mackenzie)	4–5 billion bbl of oil (Source: IHS)	Up to 1.3 billion bbl (Source: IHS)	East Schmidt (2.98 billion bbl), K-V (8.5 billion bbl) (Source: IHS)	600 million bbl
Gas reserve estimate	16.9 Tcf (Source: Wood Mackenzie)	14.2 Tcf (Source: Wood Mackenzie)	49 Tcf (Source: Gazprom)	19 Tcf, 1 Tcf in west Schmidt (Source: Roseneft)	15.2–17.7 Tcf	n/a
Primary project developers	Exxon Nefttegaz (30%), SODECO (30%), ONGC Videsh (20%), Sakhalinmo meftegaz (11.5%), and RN Astra (8.5%)	Gazprom (50%), Sakhalin Energy Investment Company:Shell (27.5%), Mitsui (12.5%), Mitsubishi(10%)	Roseneft(49.8%), Chinese Sinopec (25.1%) and Sakhaliskaya Neftyanaya Kompaniya(25.1%)	BP (49%), Rosneft (51%)	Elvary Neftegas: BP (49%), Rosneft (51%)	Urals Energy (via Petrosakh), Alfa Eco

Source: US EIA

transportation contracts it enters into with Gazprom. The UGSS transports substantially all of the natural gas sold in Russia and is owned and operated by Gazprom. UGSS transportation tariffs are set by the FTS.[44]

In June 2009, Novatek acquired 51 percent of the ordinary shares in Yamal LNG LLC, which holds the exploration and production license for the South-Tambeyskoye field. The South-Tambeyskoye field was discovered in 1974 and is located in the northeastern portion of the Yamal peninsula. As of 31 December 2009, the field contained 380.0 bcm of proved natural gas reserves and 13.7 mmt of proved liquid hydrocarbon reserves. In 2009 and 2010, various 3D and 2D seismic data were collected and a number of exploratory wells were drilled.[45]

In May 2010, Yamal LNG LLC announced that it had awarded a contract to CB&I for the concept development services for the Yamal LNG integrated project to be completed in first-half 2011. The project consists of the production, treatment, transportation, liquefaction, and shipping of natural gas and NGLs from the South Tambey field on the Yamal Peninsula in northwestern Siberia. Reserves in the South Tambey field are estimated at more than 1 trillion cu m (35.3 tcf).[46]

CB&I's project scope includes concept development of the 15–16 million tonne/year LNG liquefaction plant, including LNG storage and loading, as well as Arctic shipping and ice management, a gas transmission pipeline, central production for gas and condensate treatment, and associated well sites and gas gathering, said the announcement. Concept development will "address the technical, economic, and execution feasibility of the remote arctic project and provide a project schedule and cost estimates," in addition to the design basis for front-end engineering and design.[47]

In June 2010, Novatek entered into a cooperation agreement with Gazprom setting out the key parameters for joint activity between the companies in terms of implementing and developing the Yamal LNG project. The terms of the agreement with Gazprom set out the scope of the construction, development and subsequent utilization of related infrastructure, including energy and transportation systems and LNG production facilities.[48]

---

44 Ibid.

45 Novatek, *South-Tambeyskoye Field*, available at http://www.novatek.ru/en/business/projects/stambeyskoye/.

46 O&GJ Editors, *Yamal LNG Awards Concept Contract*, Oil & Gas Journal, May 5, 2010, http://www.ogj.com/articles/2010/05/yamal-lng-awards-concept.html.

47 Ibid.

48 Novatek, *South-Tambeyskoye Field*, available at http://www.novatek.ru/en/business/projects/stambeyskoye/.

In October 2010, the Russian government officially outlined its position on tax concessions planned for the development of the Yamal peninsula. Following this announcement and the completion of its final investment decision and feasibility studies, Novatek plans to finalize its total capital expenditure plans, including construction expenses for the Yamal LNG plant.[49]

Within the project's framework Novatek is planning further exploration and development activities at the South-Tambeyskoye field including production drilling and infrastructure development. Infrastructure plans are expected to include the construction of a gas gathering system, a gas complex processing facility, a gas condensate processing unit, an LNG plant, pipelines which enable gas transportation from the field to the plant, an offshore shipping terminal and a transport infrastructure system (including an airport, terminal and highways).

In September 2011, Novatek announced that it had exercised an option to buy a 49 percent stake in the Yamal LNG, taking full ownership of the project. While few details are available, Novatek indicated that it had exercised two call options for 23.9 percent and 25.1 percent, which will be paid in installments through June 30, 2012. The company did not provide any financial details, but said earlier it held an option to buy the 25.1 percent stake for $526 million.[50]

In October 2011, Novatek completed the sale of a 20 percent stake of Yamal LNG to French oil and gas group Total. Novatek indicated that the deal was concluded through a subsidiary, Novatek North-West, and that Total will provide financing for project development.[51] As of November 2011, Yamal LNG was in active talks with Qatar to add Qatar as a partner to the project. The project is expected to produce 15 mtpa of LNG by 2018.[52]

9.2.8 The Shtokman Project

The giant Shtokman gas field is estimated to be one of the world's largest natural gas deposits with reserve estimates of 3.8 trillion cubic meters of

---

[49] Ibid.

[50] Jacob Gronholt-Pedersen, *Novatek Exercises Option to Buy 49% Stake in Yamal LNG*, Dow Jones Newswires, Sept. 30, 2011.

[51] O&GJ Editors, *Total Novatek Sign Yamal LNG Agreement*, Oil & Gas Journal, Oct. 6, 2011, available at www.ogj.com/articles/2011/10/total-novatek-sign-yamal-lng-agreement.html.

[52] Regan Doherty, *Qatar in Active Talks to Take Stake in Yamal LNG Project*, Reuters, Nov. 13, 2011, http://www.reuters.com/article/2011/11/13/yamal-qatar-idUSL5E7MD0GI20111113.

Source: Gazprom

Figure 9.4 Shtokman field location map

natural gas and about 37m tonnes of gas condensate.[53] The Shtokman field is located in the Russian-controlled sector of the Barents Sea, about 370 miles (500 km) north of the Kola Peninsula (Figure 9.4).

Although the Shtokman gas field was identified in 1988, it was not developed due to the extreme Arctic conditions where the field is located.[54] The Arctic climate and the harsh, stormy environment leave a

---

[53] *Shtokman Gas Project*, available at http://www.hydrocarbons-technology.com/projects/shtokman_gas_project/.

[54] According to the project developers, the Shtokman gas field was identified in 1988 from offshore geophysical surveys performed by specialists on board the research vessel *Professor Shtokman*, from which the field was named. Shtokman Project, http://shtokman.ru/en/project/gasfield.

narrow window of time favorable for the actual development operations. Additional challenges include the fact that the Shtokman gas deposits lie in four main layers with the reservoir depth varying between 1,900m and 2,300m. Due to the extreme cold conditions of the Arctic, one of the greatest hazards for production platforms at these depths will be from an iceberg collision. This potential hazard will be mitigated by the use of floating removable platforms, which are technologically and economically challenging to build.[55] "The scale and complexity of the operations, harsh climate in the areas where the gas is produced and transported, and the need to apply fundamentally new engineering and process solutions during the course of development all demonstrate the unique nature of the Shtokman project."[56]

Gazprom's subsidiary, Sevmorneftegaz, has the rights to explore for and produce gas from the Shtokman field but it needed the expertise and investment of foreign oil companies to do so. In 2008, a special purpose vehicle (SPV), called the Shtokman Development Company (SDC), was set up to develop and exploit the field. Ownership of the SDC is divided between Gazprom (51 percent), Total (25 percent) and StatoilHydro (24 percent).[57] Development of the Shtokman field will be done in four phases, with the full field development expected to deliver a massive 71.1 bcm of gas per year, 30 MTA of LNG and 600,000 t of gas condensate per year over a 50-year period.[58]

According to Gazprom, the Shtokman project is of "strategic significance" and is expected to "become a pivotal point to form a new gas producing region on the Russian Arctic shelf."[59] The Shtokman Development Company has also highlighted the project's significance on a dynamic website that notes that the Shtokman field "will create a basis for further harnessing the Arctic's offshore hydrocarbon potential, simultaneously strengthening Russia's position as a key player not only on the European gas market but also on the global energy arena. The gas produced by Shtokman will play a major part in natural gas supplies to the European and world markets."[60] The Shtokman project is also viewed

---

55 *Shtokman Gas Project*, available at http://www.hydrocarbons-technology.com/projects/shtokman_gas_project/.

56 Shtokman Project, http://shtokman.ru/en/.

57 *Shtokman Gas Project*, available at http://www.hydrocarbons-technology.com/projects/shtokman_gas_project/.

58 Ibid.

59 Gazprom, Shtokman Project, available at http://www.gazprom.com/production/projects/deposits/shp/.

60 Shtokman Project, http://shtokman.ru/en/.

as a resource base to supply natural gas via the new Nord Stream pipeline to Western Europe, as well as for the production of Russian LNG for overseas markets.[61] It should be noted that the US was one of the markets the Shtokman project was targeting but, as noted in other chapters, with the production of shale gas, it now appears that the US will not be a major LNG importer.

The Shtokman field was expected to come on-stream by 2013–15, with estimated development costs between $12 and $25 billion for the first phase and a $50 billion overall investment. The first gas was expected in 2013 at 11 bcm a day along with 205,000 t of gas condensate per year and the first LNG was expected in 2014 at 7.5 MTA.[62] However, in February 2010, the project development schedule was postponed for three years due to a decrease in demand for gas in European markets. It is now expected that gas production from Shtokman will start in 2016. The first LNG production has been postponed until 2017 due to changes in the targeted US LNG market.[63]

Reaching a final investment decision (FID) for Shtokman has always been viewed as a difficult aim as the partners weigh uncertainty about demand and prices and, as this book goes to print, the FID for the Shtokman project has yet to be made after several delays.[64] Among the many challenges is the problem of where the gas will be marketed and how to transport it. Sevmorneftegaz and the Shtokman Development Company are working on the phase I development (design and construction) of the Shtokman infrastructure. This will include a production complex, a pipeline network and an LNG plant. The gas was initially destined for the US market as LNG export but Gazprom has also indicated that it wants to export some of the produced gas to Europe via the Nord Stream pipeline. However, this will require a pipeline spur from the Shtokman field to the Murmansk Oblast and then via Kola Peninsula to Volkhov in the Leningrad Oblast.

A Gazprom representative indicated at a major energy conference in November 2011 that "Shtokman will go ahead. However, it is a matter of timing. It will be underpinned by long-term contracts. We are working

---

61 Ibid.

62 hydrocarbons-technology.com, *Shtokman Gas Project*, available at http://www.hydrocarbons-technology.com/projects/shtokman_gas_project/.

63 Ibid.

64 See Josh Lewis, *Shtokman FID Delayed Again*, Upstream, March 30, 2012, http://www.upstreamonline.com/live/article1243430.ece, noting that the Shtokman partners have pushed back FID on the first phase of development until July 1, 2012.

hard to make that timing as soon as possible. For the time being we remain positive. However, challenges remain for the project."[65]

Other challenges include the fact that Shtokman's foreign shareholders, Total and Statoil, have said repeatedly that the project needs tax breaks similar to the Yamal LNG project, which received breaks on the natural resource extraction tax and local taxes, as well as a zero percent export duty.[66] Analysts have questioned whether government financial aid will be available for both Yamal and Shtokman, since the cost of Novatek's Yamal LNG project alone is estimated to reach €24 billion ($33 billion).[67] Statoil CEO Helge Lund has said repeatedly that Statoil cannot arrive at an FID without a clear definition of the fiscal regime for the project, which may not be economically viable without significant tax breaks.[68]

9.3 PERU – THE WORLD'S NEWEST LNG EXPORTER

Peru LNG[69] became the world's newest LNG exporter in June 2010, and joined the club of LNG exporters as its 18th member.[70] The Peru LNG pipeline and liquefaction project located at Pampa Melchorita is the largest industrial project ever undertaken in Peru's history.[71] It is also the

---

65 Leigh Elston and Ahmed Mehdi, *"Challenges Remain" for Giant Shtokman Project*, Interfax Energy, Nov. 2011, available at http://interfaxenergy.com/natural-gas-news-analysis/exploration-and-production/'challenges-remain'-for-giant-shtokman-project, citing comments made by Frederic Barnaud, president and managing director of Gazprom Global LNG, at the CWC 12th Annual LNG Summit in Rome.

66 Leigh Elston and Ahmed Mehdi, *"Challenges Remain" for Giant Shtokman Project*, Interfax Energy, Nov. 2011, available at http://interfaxenergy.com/natural-gas-news-analysis/exploration-and-production/'challenges-remain'-for-giant-shtokman-project.

67 Ibid.

68 The Moscow Times, *No Shtokman Clarity This Year*, Nov. 28, 2011, available at http://www.themoscowtimes.com/business/article/no-shtokman-clarity-this-year/448685.html#ixzz1fGrC8JEL.

69 Peru LNG, www.perulng.com.

70 Peru LNG was officially inaugurated on June 10, 2010, when Repsol delivered its first LNG shipment, onto the *Barcelona Knutsen* tanker. Repsol Press Release, *Repsol Delivers its First Cargo from the Peru LNG Plant*, June 24, 2010, http://www.repsol.com/es_en/corporacion/prensa/notas-de-prensa/ultimas-notas/24062010-repsol-primer-cargamento-gnl-peru.aspx.

71 CB&I, a leading engineering firm, engineered the liquefaction plant and jetty topsides. CB&I, www.CBI.com.

first LNG plant in South America and was the only LNG project worldwide that took an affirmative final investment decision in 2006.[72]

9.3.1 Project Overview

The Peru LNG project is made up of two major segments. The first segment, and a major element of the project, was the construction of a 34-inch diameter pipeline for the transportation of natural gas from the Camisea gas field located in the Andes Mountains east of Ayacucho to the LNG plant at Pampa Melchorita on the coast.[73] Development of the 408 km pipeline began in 2005 and faced numerous challenges including difficult terrain conditions, landslides, altitude considerations, government regulations, and difficult environmental and ecological conditions. The ultimate route evaluated by Gulf Interstate Engineering and various specialists "follows an undulating route, winding along narrow mountain ridges through the Andes Mountains with near inaccessible locations and terrain conditions that are considered the toughest in the world when engineering and constructing a pipeline project of this sort."[74] The pipeline ends at the LNG Plant at Pampa Melchorita, located on the coast. (See Figure 9.5.)

9.3.2 The LNG Plant

The LNG plant is a single-train facility with a capacity of 4.4 million tonnes a year. The equipment and utilities include gas dehydration and carbon absorption units, facilities for refrigeration and liquefaction, two storage tanks of 130,000 m^3 LNG capacity each and refrigerant storage.[75]

Initially 17 sites were evaluated for the LNG facility and then a more detailed assessment was applied to the selected few. Although the ultimate site at Pampa Melchorita is located on a steep cliff with a 140 m drop to the beach, it was chosen as the best option based on a number of environ-

---

72 *Introducing Peru: World's Newest LNG Exporter*, Facts Global Energy, Gas Advisory, Issue no. 50, June 2010.

73 *Gulf Interstate Engineering Case History, Peru LNG Pipeline Project* (2009), www.gie.com. Gulf Interstate Engineering (Gulf) was selected to provide basic design services (FEED) for Peru LNG, including assessment of the route selection and constructability analysis and the development of a capital cost estimate and project schedule. Gulf was subsequently awarded the detailed engineering, design and procurement services on the Peru LNG Pipeline Project.

74 Ibid.

75 *Peru LNG*, http://www.hydrocarbons-technology.com/projects/peru-lng/.

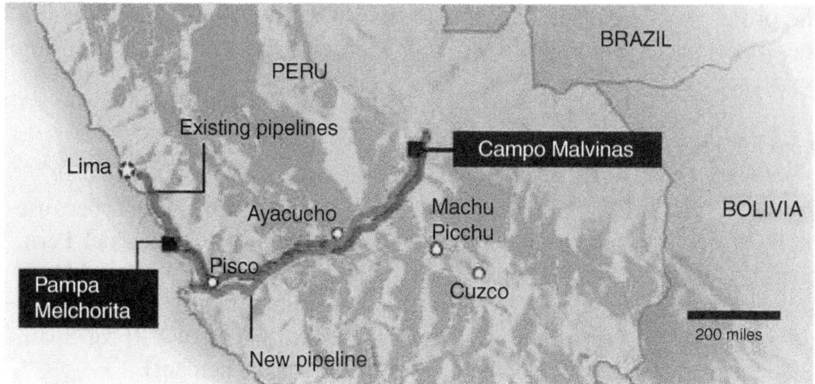

Source: www.hydrocarbons-technology.com

Figure 9.5 Peru LNG project

mental, technical and economic considerations including stable soils to prevent damage from Peru's high level of seismic activity.[76]

9.3.3 Marine facilities[77]

The marine facilities consist of various projects including a 1.35 km-long trestle from the shore to the loading platform. A 1 km-long marine terminal was constructed to facilitate tankers of capacities ranging from 90,000 to 173,000 m^3. Because the Peruvian coastline is exposed to long period Pacific swells during parts of the year, a breakwater will provide for safe berthing and will allow the marine facilities to remain accessible all year round.

An LNG tanker navigational channel was dredged to provide access in and out of the berth area. The channel has a water depth of 15 m, a width of 250 m and a total length of 3,600 m. The channel is 18 m deep where the LNG carriers make turns outside the protection of the breakwater.

The engineering, procurement and construction (EPC) contract for the project was awarded to CB&I and the entire project took four years to complete.[78] One interesting construction detail to note is that

---

[76] Ibid.

[77] Ibid.

[78] CB&I Factsheet, *LNG Production Facility, Pampa Melchorita, Peru*. CB&I is a leading engineering firm with extensive experience in the construction of LNG facilities. CB&I, www.CBI.com.

the project was designed and constructed for a high seismic region and the project site experienced an 8.0 Richter scale earthquake during construction.[79]

9.3.4 Project Partners and Financing

Peru LNG is somewhat unique among global liquefaction projects because the host country's state-owned petroleum company, Petroleos del Peru, plays no role in its upstream or downstream business. The US-based Hunt Oil company is the majority (50 percent) owner and operator of the project along with SK Corporation of Korea (20 percent), Repsol YPF of Spain (20 percent) and Marubeni Corporation of Japan (10 percent).

Repsol has an exclusive rights contract with Peru LNG to market the entire output of the plant, in accordance with an agreement signed with Peru LNG in 2005 for a term of 18 years from the start of commercial operations. In terms of volume, this will be the biggest LNG acquisition ever made by Repsol.[80]

Repsol also has an LNG supply contract with the Federal Commission of Electricity (CFE), Mexico, for the natural gas terminal at Puerto de Manzanillo, on the Mexican Pacific coast. This contract provides for the supply of LNG to the Mexican plant for a period of 15 years with a volume of at least 67 bcm, which is due to be operative during the fourth quarter of 2011. The Manzanillo plant, which will supply power plants of the Federal Commission of Electricity (CFE) in West Central Mexico, will source the gas from the Peru LNG plant.

The total investment for the project, including the liquefaction plant ($1.5 billion), related marine and pipeline facilities, and development and financing costs, was $3.8 billion.[81] The project was financed by a variety of sources, including approximately $2.25 billion secured by Hunt Oil, Repsol, SK Energy and Marubeni through third parties including export credit agencies, multilateral institutions and commercial banks in June 2008. The Inter American Development Bank (IDB) and other multilateral institutions signed loans of $800 million for the project. About $1.6 billion was secured through equity. A sum of $200 million was raised through bond issue in Peru's capital market in 2009. In July 2010 Peru

---

[79] CB&I Factsheet, *LNG Production Facility, Pampa Melchorita, Peru.*

[80] Repsol Press Release, *Repsol Delivers its First Cargo from the Peru LNG Plant*, June 24, 2010, http://www.repsol.com/es_en/corporacion/prensa/notas-de-prensa/ultimas-notas/24062010-repsol-primer-cargamento-gnl-peru.aspx.

[81] CB&I engineered the liquefaction plant and jetty topsides. CB&I, www.CBI.com.

LNG received a two-year credit line of $75 million for short-term financial needs to export the LNG.[82]

In addition, the International Finance Corporation (IFC) approved a $300 million loan for the Peru LNG project in June 2008. The IFC also advised Peru LNG on numerous environmental issues to ensure that communities surrounding the project site were protected.[83] According to the IFC, Peru LNG will enhance opportunities for local businesses to sell goods and services to the project and raise people's incomes and job prospects. IFC is developing programs to engage local communities in monitoring the effects the project will have on their lives and is also establishing training for nearby municipalities to make best use of the significant new revenues they will receive as a result of the project. These programs build on IFC's experience with similar initiatives in some of Peru's poorest regions. The Peru LNG project is expected to help generate significant tax and incremental royalty payments to the government, equivalent to over 1.5 percent of current state revenues, and the project is expected to make Peru a net gas exporter after operations begin in 2010.[84]

9.4 YEMEN AND YEMEN LNG

Located at the southwestern tip of the Arabian Peninsula, Yemen is "an impoverished Arab country that faces an array of daunting development challenges that some observers believe make it at risk for becoming a failed state."[85] Moreover, the country's rugged terrain and geographic isolation, strong tribal social structure, and sparsely settled population have historically made Yemen difficult to centrally govern.

According to the US EIA, Yemen is located along one of the world's most strategic oil shipping routes – the Bab al-Mandab – from which

---

[82] hydrocarbons-technology.com, *Peru LNG*, http://www.hydrocarbons-technology.com/projects/peru-lng/.

[83] International Finance Corporation, World Bank Group, *Peru LNG Project*, http://www.ifc.org/ifcext/plng.nsf/content/Home.

[84] International Finance Corporation, World Bank Group, *Peru LNG Project*, http://www.ifc.org/ifcext/plng.nsf/content/Home.

[85] Jeremy M. Sharp, CRS Report for Congress (RL34170), *Yemen: Background and U.S. Relations*, Jan 13, 2010, available at http://opencrs.com/document/RL34170/2010-01-13/. In 2009, Yemen ranked 140 out of 182 countries on the United Nations Development Program's Human Development Index, a score comparable to the poorest sub-Saharan African countries. Over 43% of the population of nearly 24 million people live below the poverty line, and per capita GDP is estimated to be between $650 and $800. Ibid.

an estimated 3.5 million barrels of oil passed daily in 2010. Disruption to shipping in the Bab al-Mandab could prevent tankers in the Persian Gulf and the Gulf of Aden from reaching the Suez Canal/Sumed pipeline complex, thus requiring a costly diversion around the southern tip of Africa to reach western markets.[86] (See Figure 9.6.)

Yemen's economy is heavily dependent on hydrocarbons, which account for 30 percent of GDP, nearly 75 percent of government revenues, and over 90 percent of foreign exchange earnings. In more recent years, Yemen has sought to increase foreign investment in its energy sector since declining oil revenues are weakening the government's ability to provide basic services.[87] However, foreign investment in Yemen has been limited due to the myriad of problems and challenges facing Yemen, including poor governance and lack of succession plans stemming from President Saleh's 30-year reign over Yemen.[88]

In addition to Yemen's development challenges, security concerns over actual and potential attacks, involving militant and terrorist groups such as Al Qaeda, have also deterred investment, with numerous attacks to energy infrastructure slowing production and increasing costs. Tribal conflicts have also resulted in attacks on pipelines in the north of the country.[89] Adding to Yemen's internal struggles is the fact that in recent years the region surrounding Yemen has seen rising piracy off the Somali coast in the Gulf of Aden and southern Red Sea, reaching further out into the Indian Ocean.[90]

9.4.1 Yemen LNG Project

One of the few bright spots in Yemen is the Yemen LNG project.[91] Launched in 2005, the $4.5 billion Yemen LNG project is the largest

---

86 EIA Country Analysis Briefs, *Yemen* (last updated Feb. 2011), available at http://www.eia.gov/emeu/cabs/Yemen/pdf.pdf.

87 EIA Country Analysis Brief, *Yemen*.

88 Jeremy M. Sharp, CRS Report for Congress (RL34170), *Yemen: Background and U.S. Relations*, Mar. 22, 2011, available at http://assets.opencrs.com/rpts/RL34170_20110322.pdf, hereineafter CRS Report RL34170, Yemen.

89 EIA Country Analysis Brief, *Yemen*. According to experts, "Al Qaeda's attack against the USS Cole in 2000 (while docked at a Yemeni port) coupled with the attacks of September 11, 2001, a year later officially made Yemen a front in the so-called war on terror." CRS Report RL34170, Yemen.

90 EIA Country Analysis Brief, *Yemen*.

91 According to information posted on Yemen LNG's website, Yemen LNG produced a World Bank compliant Environmental and Social Impact Assessment which was revised and updated in February 2006 to provide a full description of the environmental and socio-economic impacts of the Yemen LNG project and

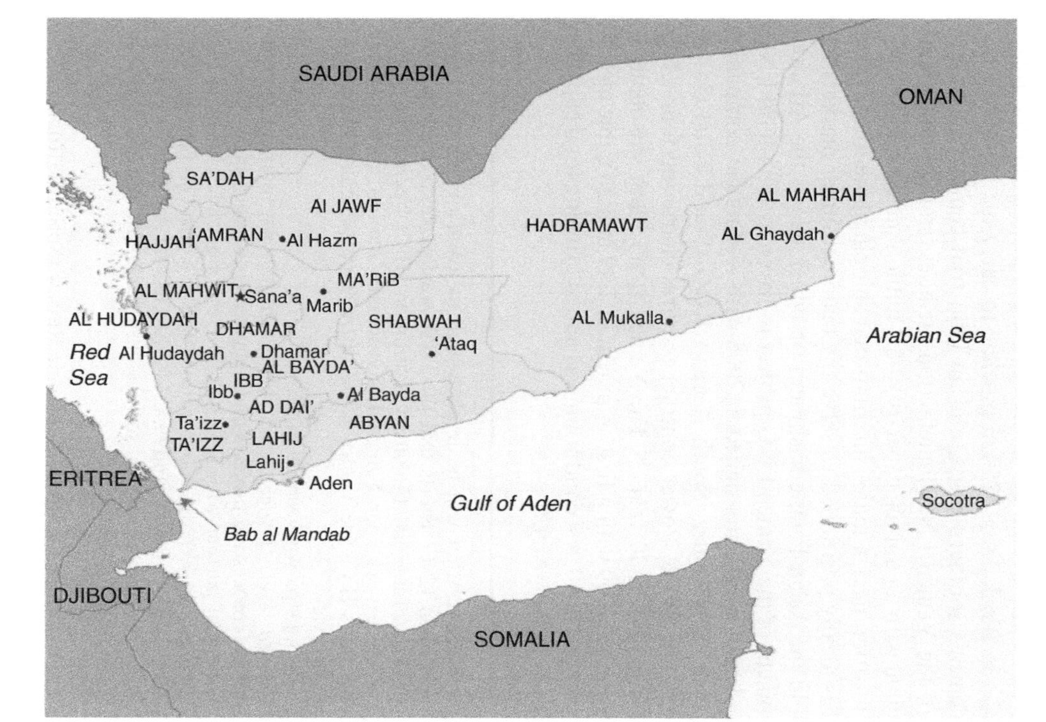

Source: CRS Report for Congress, RL34170

Figure 9.6 Map of Yemen

industrial project in Yemen's history.[92] While Yemen had been producing oil since 1986, it had not sought to leverage its gas reserves until 2005 when the Yemeni government approved plans by Total and partners to develop the majority of the 260 billion cubic metres of proved gas resources located in Block 18 in the Marib region, east of the Yemeni capital of Sana'a.[93]

The project has three main components: the gas production and treatment facilities in Block 18, a 325 km pipeline to carry the gas across the desert to Yemen's southern coast, and an industrial complex in Balhaf consisting of the Yemen LNG liquefaction facilities, storage tanks, and export terminal with an annual capacity of 6.7 million metric tons.[94]

With its location near a major international shipping lane, the Yemen LNG plant is ideally positioned to serve the main consumer markets for LNG, both in the Asian-Pacific basin and on either side of the Atlantic. The development of the Yemen LNG project is driven by long-term LNG purchase and sale agreements signed in 2005 with leading gas companies such as France's GDF Suez, South Korea's KOGAS and Total Gas & Power. The Total Gas & Power contract requires chartering four LNG carriers for a period of 20 years.[95] The Yemen LNG Plant was inaugurated in November 2009 with a first shipment of 1.49 million cubic meters of LNG.[96] In April 2010, a second train was added bringing total capacity to 6.7 mmtpa (326 Bcf/year).[97]

---

the mitigation measures that will be implemented to avoid or minimize these impacts. While Yemen LNG asserts that the ESIA was prepared in accordance with World Bank/International Finance Corporation (IFC) operational policy guidelines, it is unclear whether Yemen LNG actually received any financing from the World Bank/IFC. See Yemen LNG, http://www.yemenlng.com/ws/en/go.aspx?c=soc_overview.

92 Yemen LNG, http://www.yemen.lng.com.

93 Total, *Yemen LNG Project*, available at http://www.total.com/en/our-energies/natural-gas-/processing/projects-and-achievements/yemen-lng-940894.html. Total is the lead partner with a 39.62% interest, alongside Hunt, 17.22%; Yemen Gas Company, 16.73%; SK Corporation, 9.55%; Korea Gas Corporation, 6%; Hyundai Corporation, 5.88%; and GASSP, the General Authority for Social Security and Pensions, Yemen's largest social security organization, 5%.

94 Total, *Yemen LNG Project*.

95 Ibid.

96 World Bank, Yemen Quarterly Economic Review (Spring 2010), available at http://siteresources.worldbank.org/INTYEMEN/Resources/2010_YEU.pdf. *Yemen Quarterly Economic Review* is a quarterly report produced by the World Bank Country Office in Sana'a. It consists of several sections covering major political, social and economic developments. It also provides information on ongoing World Bank operations in Yemen, key indicators in Yemen, and a list of conferences and donor activities.

97 EIA, Country Analysis Brief, *Yemen*.

The Yemeni Ministry of Finance expects an estimated $233 million in revenues from the LNG and gas export in 2010. Once the production capacity reaches its target capacity of 6.7 mmtpa, the annual revenue flow is expected to amount to about $370 million. However, government net revenues from LNG sales will not reach their full level for the next first six to seven years, as cost recovery by the companies will weigh in to compensate for the initial investment.[98]

According to the World Bank, fiscal revenues from the Yemeni LNG project are derived from two sources: (1) directly, from taxation, bounces and royalty, and (2) indirectly, from the profit share agreement of the consortium signed with state-owned agencies (YGC, GASSP). Current estimates of the overall revenue potential over the estimated lifespan of the 20-year sales and purchase agreement (SPA) for the LNG production and export vary between $8 billion (government estimates) and $20 billion (industry estimates).[99]

9.4.2 Prospects Going Forward for Yemen LNG Exports

According to recent estimates by Cedigaz, Yemen exported a total of 243 Bcf of LNG in 2010, comprising 85 cargoes to markets in North America[100] and Asia, principally South Korea and China. Due to a recent shift in relative demand growth from the Atlantic Basin to the Asia-Pacific region, Yemen LNG is seeking to divert as much as 560 million cubic feet per day (MMcf/d) to Asian markets through the negotiation of mid-term supply contracts.[101]

Earnings from Yemen LNG's exports are projected to provide a partial offset of Yemen's falling oil export revenues in 2011, when the Yemen LNG project reaches full production capacity. However, due to the low price previously negotiated with Korea Gas for its LNG purchases, natural gas exports can only mitigate a downward trend in government

---

98 World Bank, Yemen Quarterly Economic Review (Spring 2010), available at http://siteresources.worldbank.org/INTYEMEN/Resources/2010_YEU.pdf.

99 Ibid.

100 In February 2010, the first LNG tanker from Yemen arrived in the US's Boston Harbor despite protests from Boston's Mayor that LNG from Yemen should not be allowed due to security risks. According to the US Coast Guard, extra security measures for the tanker were put in place due to Yemen's strong ties to Al Qaeda, including boarding the tanker out at sea to inspect it for proper documentation and possible stowaways. Kathleen McNerney, *Yemeni LNG Tanker Arrives Safely in Boston Harbor*, WBUR, Boston NPR, Feb 23, 2010, available at http://www.wbur.org/2010/02/23/lng-tanker.

101 EIA Country Analysis Brief, *Yemen*.

revenues. Yemen also signed 20-year contracts with GDF Suez Company and lead developer, Total.[102]

More recently, the unrest in the Arab World (commonly referred to as the "Arab Spring") in early 2011 focused increased international attention on Yemen as sustained mass protests and an assassination attempt on President Saleh caused him to flee to Saudi Arabia,[103] and later led to his official ousting, creating a power vacuum in the country.[104] Despite the unrest, in March 2011, it was reported that Yemen LNG had loaded the 100th LNG carrier at Balhaf and that production capacity had reached 6.7 MT.[105]

As this book goes to print, Yemeni soldiers have been deployed to protect the gas pipeline feeding the Yemen LNG export terminal as the country tries to shield its biggest industrial asset from attacks by armed groups that have repeatedly sabotaged Yemen's oil and gas pipelines, causing fuel shortages and slashing export earnings for the impoverished country.[106]

9.5 PAPUA NEW GUINEA

In December 2009, the Export-Import Bank of the United States (Ex-Im Bank)[107] "approved the largest financing transaction in its 75-year history – $3 billion to support U.S. exports for a liquefied natural gas (LNG) project in Papua New Guinea."[108] Along with the Ex-Im Bank, five other export credit agencies and 17 commercial banks will provide financing for

---

[102] Ibid.

[103] UPI, *Yemen: Saleh and 'New Guard' Play for Time*, Nov. 16, 2011, available at http://www.upi.com/Top_News/Special/2011/11/16/Yemen-Saleh-and-new-guard-play-for-time/UPI-78121321460685/#ixzz1du1XKOxj.

[104] Sami Aboudi, *Yemen's Saleh Faces Music after 33 Years in Power*, Reuters, Feb. 12, 2012, http://www.reuters.com/article/2012/02/20/us-yemen-saleh-idUSTRE81J0IS20120220.

[105] LNG World News, *Yemen Loads 100th Cargo*, March 27, 2011, http://www.lngworldnews.com/yemen-lng-loads-100th-cargo/.

[106] Mohammed Ghobari and Daniel Fineren, *Yemen sends soldiers to protect its LNG*, Reuters (June 21, 2012), http://www.reuters.com/article/2012/06/21/energy-yemen-lng-idUSL5E8HL7DW20120621.

[107] The Ex-Im Bank is an independent, self-sustaining US federal agency that helps to create and maintain US jobs by financing the sale of US exports, primarily to emerging markets throughout the world, by providing loan guarantees, export credit insurance and direct loans. More information is available on the Bank's web site at www.exim.gov.

[108] Ex-Im Bank Press Release dated December 14, 2009, *Ex-Im Bank Financing for Papua New Guinea LNG Project to Generate Significant Revenue for Island Nation, While Employing Workers at Dozens of American Companies*,

the project, which is estimated to cost about $18.3 billion.[109] According to the lenders, the project has the potential to double the gross domestic product of Papua New Guinea.[110]

The Papua New Guinea Liquefied Natural Gas (PNG LNG) Project is an integrated development that includes gas production and processing facilities in the Southern Highlands and Western Provinces of Papua New Guinea. Liquefaction and storage facilities with capacity of 6.6 million metric tons per year will be located near Port Moresby, the capital of Papua New Guinea. Over 700 km (450 miles) of pipelines will connect the facilities, which includes an offshore pipeline connecting to the LNG terminal.[111] (See Figure 9.7.)

9.5.1 Project Timeline

The Project is to be undertaken in a series of development phases, with Phase I scheduled to commence operations in 2014, followed by additional developments in subsequent years[112] (Figure 9.8).

9.5.2 Project Customers and Sponsors

It is expected that over 9 Tcf of gas will be produced and sold over the life of the project. The Project will supply four major LNG customers in the Asia-Pacific region under long-term sales agreements: China Petroleum and Chemical Corporation (Sinopec), Osaka Gas Company Limited, the Tokyo Electric Power Company, Inc, and Chinese Petroleum Corporation.[113]

The original Project sponsors are: ExxonMobil, which, through its subsidiary Esso Highlands Limited, will be the operator of the project with a 33.2 percent interest; Oil Search Limited (29 percent), Papua New Guinea's largest oil and gas producer and operator; National Petroleum Company Papua New Guinea (16.6 percent), owned by the Papua New

---

available at http://www.exim.gov/pressrelease_print.cfm/8F2036C5-C2D5-A3DB-0E90CFB3EFBC0262/ (hereinafter Ex-Im Bank Press Release).

[109] The original cost estimate for the PNG LNG Project was $18.3 billion. More recent cost estimates indicate that "The investment for the initial phase of the Project, excluding shipping costs, is estimated at US$15.7 billion." PNG LNG, www.http://pnglng.com/project/index.htm.

[110] Ex-Im Bank Press Release.

[111] Papua New Guinea Liquefied Natural Gas Project, Quarterly Environmental and Social Report, First Quarter 2010, http://pnglng.com/media/pdfs/quarterly_reports/Q1_2010_PNG_LNG_FINAL_E&S_Report_31-05-10.pdf (hereinafter PNG 1Q 2010 Report).

[112] PNG 1Q 2010 Report.

[113] Ibid.

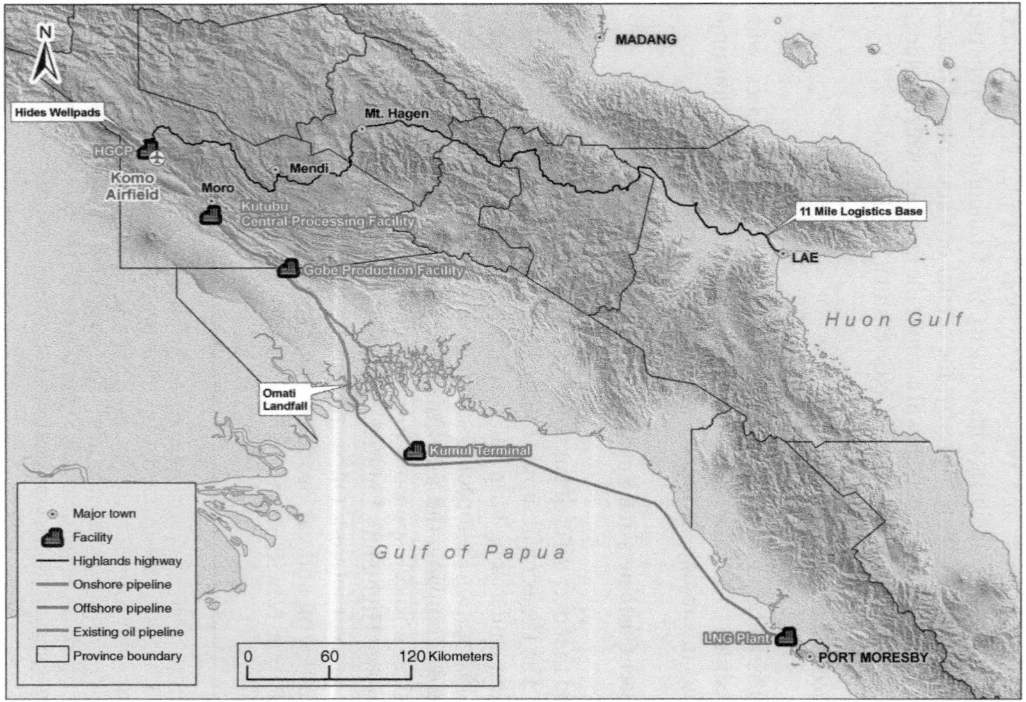

Source: PNG LNG

Figure 9.7 PNG LNG project overview map

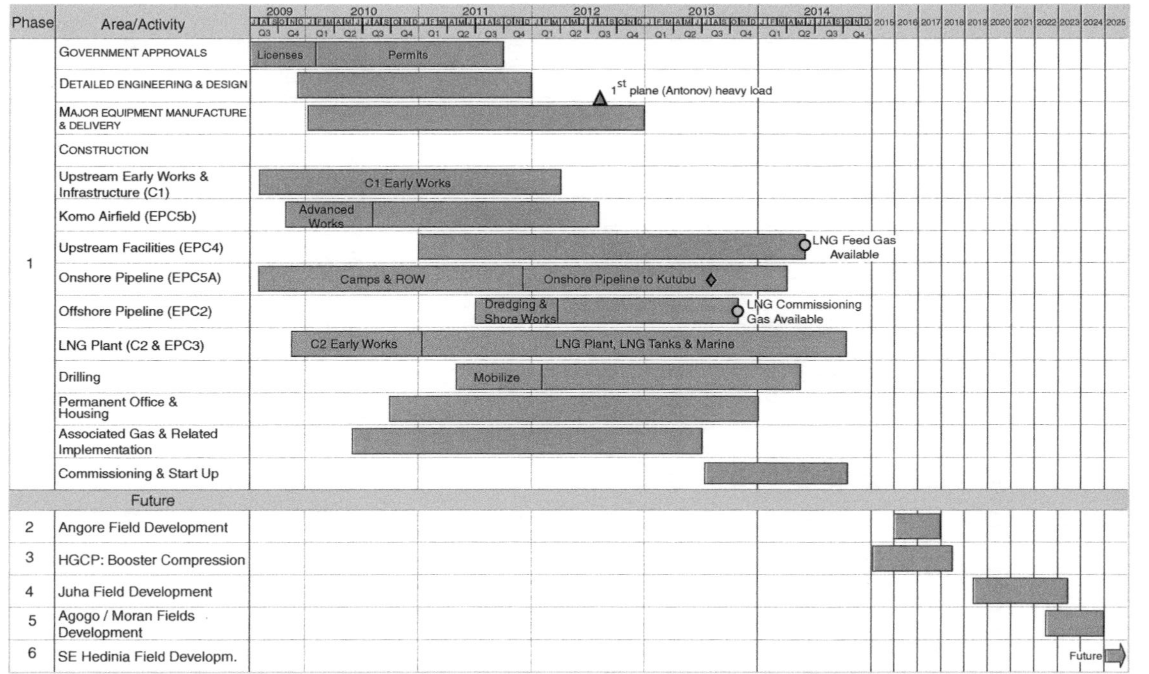

Note: This was the project schedule at the time of the 1Q 2010 Report but may not be the current schedule. It is provided here for illustrative purposes.

Source: PNG 1Q 2010 Report

Figure 9.8 Phases of development for PNG LNG

Guinean Government and holder of the majority of state-owned commercial assets; Santos (13.5 percent), Australia's largest domestic gas producer; Nippon Oil and Gas Exploration (4.7 percent), Japan's largest integrated oil company; Mineral Resources Development Company (MRDC) (2.8 percent), trustee and manager for landowner interests in PNG oil and gas; and Petromin PNG Holdings (0.2 percent).[114]

9.5.3 Economic Benefits and Revenue Sharing

The Project sponsors and lenders maintain that the Project has the potential to "transform the economy of Papua New Guinea" by, among other economic benefits, boosting GDP and export earnings, providing employment opportunities, and providing royalty payments to landowners.[115]

In fall 2009, the State of Papua New Guinea, representatives of Project area landowners, and various provincial and local level governments executed the Project License based Benefit Sharing Agreement as a first step in the allocation and sharing of the revenues associated with the Project. In general, the agreement outlines the sharing of revenue streams from royalties, development taxes and equity dividends totaling approximately US$5.6–$7.5 billion over the Project life. Under the agreement, a portion of the state's project equity will go to landowners and provincial governments. In addition, US$432 million was committed over the next ten years for infrastructure development such as roads, bridges, airports and townships.[116]

9.5.4 Environmental and Social Plans

Since Esso, a subsidiary of ExxonMobil, operates the project, ExxonMobil's management systems regarding all safety, security, health, environmental and social aspects underpin the entire Project and are quite broad and comprehensive (Figure 9.9).

9.5.5 Project Financing and Environmental Requirements

The PNG LNG Project secured financing from various export credit agencies and commercial banks and these lenders require the Project to conform to a number of environmental and social principles and

---

[114] Ibid.
[115] Ibid.
[116] Ibid.

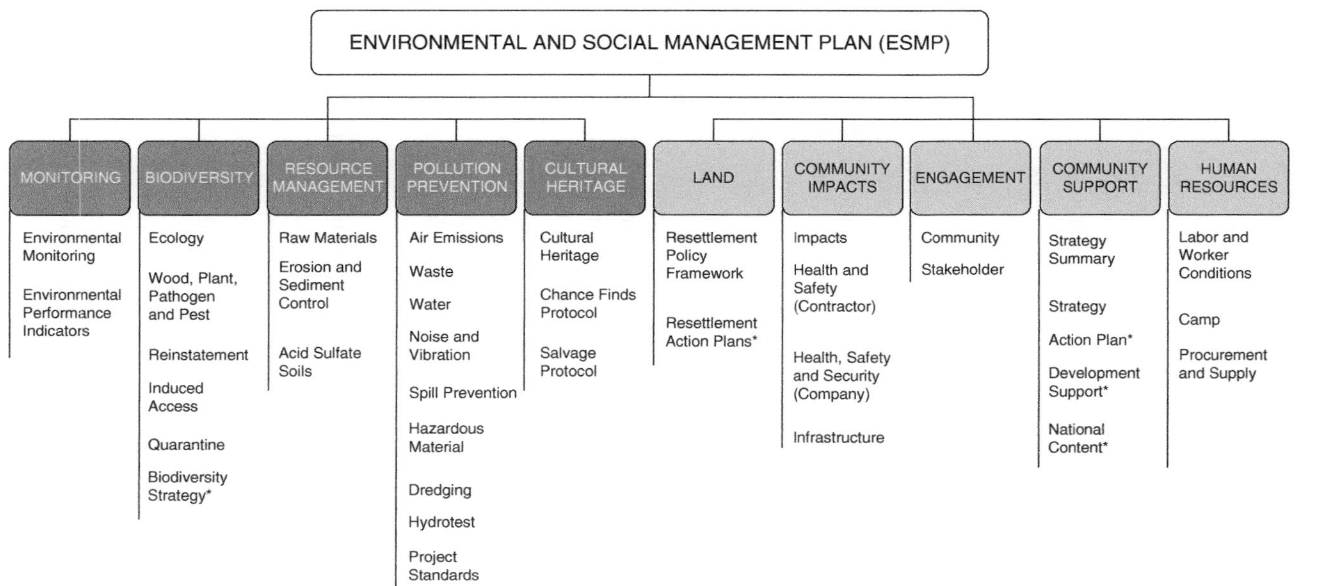

Source: PNG LNG, http://pnglng.com/commitment

Figure 9.9 Environmental and Social Management Plan

standards.[117] For example, the Project assumed numerous environmental requirements in conjunction with the Ex-Im Bank financing. The material requirements of the Ex-Im Bank are summarized below.[118]

The Project must comply in all material respects with the approved environmental and social management plan, national environmental laws and lender environmental and social standards (including the Ex-Im Bank Environmental Procedures and Guidelines). Lenders must be promptly notified of any material non-compliance with the social or environmental requirements or the occurrence of events relating to social or environmental matters that could cause significant adverse risk.

The Project must provide periodic Environmental and Social Reports on the results of monitoring of the Project's compliance with the Environmental and Social Management Plan. The Reports will be submitted quarterly until the construction is completed and on a semiannual basis for the first year following completion, and annually thereafter.

In addition to the above requirements, numerous environmental reports and plans were required to be submitted by specified due dates, including the Construction Phase Environmental and Social Management Plan, Environmental and Social Milestones Schedule, Environmental Monitoring and Reporting Plan, Social Monitoring Plan, Operations Environmental and Social Management Plan, Operation Oil Spill Response Plan, Noise and Vibrations Plan, Ambient Noise and Vibration Plan, Traffic Management Plan, Biodiversity Strategy, Biodiversity Monitoring Program, Community Support Strategy, Cultural Heritage Management Plan and various Resettlement Plans.

In addition to the above required reports, the Project elected to make publicly available the Environmental and Social Compliance Reports prepared by the Independent Environmental and Social Consultants by posting the reports on the Project website.[119]

9.5.6 The Role of the IFC Guidelines in Promoting Better Environmental Practices

The PNG LNG Project is a great resource for learning more about environmental issues associated with LNG projects and numerous environmental documents have been made publicly available and are easily

---

[117] Ibid.

[118] Ex-Im Bank, *Papua New Guinea LNG Project AP084099XX Environmental Requirements Summary*, http://www.exim.gov/products/policies/environment/png_liquified_natural_gas_project.cfm.

[119] PNG LNG, http://www.pnglng.com/quarterly_reports/.

accessed at the PNG LNG Project website.[120] As an IFC financed project, the PNG LNG Project also illustrates the role that the IFC Guidelines (discussed in more detail in Chapter 7) can play in LNG Projects. For example, the contractors in the PNG LNG Project are required to comply with all applicable laws of Papua New Guinea as well as the applicable regulations of the IFC.

Thus, although PNG has no specific air emissions or ambient air quality standards, the IFC Performance Standards apply to air emissions and ambient air quality during construction. These standards generally establish requirements for the assessment, management, and abatement of pollution from project activities with the goal of promoting the reduction of emissions.[121] This is just one example of the many areas where the IFC Guidelines might play a role in promoting better environmental practices in the construction of LNG projects.

9.5.7 The Environmental Impact Statement (EIS)

Chapter 26 of the Environmental Impact Statement (EIS) in the PNG LNG Project provides a detailed analysis and assessment of the greenhouse gas emissions associated with the construction and operation of the Project.[122] The chapter begins by outlining the context of greenhouse gas emissions in Papua New Guinea and its contribution to future global carbon reductions (Section 26.1), then outlines the methods of assessment (Section 26.2), describes the predicted emissions from the Project during its life-cycle (Section 26.3), benchmarks the emissions against other liquefied natural gas facilities (Section 26.4), outlines the Project design features to maximize energy efficiency (Section 26.5) and then provides an assessment of the residual impact (Section 26.6).

At the outset, the EIS notes that greenhouse gas intensity refers to the volume of greenhouse gases emitted per unit of energy or economic output. The greenhouse gas intensity of the facilities is influenced by a range of internal (technology) and external (environmental/geological) factors. The main factors that have a significant impact on greenhouse

---

120 PNG LNG, http://www.pnglng.com/commitment/.

121 PNG LNG, Environmental and Social Management Plan, Appendix 2: *Air Emission Management Plan*, http://www.pnglng.com/media/pdfs/commitment/Appendix_2_PGGP-EH-SPENV-000018-004_Air_Rev_2_Nov_22.pdf.

122 PNG LNG Project, environment impact statement (EIS), Chapter 26, *Greenhouse Gases and Climate Change*, http://www.pnglng.com/media/pdfs/environment/eis_chapter26.pdf.

efficiency are the reservoir CO_2 content (a higher CO_2 content equates to more CO_2 removal prior to liquefaction) and ambient temperature (compressor efficiency is favored by cooler temperatures). Other technological and process factors that influence greenhouse gas intensity include: the choice of liquefaction technology; the power generation source, technology and configuration; waste heat recovery; and the acid gas removal process.

9.5.8 Project Design[123]

The proposed PNG LNG Project includes a number of design features and industry good practice initiatives that will ensure that energy efficiency is maximized and greenhouse gas emissions are minimized. For example, waste heat will be recovered and used as an energy source at the Hides Gas Conditioning Plant and the LNG Plant, which will reduce fuel gas consumption and greenhouse gas emissions that would otherwise be associated with providing an alternative heat source. At the Hides Gas Conditioning Plant, waste heat from the exhaust of the pipeline compressor gas turbines will be used to provide heat to the thermal-fluid-based, hot-oil system. Similarly at the LNG Plant, waste heat recovery units will utilize heat from the exhaust from the aero-derivative turbines driving the two propane refrigeration compressors to provide the main source of heat to the hot-oil system.

9.5.9 Greenhouse Gas Emissions[124]

In assessing the greenhouse gas emissions, the EIS acknowledged that the Project would contribute to Papua New Guinea's total emissions. The PNG LNG Project is expected to generate total annual greenhouse gas emissions of 3.1 Mt CO_2–e. Over the 30-year life of the Project, total emissions are estimated at 77 Mt CO_2-e. The LNG Plant accounts for 75 percent of total emissions, the Hides Gas Conditioning Plant accounts for 20 percent of total emissions, and the balance of activities (shipping/Juha production facility) accounts for the remaining 5 percent of total emissions. According to the analysis in the EIS, the low level of CO_2 in the feed gas and the choice of aero-derivative technology have resulted in the overall greenhouse gas emissions for the Project comparing favorably with other similar LNG operations around the world.

---

123 EIS, Chapter 26.5.
124 EIS, Chapter 26.3, 26.4, 26.6.

Placed in context, the EIS noted that PNG contributes only a small proportion (less than 0.2 percent) of worldwide carbon dioxide emissions. In general, emissions in PNG have increased from 2.58 Mt in 2001 to 4.35 Mt in 2005. As a comparison, Australia's CO_2 emissions from fossil fuel combustion increased from 367 Mt in 2001 to 407 Mt in 2005.[125]

PNG ratified the United Nations Framework Convention on Climate Change (UNFCC) in 1993 and the Kyoto Protocol in 2002. In March 2008, PNG entered into a cooperative agreement with Australia to reduce greenhouse gas emissions from deforestation: the Papua New Guinea–Australia Forest Carbon Partnership. Nearly two-thirds of PNG's land area is forested (more than 29 million hectares).

Additionally, Papua New Guinea's rainforests are being targeted for carbon emission reduction schemes under the reduced emissions from deforestation and degradation (REDD) mechanism provided for under the UNFCC and Kyoto. The REDD mechanism is aimed at offsetting carbon emissions by protection of forest that would otherwise have been degraded by logging or other means.

9.5.10 Should Greenhouse Gas Emissions be Assessed on a Global Basis?

The EIS for the PNG LNG Project found that while the Project will be a contributor to Papua New Guinea's total emissions, "in a global context, the production and export of LNG from this project will represent a reduction in global greenhouse gas emissions compared to if customers were to use other fossil fuel sources for their energy requirements (e.g., coal, fuel oil or diesel)."[126] This statement is supported by the following assertions:[127]

- Gas-fired electricity generation produces less than half the greenhouse gas emissions of coal-fired generation, and uses a minute fraction of the water that coal-fired electricity requires.
- Every million tonne of LNG that replaces coal-fired power generation is equal to taking more than 500,000 cars off the road.

---

125 EIS, Chapter 26.1.

126 EIS, Chapter 26.6, "Residual Assessment".

127 PNG LNG, *Greenhouse Gas Emissions*, Factsheet, http://pnglng.com/media/pdfs/publications/PNG_LNG_Greenhouse_14.pdf.

- For every tonne of carbon dioxide emitted in LNG production, 4 tonnes of emissions from the coal alternative could be avoided in Japan. Between 5.5 and 9.5 tonnes can be avoided in China. (Source: APPEA Submission to the Carbon Pollution Reduction Scheme Green Paper.)

The analysis used in PNG LNG to justify the increase in domestic emissions by claiming that global emissions will be reduced is the same analysis used in the Australian LNG projects and the WorleyParsons study discussed in more detail in Chapter 7. How this analysis will play out over time remains to be seen and, as with the Australian projects, the PNG LNG Project will provide valuable insight into the emissions generated by these large-scale LNG projects.

It should also be noted that the Ex-Im bank also used this same analysis when reviewing the PNG LNG Project, finding that the production and export of LNG from the Project will "represent a net reduction in global greenhouse gas emissions compared to the case where customers were to meet their energy requirements by coal, fuel oil or diesel commonly used in the regional market, even though the project will add to Papua New Guinea's total emissions of greenhouse gasses."[128]

Some environmental groups have criticized the Ex-Im Bank's financing of fossil fuel projects such as PNG LNG claiming that such projects cause significant harm to the environment, contribute to global climate change, and hinder efforts to export clean technologies.[129]

In March 2010, the Ex-Im Bank announced a new carbon policy designed to ramp up financing for renewable energy and impose new reviews of large fossil fuel projects as part of a broad new climate change strategy. The announcement was part of a 2009 settlement resolving a lawsuit involving Friends of the Earth, Greenpeace and several cities that had alleged that the Ex-Im Bank had provided more than $32 billion to fossil fuel projects without considering the impacts of global warming as required under the National Environmental Policy Act (NEPA).[130]

---

[128] Ex-Im Bank Press Release.

[129] Environmental Leader, *Ex-Im Bank's Carbon Policy Criticized by Environmental Groups*, March 15, 2010, http://www.environmentalleader.com/2010/03/15/ex-im-banks-carbon-policy-criticized-by-environmental-groups/.

[130] Lisa Friedman, *Ex-Im Bank Approves New Scrutiny of Fossil Fuel Projects*, New York Times, March 10, 2010, http://www.nytimes.com/

9.5.11 Project Status and Ongoing Challenges

In May 2010, PNG LNG issued the first quarterly report[131] indicating that with the completion of the financing arrangements, the Project was now moving into the full execution phase. Managing Director of Esso Highlands, Peter Graham, commented that the Project sponsors "are developing this challenging project in a manner that reflects our high standards in business and operational integrity, and importantly in safety, security, health, environmental and social management." He stressed that "The benefits that flow from the project will support the PNG government's objective to strengthen its economy and infrastructure base for the benefit of its people. This is the first in a series of reports that details our progress in this challenging project."[132]

In September 2010, the numerous challenges the Project faces became more evident when local villagers attacked the Project, burning heavy machinery and using high-powered weapons to damage construction equipment near the site.[133]

Papua New Guinea's minister of planning and development has acknowledged that there has been "infighting" among the landowners affected by the Project, but maintained that the Project is still expected to stay within the original cost estimate and start exporting LNG in 2014. According to the minister, some of the approximately 60,000 landowners affected by the Project question the division of benefits from the Project with some LNG benefit-sharing negotiations resulting in violence as various tribal groups argue over a share of the Project.[134]

The September 2010 incident was not the first one for the Project. News accounts have indicated that in Feburary 2010 Esso Highlands stopped work after four people were killed when fighting broke out over a land dispute. In mid-August 2010, issues with landowners forced a shutdown of "some work activities."[135]

---

cwire/2010/03/10/10climatewire-ex-im-bank-approves-new-scrutiny-of-fossil-f-82557.html.

[131] PNG 1Q 2010 Report.

[132] Jane Dawson, *PNG LNG Report Shows "Progress" in Challenging Conditions*, GasWorld, June 9, 2010, www.gasworld.com.

[133] *Exxon Mobil's LNG Project in PNG Attacked*, Reuters, Sept. 27, 2010, http://www.reuters.com/article/2010/09/28/exxon-png-idUSSGE68R00C20100928.

[134] Ibid.

[135] Jonty Rushforth, *ExxonMobil Subsidiary Halts Some PNG LNG Work after*

While some have suggested that the Southern Highlands area of Papua New Guinea where most of the conflicts have been centered are too diverse for widespread insurrection, some local leaders have expressed concern that the benefits of the LNG Project will fall short of expectations, leading locals to turn their anger on the foreign energy companies building the Project. Others have called into question whether Papua New Guinea is ready to manage the potential wealth that might be generated by the Project and still others have expressed concern about possible corruption within the government that will result in a windfall for some, but not the masses.[136]

In order to stem further conflicts, the United States has offered to provide whatever help it can to help the country manage its coming resource revenues, including assisting with the creation of a sovereign wealth fund.[137] In a visit to Papua New Guinea in November 2010, US Secretary of State Hillary Clinton said, "There is a phrase 'resource curse' where . . . an abundance of natural resources like oil and gas or gold or minerals, if they are not handled right can actually (make) a country poorer instead of richer." As such, Secretary Clinton said the US was ready to help PNG translate its natural resources "into widespread prosperity." Secretary Clinton also noted that the United States would soon be breaking ground on a new embassy in PNG and that the US wants to provide technical training for the country's scientists and engineers as well as job training so that it is PNG residents "who take the jobs that are created."[138]

In the meantime, the Project is progressing from a construction standpoint but the conflicts are ongoing. In January 2011, Esso Highlands again shut down some work after a group of people entered a Project camp in the Southern Highlands area.[139] In May 2011, a group of landowners in the Highlands area threatened to shut down the Project if the PNG government did not take action on a long list of grievances. The

---

"Invasion", Platts, Jan. 24, 2011, http://www.platts.com/RSSFeedDetailedNews/RSSFeed/Oil/7963298. Singapore (Platts)--24Jan2011/.

[136] Norimitsu Onishi, *Riches May Not Help Papua New Guinea*, The New York Times, Oct. 25, 2010, http://www.nytimes.com/2010/10/26/world/asia/26papua.html?_r=1&ref=papuanewguinea&pagewanted=1.

[137] *U.S. Offers Help for Papua New Guinea LNG*, UPI, Nov. 4, 2010, http://www.upi.com/Business_News/Energy-Resources/2010/11/04/US-offers-help-for-Papua-New-Guinea-LNG/UPI-22931288892042/#ixzz1divFA2uH.

[138] Ibid.

[139] Jonty Rushforth, *ExxonMobil Subsidiary Halts Some PNG LNG Work after "invasion"*, Platts, Jan 24, 2011, http://www.platts.com/RSSFeedDetailedNews/RSSFeed/Oil/7963298. Singapore (Platts)--24Jan2011/.

impetus for this recent threat appears to have been the death of Tuguba Chief Himuni Homoko. Chief Himuni Homoko had been fighting for his Tuguba tribe over the way in which the government is handling the terms of the main revenue sharing agreement – the Kokopo agreement. While a bit unclear, it appears that the primary claim is that the government of PNG is not distributing funds quickly enough, as opposed to a grievance with the actual project.[140]

As of the time of this writing, it appears that the conflicts in Papua New Guinea can be managed so as not to become obstacles to the Project.[141] However, this will require much effort on everyone's part so that the conflicts do not turn into another "Bougainville."[142]

---

140 ABC Radio Australia, *Landowners Threaten to Shutdown PNG LNG Project*, May 26, 2011, http://www.radioaustralia.net.au/pacbeat/stories/201105/s3228006.htm.

141 In May 2012, ExxonMobil announced that the "PNG LNG Project onshore pipeline has reached a significant milestone, with 50% of the LNG gas pipeline now welded. The 292-kilometre onshore pipeline is being constructed to transport the gas from the Hides Gas Conditioning Plant, which is currently under construction, down to the Omati area where it will connect to the offshore pipeline. Elsewhere in the Project, progress is being made to meet first gas in 2014 with the first LNG tank roof installed and the 2.4km jetty trestle nearing completion at the LNG Plant site, and the offshore pipeline construction almost finished." ExxonMobil Media Release, *Onshore Pipeline Reaches Halfway Milestone*, May 22, 2012, http://pnglng.com/media/pdfs/media_releases/media_release20120523-PipelineReachesHalfwayMilestone.pdf.

142 Australia Network News, *In-depth with the Chairman of Bougainville Copper Ltd*, Feb. 21, 2011, http://australianetworknews.com/story.htm?id=37446. The Panguna copper mine on Papua New Guinea' island of Bougainville was closed in 1988 after landowner anger over the share of benefits from the mine and its environmental impacts boiled over into a bloody ten-year civil war.

10. The role of shale gas in the Golden Age of Gas

10.1 CONVENTIONAL AND UNCONVENTIONAL NATURAL GAS RESERVOIRS

Although natural gas seems poised to enter the golden age, this may ultimately depend on whether countries around the globe develop the vast resources of unconventional gas – shale gas, tight gas and coal bed methane – that exist in almost every region of the world.[1] Since the development of unconventional gas resources is different and more challenging than conventional resource development, a basic understanding of the different types of gas reservoirs is helpful in order to appreciate the difficulties involved in extracting natural gas from certain types of reservoirs. In general, gas reservoirs are classified as conventional or unconventional based on the following.[2]

Conventional reservoirs In a conventional reservoir, natural gas has migrated from a source rock into a "trap" that is capped by an impermeable layer of rock.[3] Conventional gas reservoirs are often associated with deposits of oil and are often developed in conjunction with oil.[4] In conventional gas reservoirs, a traditional well may simply be drilled directly into the reservoir.[5] Because the sands or rock that contain the gas have

---

[1] IEA, *Golden Rules for a Golden Age of Gas*, World Energy Outlook Special Report on Unconventional Gas, May 29, 2012, http://www.worldenergyoutlook.org/media/weowebsite/2012/goldenrules/WEO2012_GoldenRulesReport.pdf (hereinafter IEA Golden Rules Report).

[2] Ground Water Protection Council & ALL Consulting, MODERN SHALE GAS DEVELOPMENT IN THE UNITED STATES: A PRIMER 15 (2009), http://www.netl.doe.gov/technologies/oil-gas/publications/EPreports/Shale_Gas_Primer_2009.pdf (hereinafter Ground Water Prot Council, MODERN SHALE GAS PRIMER).

[3] See Jacqueline Lang Weaver, TEXAS OIL AND GAS LAW: CASES AND MATERIAL 1–7 (2009) (discussing conventional geology and methodology of oil and gas production).

[4] Ibid.

[5] Ibid.

interconnected pore spaces and are thus permeable in nature, the gas flows naturally to the wellbore.[6]

Unconventional reservoirs In an unconventional reservoir, natural gas must be extracted from the source rock itself using a variety of production techniques including hydraulic fracturing and horizontal drilling.[7] Because of the low permeability of unconventional reservoirs, these techniques are used to stimulate the reservoir – by creating fissures in the rock, the gas flows more easily through it, enhancing production.[8] There are three types of unconventional gas reservoirs:

1. *Tight gas* Tight gas commonly refers to natural gas that is trapped in sandstones.[9]
2. *Coal bed methane (CBM)* CBM is natural gas that is produced from coal seams, which act as the source and reservoir for the natural gas.[10]
3. *Shale gas* "Shale gas is natural gas produced from shale formations that typically function as both the reservoir and source for the natural gas."[11] The economic potential of a particular shale formation can be evaluated by identifying specific source rock characteristics.[12] These characteristics are used to predict whether commercial volumes can be produced from the shale formation.[13] A number of wells may need to be drilled and analyzed in order to sufficiently determine the potential of the shale formation, especially if the basin is large and the targeted zones varied.[14]

10.2 OVERVIEW OF THE SHALE GAS "REVOLUTION"

In recent years, there has been a dramatic increase in both the production of shale gas in the United States and the potential for worldwide shale

---

[6] Ibid.

[7] See Ground Water Prot. Council, MODERN SHALE GAS PRIMER, at 15.

[8] Ibid. see Chesapeake Energy, *Hydraulic Fracturing Fact Sheet* 1 (2010), http://www.chk.com/Media/CorpMediaKits/*Hydraulic_Fracturing_Fact_Sheet*.pdf (hereinafter Hydraulic Fracturing Fact Sheet).

[9] Enerdynamics, *The Rise of Unconventional Gas*, The Energy Insider, Sept. 18, 2007, at 4.

[10] Ground Water Prot. Council, MODERN SHALE GAS PRIMER, at 15.

[11] Ibid., at 14.

[12] Ibid. at 16.

[13] Ibid.

[14] Ibid.

gas production.[15] As a result, shale gas has been referred to as a "game changer," and for this reason this chapter focuses on shale gas exclusively although tight gas and coal bed methane are also important potential future sources of natural gas supplies.

Although experts have known for years about the vast deposits of shale gas found throughout the world, technological difficulties and the high costs of producing shale gas historically made it impractical to consider as a serious energy source.[16] More recently, however, technological innovations combining hydraulic fracturing and horizontal drilling technologies[17] have resulted in a tremendous boom in shale gas production in the United States since about 2007.[18] This boom seems likely to continue with leading energy experts proclaiming shale gas an energy "game changer" that will "revolutionize" global gas markets and help bridge the gap between conventional resources and the development of renewable energy sources.[19]

Thus far, the United States has been the undisputed leader in unlocking the vast tracts of gas-bearing shale found throughout the lower 48 states, but the so-called "shale gale," the strong wind blown by the technological advances in hydraulic fracturing and horizontal drilling, is not limited to only North America. Because shale formations exist in almost every region of the world, the potential for shale gas development is enormous and global in scope.[20]

---

15 See Am. Petroleum Inst., *Facts about Shale*, http://www.api.org/policy/exploration/hydraulicfracturing/shale_gas.cfm.

16 See Halliburton, *U.S. Shale Gas: an Unconventional Resource, Unconventional Challenges* 1 (2008), http://www.halliburton.com/public/solutions/contents/Shale/related_docs/H063771.pdf.

17 The hydraulic fracturing technology has been so successful that energy experts have called this the "most significant energy innovation so far of this century." Mary Lashley Barcella and David Hobbs, *Fueling North America's Energy Future*, Wall St J., Mar. 10, 2010, at A10.

18 See Am. Petroleum Inst., *Hydraulic Fracturing*, http://www.api.org/policy/exploration/hydraulicfracturing; see also Am. Petroleum Inst., *Advanced Drilling Techniques*, http://www.api.org/aboutoilgas/natgas/drilling_techniques.cfm (explaining "horizontal drilling" techniques).

19 See Tom Fowler, *Energy Game-Changer?*, Hous. Chron., Nov. 1, 2009, at A1.

20 See Leta Smith and Peter Jackson, *Is Unconventional Gas Going Global?*, Wall St J., Mar. 10, 2010, at A14, available at www2.cera.com/ceraweek2010/NAm2010-03-10.pdf. A new EIA sponsored study on global shale gas resources reports an initial assessment "of 5,760 trillion cubic feet (Tcf) of technically recoverable shale gas resources in 32 foreign countries, compared with 862 Tcf in the United States." Today in Energy, *Shale Gas is a Global Phenomenon*, US Energy Info. Admin. (2011), http://www.eia.doe.gov/todayinenergy/detail.cfm?id=811.

Since hydraulic fracturing is an essential part of developing global shale gas resources,[21] it is imperative that the industry ensures that the process is safe and environmentally sound before applying the technology to more areas of the world.[22] In the United States and elsewhere, numerous concerns have been raised about the potential environmental impacts of hydraulic fracturing and shale gas development with the IEA recently stating that although there are numerous factors that will impact the development of unconventional gas resources, "there is a critical link between the way that governments and industry respond to [the] social and environmental challenges" associated with shale gas development and the prospects for increased production.[23]

10.3 THE TRICK TO UNLOCKING GAS FROM SHALE: HORIZONTAL DRILLING AND HYDRAULIC FRACTURING

The trick that has enabled companies to unlock natural gas from shale rock involves the combination of two production technologies – horizontal drilling and hydraulic fracturing.[24] Although these two technologies have been around for decades, the combination of the two, coupled with technological advances in equipment and cost reductions, was the key to unlocking the vast reserves of shale gas in North America.[25]

Horizontal drilling has been instrumental in increasing production volumes from all forms of natural gas and oil wells and is used extensively in shale gas production.[26] Horizontal drilling involves drilling a vertical well to intersect the shale formations found at various depths ranging from 1,000 to more than 13,000 feet. Before the targeted depth is reached, the well is deviated, or turned to achieve a horizontal wellbore within the shale formation. Wells can be oriented in a direction that is designed to maximize

---

21 See Halliburton, at 1; see also Am. Petroleum Inst., *Hydraulic Fracturing*, at 3.

22 See Hannah Wiseman, *Untested Waters: The Rise of Hydraulic Fracturing in Oil and Gas Production and the Need to Revisit Regulation*, 20 Fordham Envtl L. Rev. 115, 116 (2009).

23 IEA Golden Rules Report at p. 9.

24 See *Coastal Oil & Gas Corp. v. Garza Energy Trust*, 268 S.W. 3d 1, 6 (Tex. 2008) (Texas Supreme Court describing the "fracking" process); see also Hydraulic Fracturing Fact Sheet, at 27.

25 See US Energy Info. Admin., Annual Energy Outlook 2010, http://www.eia.doe.gov/oiaf/aeo/pdf/trend_4.pdf.

26 Ground Water Prot. Council, MODERN SHALE GAS PRIMER at ES-3.

the number of natural fractures present in the shale and these natural fractures can provide pathways for the gas to flow into the wellbore. Horizontal drilling provides greater exposure to the shale formation, which in turn optimizes natural gas recovery and improves well economics.[27]

Once the targeted area is reached by horizontal drilling, hydraulic fracturing is then used to help produce the gas reservoir. Beginning at the tow of the long horizontal section of the well, segments of the wellbore are isolated, the casing is perforated and hydraulic fracturing fluid is pumped under high pressure (thousands of pounds per square inch) through the perforations, cracking the shale and creating one or more fractures that extend out into the surrounding rock.[28] These fractures continue to propagate for hundreds of feet or so until the pumping ceases. Proppants, the most common being sand, are carried along in the hydraulic fracturing fluids and these proppants open the fracture after pumping stops and the pressure is relieved. The propped fracture is only a fraction of an inch wide, held open by the grains of sand and hydraulic fracturing fluids. The hydraulic fracturing process and the resulting fracturing creates the pathways for the oil and gas to enter the wellbore so the fluids can be pumped to the surface.[29] (See Figure 10.1.)

The hydraulic fracturing of shale gas wells is performed in numerous stages with each stage using a series of different volumes and compositions of fracturing fluids.[30] A typical shale gas well may involve four or more stages that use millions of gallons of water-based fracturing fluids mixed with a variety of proppant materials and chemical additives.[31] This process raises a number of issues in terms of water usage and availability and chemical composition.

10.3.1 Hydraulic Fracturing Fluids

A key component of hydraulic fracturing is the high-pressure injection of hydraulic fracturing fluids[32] that increases the permeability of the rock by

---

27 See *Advanced Drilling Techniques*.

28 Am. Petroleum Inst., *Freeing Up Energy, Hydraulic Fracturing: Unlocking America's Natural Gas Resources*, 5 (2010), http://www.api.org/~/media/Files/Policy/Exploration/HYDRAULIC_FRACTURING_PRIMER.ashx (hereinafter API Freeing Up Energy).

29 Ibid.

30 Ground Water Prot. Council, MODERN SHALE GAS PRIMER at p. 58.

31 Ibid. at 60–61.

32 See Hydraulic Fracturing Fact Sheet, *supra* note 8; see also Envtl Prot. Agency, EVALUATION OF IMPACTS TO UNDERGROUND SOURCES OF DRINKING WATER BY HYDRAULIC FRACTURING OF COALBED METHANE RESERVOIRS STUDY, at

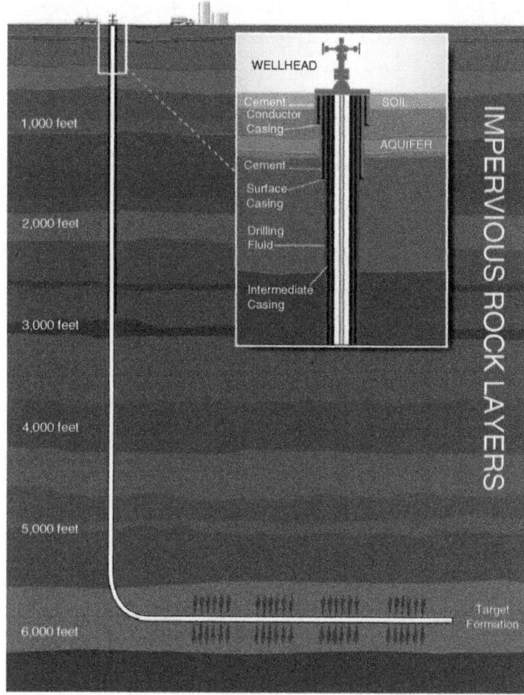

Typically, steel pipe known as surface casing is cemented into place at the uppermost portion of a well for the explicit purpose of protecting the groundwater. The depth of the surface casing is generally determined based on groundwater protection, among other factors. As the well is drilled deeper, additional casing is installed to isolate the formation(s) from which oil or natural gas is to be produced, which further protects groundwater from the producing formations in the well.

Casing and cementing are critical parts of the well construction that not only protect any water zones but are also important to successful oil or natural gas production from hydrocarbon bearing zones.

Industry well design practices protect sources of drinking water from the other geologic zone of an oil and natural gas well with multiple layers of impervious rock.[1]

[1] Industry has developed equipment-specific and operating practices for use in drilling and production activities. Examples include: API 5 Series Publications: Tubular Goods; API 7 Series Publications: Drilling Equipment; API 10 Series Publications: Oil Well Cements; API 11 Series Publications: Production Equipment; API 13 Series Publications: Drilling Fluid Material.

Source: API Freeing Up Energy

Figure 10.1 Hydraulic fracturing for shale development

"propping up" or holding open the fractures.[33] According to the industry, fracturing fluid is a mixture of about 90 percent water, 9.5 percent sand, and 0.5 percent other chemicals (Figure 10.2).[34]

Although water is the main component of hydraulic fracturing fluids, a number of additives and chemicals are also used, the number varying based on the conditions of the specific well being fractured and thus no "one-size fits all formula for the volumes for each additive."[35] The chemical additives used include "common chemicals which people regularly encounter in everyday life" as well as "chemical additives that could be

4-1 (2004), http://water.epa.gov/type/groundwater/uic/class2/hydraulicfracturing/wells_coalbedmethanestudy.cfm (hereinafter DRINKING WATER IMPACT STUDY).

[33] Ibid.

[34] API Freeing Up Energy.

[35] Ibid. at 61.

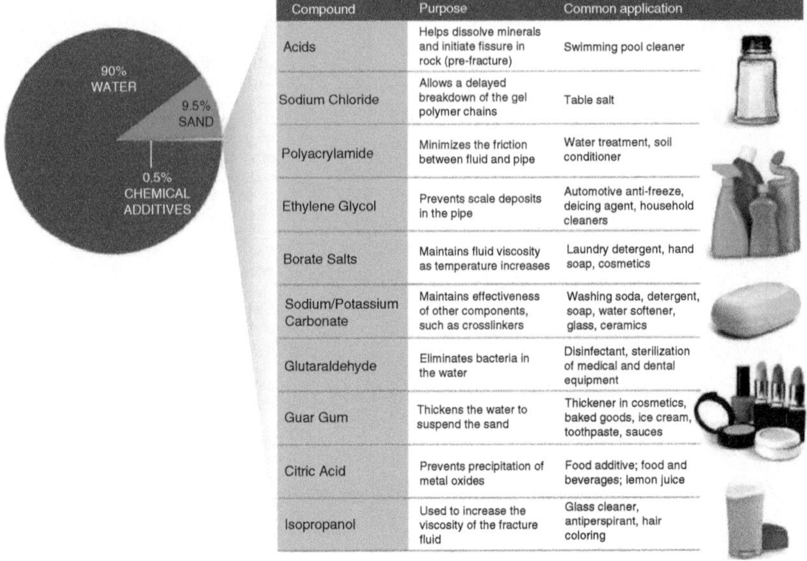

Compound	Purpose	Common application
Acids	Helps dissolve minerals and initiate fissure in rock (pre-fracture)	Swimming pool cleaner
Sodium Chloride	Allows a delayed breakdown of the gel polymer chains	Table salt
Polyacrylamide	Minimizes the friction between fluid and pipe	Water treatment, soil conditioner
Ethylene Glycol	Prevents scale deposits in the pipe	Automotive anti-freeze, deicing agent, household cleaners
Borate Salts	Maintains fluid viscosity as temperature increases	Laundry detergent, hand soap, cosmetics
Sodium/Potassium Carbonate	Maintains effectiveness of other components, such as crosslinkers	Washing soda, detergent, soap, water softener, glass, ceramics
Glutaraldehyde	Eliminates bacteria in the water	Disinfectant, sterilization of medical and dental equipment
Guar Gum	Thickens the water to suspend the sand	Thickener in cosmetics, baked goods, ice cream, toothpaste, sauces
Citric Acid	Prevents precipitation of metal oxides	Food additive; food and beverages; lemon juice
Isopropanol	Used to increase the viscosity of the fracture fluid	Glass cleaner, antiperspirant, hair coloring

Source: API Freeing Up Energy

Figure 10.2 Typical shale fracturing fluid makeup and chemicals

hazardous, but are safe when properly handled."[36] The service companies that provide these additives have developed a number of different combinations to be used depending on the well characteristics.[37]

10.4 SHALE GAS DEVELOPMENT AND RESOURCES IN THE UNITED STATES

The production of shale gas has expanded particularly rapidly in the United States[38] with natural gas production from shale gas formations rising almost 65 percent from 2007 to 2008 alone.[39] According to the US EIA, shale gas represents the largest source of growth in US natural gas

---

36 Ibid. at 62.

37 Ibid.

38 US Dep't of State, *Global Shale Gas Initiative (GSGI)*, http://www.state.gov/s/ciea/gsgi/index.htm.

39 American Clean Skies Foundation, http://www.cleanskies.org/resources-supply.html.

production for the coming decades. The rapid development of US shale gas resources has dramatically transformed the global gas markets and led many experts to proclaim shale gas an energy "game-changer."[40]

The game-changing nature of shale gas in North America is not just due to increased production but is also due to significant increases in the estimated natural gas resource base. An influential study done in 2008 estimated that North America has 2,247 Tcf of natural gas resources, which is "about 100 years of production at current levels."[41] In June 2009, the Potential Gas Committee established by the University of Colorado School of Mines estimated the US natural gas resource base at 2,074 Tcf, the highest estimate ever released by that group.[42]

In terms of technically recoverable resources, the US EIA has estimated that the US has 827 Tcf of technically recoverable shale gas resources, although this number may change as more information becomes available from developing shale plays.[43] More recently, the US EIA has noted that there is uncertainty regarding the ultimate size of technically recoverable shale gas due to a number of factors. First, because many shale gas wells are only a few years old, their long-term productivity is untested and consequently the long-term production profiles of shale wells and their estimated ultimate recovery of oil and natural gas are uncertain. Second, in emerging shale plays, production has been confined largely to those areas known as "sweet spots" that have the highest known production rates for the play. If the production rates for the sweet spots are used to infer the productive potential of entire plays, their productive potential probably will be overstated. Third, many shale plays are so large (for example, the Marcellus shale) that only portions have been extensively production tested. Fourth,

---

40 Int'l Energy Agency, Press Release, *The Time Has Come to Make the Hard Choices Needed to Combat Climate Change and Enhance Global Energy Security, Says the Latest IEA World Energy Outlook*, Nov. 10, 2009, http://www.iea.org/press/pressdetail.asp?PRESS_REL_ID=294. See also Amy Myers Jaffe, *Shale Gas Will Rock the World*, Wall St J., May 10, 2010, available at http://online.wsj.com/article/SB10001424052702303491304575187880596301668.html.

41 Ibid., citing the July 2008 study, *North American Natural Gas Supply Assessment*, done by Navigant Consulting.

42 *Potential Gas Committee Reports Unprecedented Increase in magnitude of U.S. Natural Gas Resource Base*, June 18, 2009, available at http://www.mines.edu/Potential-Gas-Committee-reports-unprecedented-increase-in-magnitude-of-U.S.-natural-gas-resource-base.

43 US Energy Information Admin., Annual Energy Outlook (AEO) 2011 (US EIA AEO 2011), available at http://www.eia.gov/forecasts/aeo/pdf/0383(2011).pdf.

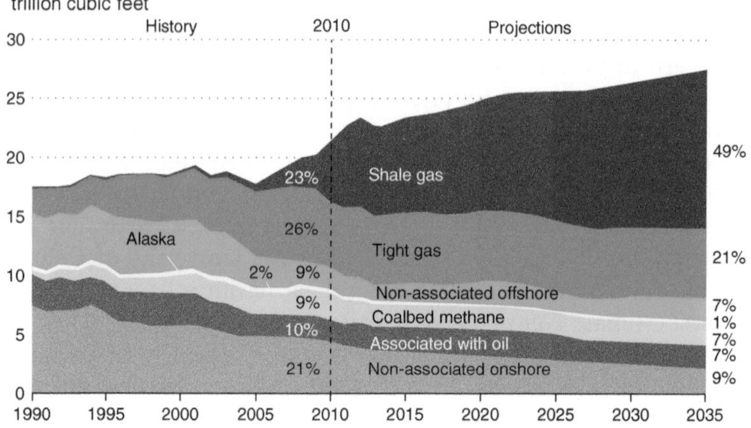

Source: US Energy Information Administration, AEO 2012, Early Release Overview, January 23, 2012

Figure 10.3 US natural gas production, 1990–2035

technical advancements could lead to more productive and less costly well drilling and completion. And lastly, currently untested shale plays, such as thin-seam plays or untested portions of existing plays, could prove to be highly productive.[44] As a result of these factors, estimating the technically recoverable shale gas resource base in the United States is an evolving process that is likely to continue for some time. Nonetheless, the general consensus is that the US shale gas resource base is significant, with the US EIA estimating that US production of shale gas is expected to increase and constitute 49 percent of total US natural gas supply by 2035.[45] (See Figure 10.3.)

10.4.1 Overview of Major US Shale Plays

In the United States, shale gas exists in most of the lower 48 states.[46] The most active shale basins to date are the Barnett Shale, the Haynesville/

---

[44] US EIA, *Review of Emerging Resources: U.S. Shale Gas and Shale Oil Plays*, Release date: July 8, 2011, available at http://www.eia.gov/analysis/studies/usshalegas/.

[45] US EIA, Annual Energy Outlook (AEO) 2012, Early Release Overview, Jan. 23, 2012, http://www.eia.gov/forecasts/aeo/er/.

[46] See generally US Energy Info. Admin., *Shale Gas Plays, Lower 48 States*, http://www.eia.doe.gov/oil_gas/rpd/shale_gas.pdf.

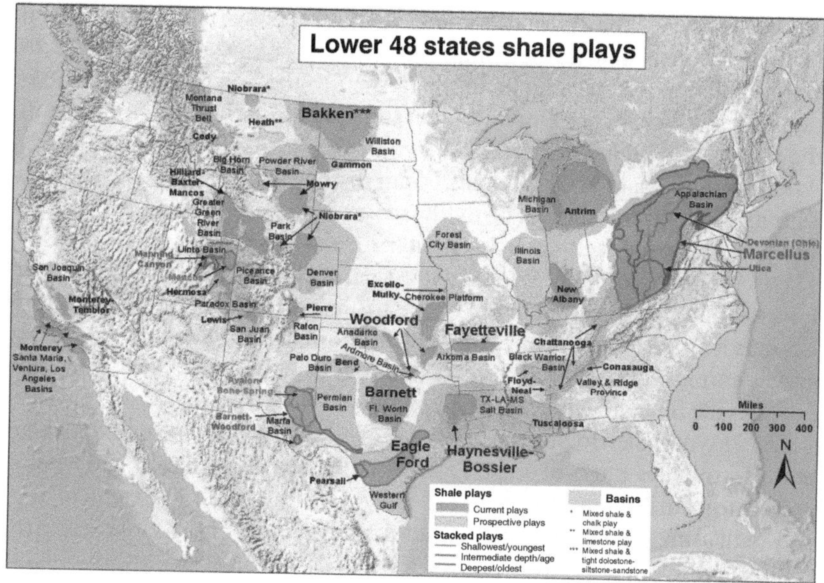

Source: US EIA

Figure 10.4 *Map of US shale basins*

Bossier Shale, the Antrim Shale, the Fayetteville Shale, the Marcellus Shale, the Eagle Ford, and the New Albany Shale.[47] (See Figure 10.4.)

The Barnett Shale is located in the Fort Worth Basin of north central Texas and was the first major shale play in the United States.[48] The prominence of the Barnett Shale and its record as one the busiest shale gas plays (home to more than 10,000 wells) in the United States is undisputed.[49] As one of the first of the modern shale plays, the Barnett Shale was the testing grounds for proving that the combined technologies of horizontal drilling and hydraulic fracturing could lead to the successful and economical development of shale gas.[50] With over 10,000 wells drilled to date, the Barnett Shale is widely considered the "gold standard" of US shale plays, having produced over 5 Tcf of natural gas.[51]

---

[47] Ground Water Prot. Council, MODERN SHALE GAS PRIMER, at ES-2.

[48] Ground Water Prot. Council, MODERN SHALE GAS PRIMER, at 13, 18.

[49] Ibid.

[50] Ibid. at 13.

[51] Hart Energy, Unconventional Oil & Gas Center, *Snapshot: Barnett Shale*, http://www.ugcenter.com/Barnett/Snapshot-Barnett-Shale_8105.

The success of the Barnett Shale grabbed the industry's attention and as production in the Barnett matured, natural gas producers looked to extrapolate the lessons learned in the Barnett to the other shale gas formations present across the United States and Canada.[52] The development of the Fayetteville Shale, which is situated in the Arkoma Basin of northern Arkansas and eastern Oklahoma, began in the early 2000s.[53] Companies who had reaped the success of the Barnett Shale were looking forward to applying the same techniques to similar formations, or new shale plays.[54] These companies quickly recognized the parallels between the Barnett and Fayetteville Shales – similar age of the formation and geologic character. Lessons learned from the horizontal drilling and hydraulic fracturing techniques employed in the Barnett assisted in the commercial viability of the Fayetteville Shale,[55] where more than 1,000 wells now exist.[56]

The Haynesville/Bossier shale play is mainly found in North Louisiana but also touches parts of East Texas.[57] Although there has already been exploratory drilling and testing for several years, "the full extent of the play will only be known after several more years of development are completed."[58]

The Marcellus Shale is "the most expansive shale gas play"[59] covering six states in the northeastern United States: New York, Pennsylvania, Ohio, Maryland, West Virginia, and Virginia.[60] Range Resources Corporation was the first company to drill economically producing wells in the Marcellus formation, with their success attributable to the use of horizontal drilling and hydraulic fracturing techniques, the same techniques used in the Barnett Shale in Texas.[61] Widely considered one of the most promising shale plays in the US, the Marcellus Shale has also been the center of environmental controversy, with New York having an effective moratorium on shale gas development pending further environmental review.

Other shale plays in the US include the Woodford Shale in south central Oklahoma, which is at "an early stage of development"[62] and the New Albany Shale located in the Illinois Basin and covering portions of Illinois,

---

52 Ground Water Prot. Council, MODERN SHALE GAS PRIMER.
53 Ibid. at 19.
54 Ibid.
55 Ibid.
56 Ibid.
57 Ibid. at 20.
58 Ibid.
59 Ibid. at 21.
60 Ibid.
61 Ibid.
62 Ibid. at 22.

Indiana and Kentucky.[63] The Antrim Shale is in the Michigan Basin[64] and next to the Barnett Shale; "the Antrim Shale has been one of the most actively developed shale gas plays."[65] Most of its expansion took place in the late 1980s.[66] As opposed to other gas shale plays in the United States, the Antrim has a shallow depth and "small stratisgraphic thickness."[67]

More recent developments in the US include a focus on "liquids-rich" shale plays such as the Eagle Ford as US natural gas producers seek to leverage liquids and light oil reserves to maximize the value of the production stream in gas shales. With the US Henry Hub price remaining persistently low (less than $4.00/Mcf), liquids-rich hybrid plays offer higher rates of return. For example, in some cases, there may be oil molecules valued at $15.00/Mcf versus $4.00/Mcf for gas with NGLs being valued somewhere in between. For this reason, the focus on liquids-rich plays in the US, and in particular the Eagle Ford, seems likely to continue into the future.[68]

10.5 SHALE GAS DEVELOPMENT AND RESOURCES IN CANADA

Canada has significant petroleum, natural gas, and coal reserves,[69] and along with Australia and Norway it is one of only three member-states of the Organisation for Economic Co-operation and Development (OECD) that are net energy exporters.[70] Canada is the largest source of US energy imports, as nearly all of its oil and gas exports go to the United States.[71] Recognizing the importance of energy trade, both the US and Canada,

---

63 Ibid. at 24.

64 Ibid. at 23.

65 Ibid.

66 Ibid.

67 Ibid.

68 Paula Dittrick, Focus: Unconventional Oil & Gas: *Industry Expects Rapid Gains in Eagle Ford Shale Output*, Oil & Gas Journal, July 4, 2011, available at http://www.ogj.com/articles/print/volume-109/issue-27/general-interest/focus-unconventional-oil-gas-industry-expects.html, noting that energy analysts Bentek Energy LLC believes the Eagle Ford "stands above the crowd" of multiple US unconventional oil and gas plays because of its high liquids potential and access to crude and natural gas liquids markets.

69 See Country Analysis Briefs: *Canada, infra* note 71.

70 Martin Ferguson, Austl. Minister for Res. and Energy and Minister for Tourism, *Australia's Energy and Resources Future* (June 23, 2010), available at http://minister.ret.gov.au/MediaCentre/Speeches/Pages/Australia%27sEnergyandResourcesFuture.aspx.

71 Energy Info. Admin., Country Analysis Briefs, *Canada*, http://www.eia.

along with Mexico, participate in the North American Energy Working Group, which seeks to improve energy integration and cooperation between the countries in the region.[72]

Although Canada is a major producer of conventional natural gas, in recent years the country has increasingly focused on developing natural gas from unconventional resources such as shale gas.[73] This is largely due to the view that production of conventional gas has peaked and new gas finds are needed to offset the decline.[74] The Canadian gas industry is currently undergoing a transformation similar to that of the United States through its increased focus on shale gas production.[75] The most significant shale basins are located in northeastern British Columbia, while some shale basins in Alberta, Ontario, Quebec, and the Maritimes also have some potential.[76] Although large-scale commercial production of shale gas has not yet occurred in Canada, this might change in the coming years.[77] More than $2 billion has been invested in northeast British Columbia to establish land positions in the Horn River Basin and the Montney Trend.[78] (See Figure 10.5.)

In terms of the potential resource base of Canadian natural gas resources, estimates show a dramatic increase in Canada's natural gas reserve potential and put Canada's natural gas in place (GIP) at almost 4,000 TCF.[79] Such a dramatic increase in the reserve estimates results from the large contribution unconventional gas resources make to the reserves, dramatically changing the picture of Canada's gas potential.

---

doe.gov/emeu/cabs/Canada/Background.html (hereinafter Country Analysis Briefs: *Canada*).

72 US Dep't of Energy, *North American Energy Working Group (NAEWG)*, http://www.pi.energy.gov/naewg.htm. (The North American Energy Working Group (NAEWG) was established in 2001 by the US Secretary of Energy, the Secretary of Energy of Mexico, and the Canadian Minister of Natural Resources.)

73 Can. Soc'y for Unconventional Gas, *Unconventional Gas Facts*, http://www.csug.ca/index.php?option=com_content&task=view&id=60&Itemid=66.

74 Ibid.

75 See Country Analysis Briefs: *Canada*, *supra* note 71; see also Gary Park, *Gas Revolution No. 2: Canadian Shale*, Pipeline & Gas J., May 2010, available at http://pipelineandgasjournal.com/gas-revolution-no-2-canadian-shale?page=show.

76 See *Unconventional Gas Facts*.

77 Can. Soc'y for Unconventional Gas, *Shale Gas*, http://www.csug.ca/index.php?option=com_content&task=view&id=60&Itemid=66#shale.

78 Ibid.

79 F.M. Dawson, *Cross Canada Check Up: Unconventional Gas Emerging Opportunities and Status of Activity* 3 (2010), http://www.csug.ca/images/Technical_Luncheons/Presentations/2010/MDawson_AGM2010.pdf (hereinafter Cross-Canada Check Up).

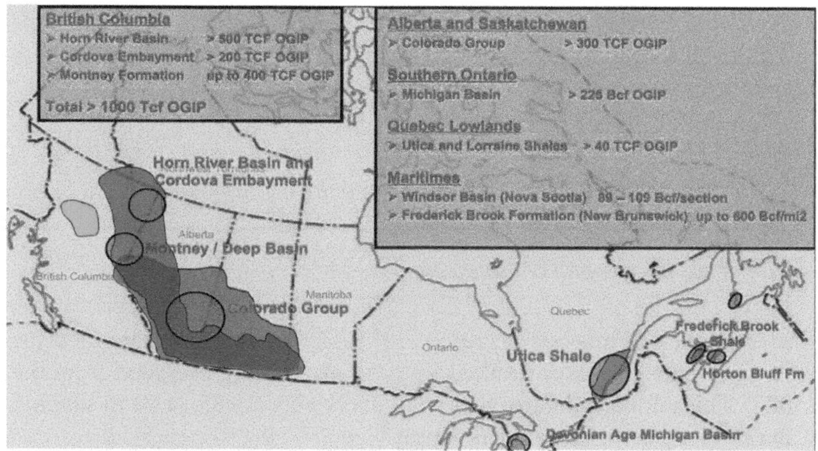

Source: Michael Dawson, President, Can. Soc'y for Unconventional Gas, *Shale Gas Plays in Canada: Opportunities from Coast to Coast*, Lecture at the Developing Natural Gas Conference April 7, 2009, available at http://www.csug.ca/images/CSUG_presentations/2009/Hart_energy_conf_Presentation_final.pdf.

Figure 10.5 Map of Canada's shale gas plays

Table 10.1 Canada's gas in place resources (TCF)

Conventional (GIP)	692
Natural gas from coal/coal bed methane	801
Tight gas	1,311
Shale gas	1,111
Total	3,915

Source: F.M. Dawson, "Cross Canada Check Up," 2010

(See Table 10.1.) The marketable portion is between 700 and 1300 TCF, of which 357 TCF are conventional and between 376 (low case) and 947 TCF (high case) are unconventional.[80]

This estimate is significantly higher than prior estimates that did not include potential unconventional resources, but it may still underestimate

80 Paul Wells, *CSUG Report Pegs Canada's Natural Gas in Place at Almost 4,000 tcf*, Oil & Gas Inquirer, June 2010, available at http://www.oilandgasinquirer.com/printer.asp?article=profiler%2F100610%2FPRO2010_UA0002.html.

Table 10.2 Canada's estimated marketable gas resources (TCF)

Conventional (Remanining GIP)	357
Natural gas from coal/coal bed methane	34–129
Tight gas	215–476
Shale gas	128–343
Total	733–1,304

Source: Petrel Robertson/CUSG Study, 2010

the true value of Canada's gas reserves.[81] A lack of available data on some emerging shale gas plays resulted in those plays being excluded from the total.[82] This additional natural gas will likely play a major role in shaping Canada's long-term natural gas supply.[83]

10.6 GLOBAL SHALE GAS DEVELOPMENT AND RESOURCES

The shale gas "revolution" that is transforming the North American natural gas market is not limited to that region.[84] It has been widely recognized that there is enormous unconventional gas potential in other parts of the world.[85] While shale gas appears to be the most promising type of unconventional gas around the world, some countries are also developing tight gas and CBM.[86] One of the main challenges for shale gas development globally is estimating the potential resource base.[87]

10.6.1 IEA Estimates of Global Shale Gas Resources

According to the IEA, there are only limited studies estimating global unconventional gas resources and "major work is still needed to refine and expand [the] data."[88] With few exceptions, unconventional gas resources

[81] Ibid.
[82] See ibid.
[83] See *Cross Canada Check Up*, at 3.
[84] Smith and Jackson.
[85] Ibid.
[86] Ibid.
[87] See Int'l Energy Agency, Medium-Term Oil & Gas Markets 185 (2010) (hereinafter IEA MTOGM 2010).
[88] Ibid.

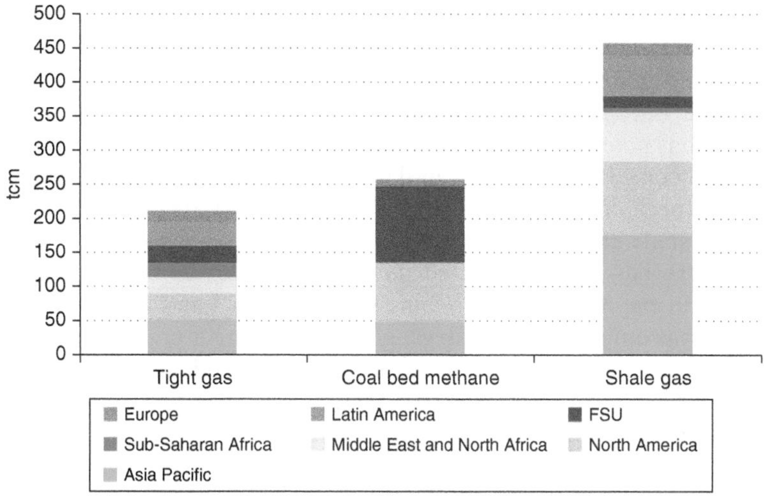

Source: IEA MTOGM 2010, p. 185

Figure 10.6 Worldwide unconventional gas resources in place

around the world have "largely been overlooked and understudied" and most "have not been appraised in any systematic way."[89]

In terms of existing regional estimates of global unconventional gas potential, Asia Pacific and North America have the highest, "with 274 TCM and 233 TCM respectively followed by [the former Soviet Union] with 155 TCM, Latin America [with] 98 TCM and [the Middle East-North Africa region with] 95 TCM."[90] Though significant attention has been devoted to Europe's potential unconventional gas resources, so far, they are estimated at only 35 TCM.[91] The IEA notes that "shale gas represents half of this global potential and is especially present in Asia and North America while CBM is mainly in [the former Soviet Union] and tight gas is quite evenly distributed between the regions"[92] (Figure 10.6). However, these numbers "should be considered with caution" as not all of this gas will be recoverable.[93]

---

[89] Ibid. at 186.

[90] Ibid. at 185; see Susan L. Sakmar, Recent Development, *The Status of the Draft Iraq Oil and Gas Law*; 30 Hous. J. Int'l L. 289, 295 n. 35 (2008) (noting Iraq's "fairly significant gas reserves").

[91] IEA MTOGM 2010 at 185.

[92] Ibid.

[93] Ibid. The IEA has estimated that "around 380 tcm would be recoverable based on current data and knowledge." Ibid. at 186.

In terms of country-specific developments, Australia ranks first among the countries able to develop their unconventional gas resources in the short term.[94] CBM has been at the "mature market stage in Australia for some time, but shale gas is still in its infancy."[95]

China has potentially significant unconventional gas resources and has expressed considerable interest in developing these. Historically China's focus has been on CBM, but recently its focus has shifted towards developing its shale gas resources. Although these are estimated at 26 TCM, the country has never appraised its shale gas reserves but is expected to do so in the near future.[96] China's Ministry of Land and Resources (MLR) "has announced a strategic goal of reaching a production target of 15–30 bcm (billion cubic meters) by 2020."[97] In this regard, it will be critical for China to acquire technology to meet these production goals. China's Sinopec has already engaged in dialogue with international oil companies in furtherance of this goal. In November 2009, China and the United States signed a Memorandum of Understanding to jointly cooperate in assessing China's shale gas resources and, consequently, promote investments in this area.[98]

Like China, India has historically focused on CBM but is now turning to shale gas, which is rapidly gaining the attention of industry players.[99] In April 2010, India's Reliance Industries Ltd invested $1.7 billion in the US Marcellus Shale play. This was viewed as an indication that Indian companies are looking to acquire expertise and technology to develop shale gas resources, both at home and abroad. The two major obstacles for India are a lack of clarity regarding upstream regulation for shale gas and a lack of data as most of India's shale gas potential remains underexplored.

Compared to Australia and India, Indonesia has been slow to develop its unconventional gas resources and foreign companies have been reluctant to invest there, largely because of the legal and regulatory uncertainty.[100] Indonesia's outlook may change, however, in light of its estimated shale gas potential of approximately 30 TCM and its plans to launch a tender of shale gas fields.

Europe has received the most attention, with many countries in the

---

94 Ibid. at 187.
95 Ibid. at 188.
96 Ibid. at 188–9.
97 Ibid. at 189.
98 Ibid. at 188.
99 Ibid. at 189.
100 Ibid. at 190.

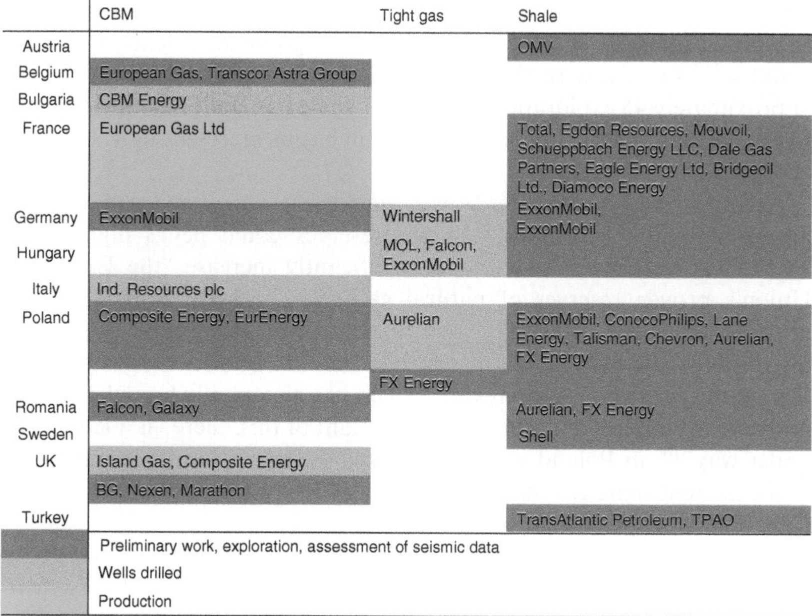

Note: The list of companies is not exhaustive.

Source: IEA MTOGM 2010, p. 191

Figure 10.7 Unconventional gas activities in Europe

region looking to replicate the US shale gas revolution. While there are "many challenges that could prevent an unconventional gas boom happening in Europe," recently there has been a lot of activity and interest in shale gas in Austria, Bulgaria, France, Germany, Italy, Poland, Romania, Spain, Sweden, and the United Kingdom. International oil companies, which were largely absent from early shale gas development in the United States, have been more proactive in Europe. Many major oil companies, including ExxonMobil, Shell, Chevron, ConocoPhillips, Marathon, and Total, are present in one or more European countries.[101] (See Figure 10.7.)

In most European countries, most of these developments are at the very early stages and seismic data are just barely being compiled.[102] The IEA notes that "only a few European countries are actually producing

[101] Ibid. at 190–91.
[102] Ibid. at 190.

unconventional gas, and then only in small quantities."[103] Of these, Poland is worth noting as shale gas has received significant attention in that country.[104] In its report, the IEA also notes that "Poland has approved approximately 45 exploration licenses for shale gas [and] ExxonMobil has five concessions in the Podlasie and Lublin basins representing 1.3 million acres."[105]

According to estimates by Wood Mackenzie, an oil and gas research group, Poland's unconventional gas reserves could be as high as 48 TCF.[106] If confirmed, this would significantly increase "the European Union's proven reserves of natural gas and . . . make Poland, which imports 72 per cent of its gas, self-sufficient for the foreseeable future."[107] Significant shale gas production in Poland could also alter the gas geopolitics for the entire European region, which has historically been dependent on Russian supplies of natural gas.[108] In light of this, there "is a land grab under way"[109] in Poland with several major energy companies investing in the nascent shale gas industry including Chevron, ConocoPhillips, and Canadian-based Talisman.[110]

France, Germany, and Hungary are also just emerging as potential shale gas players while other countries are starting to assess their potential reserves.[111] The IEA notes that "many initiatives are underway such as the Gas Shales in Europe ('GASH'), coordinated by the German GeoForschungsZentrum (GFZ) and The Institut Français du Pétrole (IFP). In other regions, [international oil companies (IOCs)] and National Oil Companies (NOCs) have been carrying out exploratory work [on unconventional resources,]" yet the results remain to be seen.[112]

---

103 Ibid.

104 Ibid. at 191.

105 Ibid.

106 Robin Pagnamenta, *Dash for Poland's Gas Could End Russian Stranglehold on Supplies*, The Times (London), Apr. 5, 2010, at 33.

107 Ibid.

108 See Ibid.; see also Kim Talus, *Access to Gas Markets: A Comparative Study on Access to LNG Terminals in the European Union and the United States*, 31 Hous. J. Int'l L. 343, 354 (2009).

109 Pagnamenta (quoting Oisin Fanning, executive chairman of San Leon Energy, a British company that has secured three license areas in Poland); see also Dinakar Sethuraman, *Exxon, Chevron "Land Grab" for Europe Shale Gas, JP Morgan Says*, Bloomberg Businessweek, Feb. 11, 2010, available at http://www.businessweek.com/news/2010-02-11/exxon-chevron-land-grab-for-europe-shale-gas-jpmorgan-says.html.

110 IEA MTOGM 2010 at 191.

111 See generally Ibid. at 192.

112 Ibid. at 186.

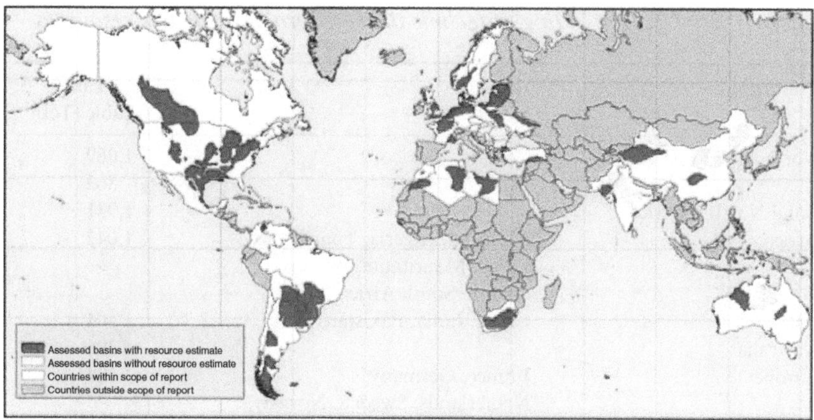

Source: EIA/ARI World Shale Gas Resource Assessment

Figure 10.8 Worldwide shale basins

10.6.2 World Shale Gas Resources: an Initial Assessment

While assessments are underway in Europe, the US EIA recently released a study commissioned by an external consultant to assess the potential of international shale gas resources.[113] The report assessed 48 shale basins in 32 countries containing almost 70 shale gas formations. Figure 10.8 indicates the location of assessed shale gas basins for which estimates of the risked gas in place and technically recoverable resources were provided.

Although the shale gas resource estimates will change over time as additional information becomes available, the report shows that the global shale gas resource base is vast. The initial assessment of technically recoverable shale gas resources in the 32 countries outside the US is 5,760 Tcf (trillion cubic feet). When the US shale gas resource base of 862 Tcf is added to this, the total global shale gas resource base is 6,622 TCF. (See Table 10.3.)

The report noted that these estimates are relatively conservative and likely to go up as more information becomes known and this has certainly been the case in the US. However, it is also important to note that the report estimated "technically recoverable" resources, which does not

[113] US EIA, *World Shale Gas Resources: An Initial Assessment of 14 Regions Outside the United States*, Release date: April 5, 2011, http://www.eia.gov/analysis/studies/worldshalegas/.

Table 10.3 Country listing of technically recoverable shale gas resources

Continent		Technically recoverable (Tcf)
North America (non-US)	Canada, Mexico	1,069
	US	862
Total North America		1,931
Africa	Morocco, Algeria, Tunisia, Libya, Mauritania, Western, Sahara, South Africa	1,042
Asia	China, India, Pakistan	1,404
Australia		396
Europe	France, Germany, Netherlands, Sweden, Norway, Denmark, UK, Poland, Lithuania, Ukraine, Turkey	624
South America	Colombia, Venezuela, Argentina, Bolivia, Brazil, Chile, Uruguay, Paraguay	1,225
Total		6,622
Total without US		5,760

Source: EIA/ARI World Shale Gas Resource Assessment

mean commercially viable resources. In other words, it may not make commercial sense to exploit all of these resources.

The report also noted that there were two country groupings where shale gas development might be most attractive. The first group consists of countries that are currently dependent upon natural gas imports and have at least some gas production infrastructure and where the estimated shale gas resources are substantial relative to the current gas consumption. This group includes France, Poland, Turkey, Ukraine, South Africa, Morocco and Chile. The second group comprises those countries where the shale gas resource estimate is large and there already exists a significant natural gas production infrastructure. In addition to the US, this group includes Canada, Mexico, China, Australia, Libya, Algeria, Argentina and Brazil.

10.7 CHALLENGES TO DEVELOPING GLOBAL UNCONVENTIONAL GAS

The IEA has recognized that there are numerous challenges to replicating the success of the US unconventional gas revolution overseas.[114] The issues raised by the IEA that may impact the development of global unconventional gas resources include:[115]

1. limited studies on unconventional gas potential around the world
2. environmental concerns
3. fiscal conditions
4. landowner acceptance
5. interference from local authorities
6. pipeline and infrastructure issues
7. availability of technology, equipment and skilled labor force
8. gas players' experience.

Of these, landowner acceptance and environmental concerns are worth noting since these two areas have been the most challenging in the development of shale gas in the US.[116]

Landowner acceptance is likely to vary depending on whether the landowner stands to gain financially from the drilling activity.[117] In the United States, landowners often stand to benefit financially from drilling on their property – if they own the underground resources, they may receive a bonus or royalties upon leasing to an oil company in order to develop the resources.[118] For example, some US landowners who own the underground mineral resources have received "up to $25,000 per acre, and sometimes up to 25% royalty" by leasing their property for shale gas development.[119] Although this financial incentive has been particularly helpful in the development of shale gas in the United States, it may not be as relevant in other areas of the world where landowners do not own the underground resources.[120]

In its report, the IEA also noted the numerous environmental concerns that have been raised in the United States.[121] These concerns include the impact hydraulic fracturing might have on local water supplies in terms of potential contamination of underground drinking water sources and surface waters as well as issues related to the quantity of water used in the process.[122] These issues are discussed in detail in the following sections.

---

114 See IEA MTOGM 2010 at 184–5.
115 Ibid.
116 See ibid. at 186–7.
117 Ibid.
118 See ibid. at 187.
119 Ibid.
120 See ibid.
121 See ibid. at 186–7.
122 See ibid.

10.8 GLOBAL SHALE GAS INITIATIVE: WILL THE UNITED STATES BE A MODEL FOR GLOBAL SHALE GAS DEVELOPMENT?

In recognition of the growing worldwide interest in developing unconventional gas resources, in April 2010 the US Department of State launched the Global Shale Gas Initiative (GSGI) "in order to help countries seeking to utilize their unconventional natural gas resources to identify and develop them safely and economically" and in an "environmentally sensitive manner." The goal of the GSGI is to assist countries seeking to develop their own unconventional gas resources with balancing energy security and environmental concerns.[123]

A country's ability to participate in the initiative depends largely on the "presence of gas-bearing shales within their borders, market potential, business climates, geopolitical synergies, and host government interest." Countries have been classified into tiers, with Tier 1 countries being those that "have the greatest potential for benefiting from GSGI opportunities" and Tier 2 countries being those "that have expressed interest and meet GSGI criteria." So far, partnerships have been arranged with China, India, and Poland.[124]

In August 2010, when the first meeting of the GSGI took place, the representatives of 17 different countries discussed "the importance of shale gas as a lower-carbon fuel option that can help reduce CO_2 emissions while ensuring energy security and economic development in the 21st century." The meeting was a "regulatory conference" designed to showcase the "umbrella of laws and regulations [in the United States] that makes sure [shale gas development] is done safely and efficiently."[125]

At the conference, the State Department noted that the United States has both federal and state laws to protect land use, water, and air as well as the capacity to monitor, regulate and enforce the laws. The conference gave US agencies, such as the EPA and the EIA, the opportunity to explain the laws and regulations pertaining to shale gas development in the

---

123 US Dept of State, *Briefing on the Global Gas Shale Initiative Conference*, Remarks by David L. Goldwyn, Special Envoy for International Energy Affairs (Aug. 24, 2010), available at http://www.state.gov/s/ciea/rmk/146249.htm.

124 *Polish Delegation Attends First Multilateral Meeting of the Global Shale Gas Initiative*, US Diplomatic Mission to Warsaw, Poland (Aug. 24, 2010), http://poland.usembassy.gov/shalegas.html

125 *Briefing on the GSGI Conference*.

United States, with particular attention being paid to issues pertaining to water protection since water is scarce in many countries.[126]

Whether the GSGI can provide a regulatory model for environmental best practices is debatable and remains to be seen.[127] In light of the growing environmental challenges and potential additional regulations facing the US shale gas industry, the usefulness of the US legal scheme as a model framework is still an open question,[128] especially as it relates to environmental issues.[129] As discussed in detail below, there is some indication that production in the US may have outpaced the ability of some states to effectively oversee the safety and environmental sustainability of shale gas development.[130] If the United States is having difficulty with the safety and environmental aspects of shale gas drilling, how can other countries keep pace with shale gas developments? This question is especially critical for those countries with less-developed laws and regulations.[131]

10.9 ENVIRONMENTAL ISSUES ASSOCIATED WITH SHALE GAS DEVELOPMENT IN THE UNITED STATES[132]

The development of shale gas in the United States has been widely recognized as one of the most promising trends in the US in terms both of job creation and economic benefits and of its resulting increase in the domestic

---

126 Ibid.

127 To date, there appears to be only limited activity related to the GSGI and it remains to be seen whether this initiative gains in prominence. The GSGI has been renamed the Unconventional Gas Technical Engagement Program (UGTEP). US Dept of State, *Unconventional Gas Technical Engagement Program (UGTEP)*, available at http://www.state.gov/s/ciea/ugtep/index.htm.

128 See IEA MTOGM 2010 at 186–7; see also Amy Westervelt, *Shale Gas Booming Globally, Despite Chemical Dangers*, Solve Climate News, Aug. 9, 2010, http://solveclimatenews.com/news/20100809/shale-gas-booming-globally-despite-chemical-dangers.

129 Ibid.

130 See Anthony Andrews et al. (2009), *Unconventional Gas Shales: Development, Technology, and Policy Issues*, pp. 33–8, http://www.fas.org/sgp/crs/misc/R40894.pdf.

131 See Laura C. Reeder, Note, *Creating a Legal Framework for Regulation of Natural Gas Extraction from the Marcellus Shale Formation*, 34 Wm & Mary Envtl L. & Pol'y Rev. 999, 1022 (2010) (describing the complex legal obstacles inherent in shale gas development).

132 Since the US has thus far been the leader in shale gas development, this section focuses on environmental issues in the US with the view that other

supplies of natural gas.[133] Many people view natural gas as a cleaner-burning fossil fuel that could enhance energy independence, reduce emissions and serve as a bridge fuel to renewable energy.[134]

While there are many proponents of shale gas, there are also many who oppose it because of the technology necessary to produce it.[135] This opposition has intensified as hydraulic fracturing has become more commonplace in wells around the country and around the world.[136] For its part, the gas industry contends that hydraulic fracturing is safe, well regulated, and has a proven track record having been used in the United States since the 1940s in drilling more than one million wells.[137]

In support of the safety of hydraulic fracturing, the industry often points to a 2004 EPA study that assessed the potential for contamination of underground sources of drinking water from the injection of hydraulic fracturing fluids into CBM wells.[138] In that study, the EPA concluded that the injection of hydraulic fracturing fluids into these wells posed "little or no threat to [underground drinking water]."[139] After reviewing incidents of drinking water well contamination, the EPA found "no confirmed cases that are linked to fracturing fluid injection into coalbed methane wells or subsequent underground movement of fracturing fluids."[140]

The industry also maintains that the continued use of hydraulic

---

countries may face similar issues should they choose to develop their own shale gas resources.

[133] IHS Report (2012), *America's New Energy Future: The Unconventional Oil and Gas Revolution and the US Economy*, http://www.ihs.com/info/eec/a/americas-new-energy-future.aspx. ("Unconventional oil and gas activity is already revolutionizing America's energy future and bringing enormous benefits to its economy. Unlocking unconventional energy will generate millions of jobs and billions in government receipts.")

[134] See *Shale Gas*, Energy Tomorrow, http://www.energytomorrow.org/Shale_Gas.aspx; *Briefing on the GSG Conference*. See generally Marianne Lavelle, *Natural Gas Stirs Hope and Fear in Pennsylvania*, National Geographic, Oct. 13, 2010, available at http://news.nationalgeographic.com/news/2010/10/101022-energy-marcellus-shale-gas-overview/.

[135] See, e.g., Wes Deweese, *Fracturing Misconceptions: A History of Effective State Regulation, Groundwater Protection, and the Ill-conceived FRAC Act*, 6 Okla. J. L. & Tech. 49, 6 (2010).

[136] As shale goes global, concerns have been raised in other countries as well. See e.g., Monique Beau Din, *Shale-gas Opposition is Growing, Survey Concludes*, The Gazette (Montreal), Feb. 16, 2011, at A6; *Exploration Ban in France Extended*, Calgary Herald (Can.), Jan. 20, 2011, at B4.

[137] API Freeing Up Energy.

[138] See ibid.

[139] Drinking Water Impact Study at 7-5.

[140] Ibid. at 7-6.

fracturing is critically important to producing the natural gas America will need in the future.[141] It is estimated that "[80%] of natural gas wells drilled in the next decade will require hydraulic fracturing"[142] and that, without it, the US could lose "45 percent of domestic natural gas production."[143]

10.9.1 Water Contamination Concerns

Despite the industry's claims that hydraulic fracturing is a safe and proven technology, environmental organizations, public health groups, and local communities have expressed numerous concerns about the potential environmental impacts of the use of hydraulic fracturing around the country.[144] There have been many allegations that hydraulic fracturing has led to the contamination of drinking water in many communities.[145] This has led to increased calls for federal regulation of hydraulic fracturing under the Safe Drinking Water Act (SDWA) that would provide a minimum federal floor for drinking water protection in the states engaged in drilling shale gas.[146]

The nonprofit, investigative journalism organization ProPublica has an extensive investigation of hydraulic fracturing underway.[147] According to that investigation, numerous states have reported cases involving spills of hazardous materials or other occurrences of water contaminated by oil or gas operations.[148] There are also hundreds of cases of water

---

141 Hydraulic Fracturing.

142 Ibid.

143 American Petroleum Institute, www.api.org, citing Global Insight, *Measuring the Economic and Energy Impacts of Proposals to Regulate Hydraulic Fracturing*, (2009).

144 See Amy Mall, *Incidents Where Hydraulic Fracturing is a Suspected Cause of Drinking Water Contamination*, Switchboard: Nat'l Res. Def. Council Staff Blog, Oct. 4, 2010, http://switchboard.nrdc.org/blogs/amall/incidents_where_hydraulic_frac.html (listing incidents of drinking water contamination and supporting regulation of hydraulic fracturing under the Safe Drinking Water Act).

145 Ibid.

146 Ibid.

147 See *Buried Secrets: Gas Drilling's Environmental Threat*, ProPublica, http://www.propublica.org/series/buried-secrets-gas-drillings-environmental-threat (containing links to various investigative pieces concerning the environmental impact of gas drilling). The New York Times is also examining the risks of shale gas drilling and efforts to regulate the rapidly growing industry in its Drilling Down series, available at http://topics.nytimes.com/top/news/us/series/drilling_down/index.html.

148 Abraham Lustgarten, *Setting the Record Straight on Hydraulic Fracturing*, ProPublica, Jan. 12, 2009, http://www.propublica.org/article/setting-

contamination in drilling areas where hydraulic fracturing is used, including some pending lawsuits alleging contamination.[149]

ProPublica has also noted the difficulty scientists face in specifically determining "which aspect of drilling – the hydraulic fracturing, the waste water that accidentally flows into the ground, the leaky pits of drilling fluids or the spills from truckloads of chemicals transported to and from the site – causes [the reported] pollution."[150] One challenge has been the refusal by the industry to make public the chemical makeup of the hydraulic fracturing fluid used on a particular well.[151] Without this information, "environmental officials say they cannot conclude with certainty when or how certain chemicals entered the water."[152]

10.9.2 Water Quantity and Flowback Concerns

Concerns have also been raised pertaining to the large volumes of water needed during the hydraulic fracturing process, and the disposal of the flowback or wastewater from fracturing operations.[153] A recent US Geological Survey (USGS) report noted these concerns with regard to water resources and gas production in the Marcellus Shale.[154] According to the USGS report, "many regional and local water management agencies [in the Marcellus Shale region] are concerned about where such large volumes of water will be obtained, and what the possible consequences might be for local water supplies."[155]

Chesapeake Energy Corp., one of the most active drillers in the Marcellus Shale,[156] candidly admits water is an essential component of its deep shale gas development.[157] According to the company, "fracturing

---

the-record-straight-on-hydraulic-fracturing-090112 (hereinafter *Setting the Record Straight on Hydraulic Fracturing*).

149 Ibid.; Abraham Lustgarten, *Pa. Residents Sue Gas Driller for Contamination, Health Concerns*, ProPublica, Nov. 20, 2009, http://www.propublica.org/article/pa-residents-sue-gas-driller-for-contamination-health-concerns-1120.

150 *Setting the Record Straight on Hydraulic Fracturing*.

151 Ibid.

152 Ibid.

153 Andrews et al.

154 Daniel J. Soeder and William M. Kappel, *Water Resources and Natural Gas Production from the Marcellus Shale* 3–4 (2009), http://pubs.usgs.gov/fs/2009/3032/pdf/FS2009-3032.pdf.

155 Ibid. at 4.

156 Chesapeake Energy, Press Release, *Chesapeake Energy Corporation Confirms Decision Not to Drill for Natural Gas in the New York City Watershed* (Oct. 28, 2009) available at http://www.chk.com/news/articles/pages/1347788.aspx.

157 Chesapeake Energy (2010), *Fact Sheet: Water Use in Marcellus Deep*

a typical Chesapeake Marcellus horizontal deep shale gas well requires an average of five and a half million gallons per well."[158] Chesapeake also maintains that water resources are protected through stringent state, regional and local permitting processes and, in comparison to other uses within the area, deep shale gas drilling and fracturing uses a small amount of water.[159]

Hydraulic fracturing also gives rise to concerns pertaining to the disposal of wastewater.[160] While some of the injected hydraulic fracturing fluids remains trapped underground, the majority – 60–80 percent – returns to the surface as "flowback."[161] The USGS has noted that because the quantity of fluids is so large, the additives in a 3 million gallon frac job would yield about 15,000 gallons of chemicals in the flowback water.[162] Some states, such as West Virginia, have noted that wastewater disposal is "perhaps the greatest challenge" in hydraulic fracturing operations.[163]

Other shale producing areas face the same challenges. In north Texas, increased water use stemming from a growing population, drought, and the Barnett Shale development has led to heightened concerns about water availability.[164] In January 2007, the Texas Water Development Board (TWDB) published a study of a 19-county area in North Texas that contains estimates of water used in the Barnett Shale development.[165] The TWDB report indicates that the fracturing of a horizontal well completion can use more than 3.5 million gallons (more than 83,000 barrels) of water.[166] In addition, the wells may be re-fractured multiple times when the natural gas flow slows after being in production for several years.[167] However, the report estimates that the amount of water used for development has been a relatively small percentage of the total water use.[168]

Shale Gas Exploration, http://www.chk.com/media/marcellusmediakits/marcellus_water_use_fact_sheet.pdf (hereinafter Chesapeake Energy, *Water Use*).

158 Ibid.

159 Ibid.

160 See Drinking Water Impact Study at 3-11.

161 Ibid.

162 Soeder and Kappel at 4.

163 Andrews et al. at 35.

164 James E. Bené and Robert Harden, Northern Trinity/Woodbine Groundwater Availability Model: Assessment of Groundwater Use in the Northern Trinity Aquifer due to Urban Growth and Barnett Shale Development 1 (2007), http://rio.twdb.state.tx.us/RWPG/rpgm_rpts/0604830613_BarnetShale.pdf.

165 Ibid.

166 Ibid. at 14.

167 Ibid. at 2-44.

168 Ibid. at 2-3.

Although growing, the report calculated water used for the Barnett Shale accounted for only 3 percent of the total groundwater used.[169] The TWDB report makes predictions of future water needs for the area, including Barnett Shale development.[170] These estimate an increase in the groundwater used from 3 percent in 2005 to 7–13 percent in 2025.[171]

10.9.3 The EXXON/XTO Merger

The Exxon/XTO merger was announced against the backdrop of increased interest and scrutiny in developing US shale gas resources.[172] In December 2009, ExxonMobil (Exxon)[173] announced plans to buy XTO Energy (XTO)[174] in an all-stock transaction worth about $41 billion (including debt of $10 billion), which would create the largest US natural gas producer and holder of gas reserves.[175] Exxon's interest in XTO was primarily due to XTO's strong unconventional gas resource focus, including XTO's resource base, as well as its technical expertise in extracting shale gas through hydraulic fracturing technology.[176]

The Exxon/XTO merger was seen by many in the oil and gas industry as a show of confidence in the future of shale gas.[177] Many praised the deal as a boost for shale gas to play a greater role in supplying the world with abundant, affordable, and cleaner-burning energy.[178] At the same time, however, the proposed merger led to greater scrutiny of the hydraulic fracturing technology, which has drawn intense criticism from environmentalists and lawmakers concerned about the

---

169 Ibid.

170 Ibid.

171 Ibid. at 3.

172 *ExxonMobil to Boost Unconventional Focus by Acquiring XTO*, Oil & Gas J., Dec. 21, 2009, at 31; see *Natural Gas Helps Exxon and Shell Lift Profits*, N.Y. Times, July 30, 2010, at B4.

173 ExxonMobil website, www.exxonmobil.com.

174 XTO Energy website, www.xtoenergy.com.

175 *ExxonMobil to Boost Unconventional Focus by Acquiring XTO*, *supra* note 172.

176 *The ExxonMobil-XTO Merger: Impact on U.S. Energy Markets*: Hearing Before the Subcomm. on Energy and Env't of the H. Comm. on Energy and Commerce, 111th Cong. 53 (2010) (statement of Rex Tillerson, CEO, ExxonMobil Corp.), available at http://energycommerce.house.gov/Press_111/20100120/transcript_01202010_ee.pdf.

177 Katie Howell, *House Panel Looks into Effects of Exxon-XTO Merger*, N.Y. Times, Jan. 19, 2010, http://www.nytimes.com/gwire/2010/01/19/19greenwire-house-panel-looks-into-effects-of-exxon-xto-me-96870.html.

178 Ibid.

potential impact of hydraulic fracturing on water supplies and the environment.[179]

At the congressional hearings related to the merger, several lawmakers expressed concern that the proposed merger would reduce competition in the oil and gas industry and also lead to an increase in the use of hydraulic fracturing and horizontal drilling.[180] Other lawmakers expressed concern that the technologies could pollute drinking water supplies.[181] Exxon Chairman Rex Tillerson defended the controversial hydraulic fracturing technology and assured lawmakers that "[w]ith recent advances in extended reach horizontal drilling, combined with the time-tested technology of hydraulic fracturing . . . we can now find and produce unconventional natural gas supplies miles below the surface in a safe, efficient and environmentally responsible manner."[182]

Mr Tillerson also indicated that continued use of hydraulic fracturing was essential for the industry and the merger.[183] Indeed, the continued use of hydraulic fracturing was so important to the combined company's success that the merger agreement provided an "opt out" provision allowing the deal to be called off if any event or action gave rise to a "Company Material Adverse Effect," which included changes in laws that made hydraulic fracturing illegal or commercially impracticable.[184] Although the XOM/XTO merger closed on June 25, 2010, without any congressional or regulatory action to limit or ban hydraulic fracturing,[185] the controversy surrounding the use of hydraulic fracturing continues to this day in the US and has spread to other continents.

---

179 Ibid.

180 Tom Doggett, *Exxon-XTO Merger Draws Scrutiny from Congress*, Reuters, Jan. 20, 2010, available at http://www.reuters.com/article/idUSTRE60J53920100120.

181 Ibid.

182 Ibid.

183 Ibid.

184 XTO Energy Inc., Agreement and Plan of Merger, Dated as of Dec. 13, 2009–Dec. 15, 2009, art. I, IX, available at http://www.faqs.org/sec-filings/091215/XTO-ENERGY-INC_8-K/dex21.htm; see also Russell Gold, *Exxon Can Stop Deal if Drilling Method is Restricted*, Wall St J., Dec. 17, 2009, at B3.

185 *ExxonMobil Announces Completion of All-Stock Transaction for XTO*, Business Wire, June 25, 2010, available at http://www.businesswire.com/portal/site/exxonmobil/index.jsp?ndmViewId=news_view&ndmConfigId=1001106&newsId=20100625005806&newsLang=en; see also Anna Driver, *Exxon Sees Greater Scrutiny After BP Spill*, Reuters, July 8, 2010, available at http://www.reuters.com/article/idUSN0821129720100708?rpc=21.

10.10 OVERVIEW OF THE REGULATORY FRAMEWORK FOR SHALE GAS DEVELOPMENT IN THE UNITED STATES

As the undisputed leader in shale gas development, the US also has the most experience with the regulatory framework to govern the shale gas industry. As other countries assess their shale gas resources, they will also need to assess whether they have the sufficient regulatory framework in place to manage shale gas development. In this regard, the US framework could prove to be a useful starting point for some countries and, for that reason, an overview of the US regulatory framework is provided below.

As previously noted, hydraulic fracturing is a water intensive technology that raises many issues related to the environmental protection of US water supplies. The gas industry believes that existing state regulations are adequate to protect water resources during the development of shale gas resources.[186] This view is shared by the Ground Water Protection Council (GWPC), which represents state groundwater protection agencies and underground injection control (UIC) program administrators.[187] However, there is a growing contingent of landowners, environmental groups and citizen groups calling for federal regulation and further investigation of hydraulic fracturing due to concerns about water usage and possible contamination issues.[188] While an analysis of individual existing state laws is beyond the scope of this chapter, there are several important federal regulations that are relevant and discussed in detail below.

---

[186] Hydraulic Fracturing Fact Sheet, *supra* note 8; see Hannah Wiseman, *Regulatory Adaptation in Fractured Appalachia*, 21 Vill. Envtl L. J. 229, 288–9 (2010); see also *Hydraulic Fracturing*, *supra* note 18 (outlining industry practices relating to hydraulic fracturing).

[187] Hydraulic Fracturing Fact Sheet; Ground Water Prot. Council, *About Us*, http://www.gwpc.org/about_us/about_us.htm.

[188] See Mireya Navarro, *8,000 People? E.P.A. Defers Hearing on Fracking*, Green: a Blog about Energy and the Env't, Aug. 10, 2010, 5:28 p.m., http://green.blogs.nytimes.com/2010/08/8000.people-e-p-a-defers-hearing-on-fracking; see also Mike Soraghan, *BP, Others Push Against Federal Regulation of Fracturing*, N.Y. Times, Mar. 23, 2010, available at http://www.nytimes.com/gwire/2010/03/23/23greenwire-bp-others-push-against-federal-regulation-of-f-95671.html.

10.10.1 The Safe Drinking Water Act

The SDWA[189] is the primary federal law for protecting public water supplies from harmful contaminants.[190] Enacted in 1974,[191] and broadly amended in 1986 and 1996,[192] the SDWA is administered through a variety of programs that regulate contaminants in public water supplies, provide funding for infrastructure projects, protect underground sources of drinking water, and promote the capacity of water systems to comply with SDWA regulations.[193]

The EPA is the federal agency responsible for administering the SDWA[194] but a federal–state structure exists in which the EPA may delegate primary enforcement and implementation authority (primacy) for the drinking water program to states and tribes.[195] The state-administered Public Water Supply Supervision (PWSS) program remains the basic program for regulating public water systems,[196] and the EPA has delegated primacy for this program to all states, except Wyoming and the District of Columbia (which SDWA defines as a state).[197] The EPA has responsibility for implementing the PWSS program in these two jurisdictions and throughout most Indian lands.[198]

A second key component of the SDWA requires the EPA to regulate the underground injection of fluids to protect underground sources of drinking water.[199] In terms of oil and gas drilling, the UIC program regulations specify siting, construction, operation, closure, financial responsibility, and other requirements for owners and operators of injection wells.[200] Thirty-three states (including West Virginia, Ohio, and

---

189 Safe Drinking Water Act, 42. U.S.C. § 300f (2005).

190 Office of Water, Envtl Prot. Agency, *Safe Drinking Water Act*, http://water.epa.gov/lawsregs/rulesregs/sdwa/index.cfm.

191 Ibid.

192 Ibid.

193 See generally Envtl Prot. Agency Office of Water, *Understanding the Safe Drinking Water Act* (2004), http://water.epa.gov/lawsregs/guidance/sdwa/upload/2009_08_28_sdwa_fs_30ann_sdwa_web.pdf (hereinafter *Understanding the Safe Drinking Water Act*).

194 Ibid.

195 See ibid.

196 Office of Water, Envtl Prot. Agency, *Public Water System Supervision (PWSS) Grant Program*, http://water.epa.gov/grants_funding/pws/index.cfm.

197 Understanding the Safe Drinking Water Act, supra note 193.

198 See ibid.

199 Andrews et al., *supra* note 130, at 37.

200 Ibid. (noting that requirements for Class II wells are found in 40 C.F.R. §§ 144–6).

Texas) have assumed primacy for the UIC program.[201] The EPA has lead implementation and enforcement authority in ten states, including New York and Pennsylvania, and authority is shared in the remainder of the states.[202]

Notwithstanding the SDWA's general mandate to control the underground injection of fluids to protect underground sources of drinking water, the law specifically states that EPA regulations for state UIC programs "may not prescribe requirements which interfere with or impede . . . any underground injection for the secondary or tertiary recovery of oil or natural gas, unless such requirements are essential to assure that underground sources of drinking water will not be endangered by such injection."[203] Consequently, the EPA has not regulated gas production wells, and historically has not considered hydraulic fracturing to fall within the regulatory definition of underground injection.[204]

10.10.2 *Leaf v. EPA*

Until 1997, it was unclear whether hydraulic fracturing was regulated under the UIC programs.[205] However, the US Court of Appeals for the 11th Circuit ruled that the hydraulic fracturing of coal beds for methane production constituted an underground injection that must be regulated.[206] Since this decision applied only in the 11th Circuit, the only state required to revise its UIC program was Alabama.[207]

---

201 Ibid.

202 See ibid. To receive primacy, a state must demonstrate to the EPA that its UIC program is at least as stringent as the federal standards. Ibid. For Class II wells, states must demonstrate that their programs are effective in preventing pollution of underground sources of drinking water. Ibid. at 37, n. 77.

203 Safe Drinking Water Act, 42 U.S.C. § 300h(b)(2) (2005).

204 Andrews et al., *supra* note 130, at 37.

205 Deweese, *supra* note 135, at 10.

206 *Legal Envtl Assistance Found. v. Envtl Prot. Agency*, 118 F.3d 1467, 1477 (11th Cir. 1997).

207 Ibid. In 2000, a second suit was filed against the EPA wherein the court approved Alabama's revised UIC program, despite several alleged deficiencies. *Legal Envtl Assistance Found. v. Envtl Prot. Agency*, 276 F.3d 1253, 1256 (11th Cir. 2001). The US Court of Appeals for the 11th Circuit directed the EPA to require Alabama to regulate hydraulic fracturing under the SDWA. Ibid. at 1477–8. The court determined that the EPA could regulate hydraulic fracturing under the SDWA's more flexible state oil and gas provisions in section 1425, rather than the more stringent underground injection control requirements of section 1422. Ibid. at 1260–61.

In response to the decision in *Leaf v. EPA*[208] and citizen complaints about water contamination attributed to hydraulic fracturing, the EPA began to study the impacts of hydraulic fracturing practices used in CBM production on drinking water sources, and to determine whether further regulation was needed.[209] In 2004, the EPA issued a final (phase I) report, based primarily on interviews and a review of the available literature, and concluded that the injection of hydraulic fracturing fluids into CBM wells posed little threat to underground sources of drinking water and required no further study.[210]

The EPA noted, however, that very little documented research had been done on the environmental impacts of injecting fracturing fluids.[211] It also noted that estimating the concentration of diesel fuel components and other fracturing fluids beyond the point of injection was beyond the scope of its study.[212] Some members of Congress and some EPA professional staff criticized the report, asserting that its findings were not scientifically founded.[213]

Ultimately, in the Energy Policy Act of 2005,[214] Congress amended SDWA Section 1421 to specify that the definition of "underground injection" excludes the injection of fluids or propping agents (other than diesel fuels) used in hydraulic fracturing operations related to oil, gas, or geothermal production activities.[215] This exclusionary language effectively removed the EPA's (unexercised) authority under the SDWA to regulate the underground injection of fluids for hydraulic fracturing purposes.[216] Environmentalists and others opposed to hydraulic fracturing commonly refer to this exclusionary language as "The Halliburton Loophole," based on a *New York Times* editorial of the same title.[217]

---

208 *Legal Envtl Assistance Found. v. Envtl Prot. Agency*, 118 F.3d 1467 (11th Cir. 1997).

209 DRINKING WATER IMPACT STUDY, at ES-1.

210 Ibid.

211 Ibid. at 4-1.

212 Ibid. at 4-12.

213 Mike Soraghan, *Natural Gas Drillers Protest Nomination of Fracking Critics for EPA Review Panel*, N.Y. Times, Sept. 30, 2010, available at http://www.nytimes.com/gwire/2010/09/30/30greenwire-natural-gas-drillers-protest-nomination-of-fra-98647.html.

214 Energy Policy Act of 2005, Pub. L. No. 109-58, 119 Stat. 594 (2005).

215 Ibid. § 322.

216 See Safe Drinking Water Act § 1421, 42 U.S.C. § 300h.

217 *The Halliburton Loophole*, New York Times Editorial, Nov. 2, 2009, http://www.nytimes.com/2009/11/03/opinion/03tue3.html?_r=3&th&emc=t.

10.10.3 The FRAC Act

As shale gas development spread across the US, so too did public concern about the safety and environmental impact of hydraulic fracturing. These concerns ultimately made their way to the US Congress where companion bills H.R. 2766 and S. 1215 were introduced in 2009 in an effort to amend the SDWA to include hydraulic fracturing.[218] Representative Diana DeGette introduced H.R. 2766 on June 9, 2009, and Senator Robert Casey Jr. introduced S. 1215 as the Fracturing Responsibility and Awareness of Chemicals Act (or "FRAC Act").[219]

The FRAC Act would amend the SDWA definition of "underground injection" to expressly include "the underground injection of fluids or propping agents" used for hydraulic fracturing in oil and gas operation and production activities.[220] The bill would also require public disclosure of the chemical constituents (but not the proprietary chemical formulas) used in the fracturing process.[221] As of October 23, 2010, H.R. 2766 had 69 co-sponsors but ultimately the FRAC Act did not reach the house floor before the 111th Congress recessed.[222] The FRAC Act was re-introduced in the 112th Congress[223] but is unlikely to pass before the 112th Congress adjourns but could be re-introduced in the 113th Congress.

10.10.4 Other Congressional Actions: Disclosure of Frac Fluid Chemicals

In addition to the FRAC Act seeking disclosure of the chemicals used in hydraulic fracturing, Congress has separately requested information from the industry about the chemicals used in hydraulic fracturing.[224] On

---

[218] Fracturing Responsibility and Awareness of Chemicals Act of 2009, S. Con. Res. 1215, 111th Cong. (2009); Fracturing Responsibility and Awareness of Chemicals (FRAC) Act, H.R. Con. Res. 2766, 111th Cong. (2009).

[219] S. 1215; H.R. 2766.

[220] S. 1215 § 2(a); H.R. 2766 § 2(a).

[221] S. 1215 § 2(b).; H.R. 2766 § 2(b).

[222] Bill Summary and Status, H.R. 2766, 111th Congress (2009), The Library of Congress, Thomas, http://thomas.loc.gov/cgi-bin (follow "Bills, resolutions" hyperlink; then follow "Bill summary and status" hyperlink; then search "Fracturing Responsibility and Awareness of Chemicals Act").

[223] S. 587, 112th Cong. (2011); H.R. 1084, 112th Cong. (2011).

[224] Press Release, Comm. on Energy and Commerce, *Energy & Commerce Committee Investigates Potential Impacts of Hydraulic Fracturing*, Feb. 18, 2010, available at http://energycommerce.house.gov/index.php?option=com_content&view=article&id=1896:energy-a-commerce-committee-investigates-potential-impacts-of-hydraulic-fracturing&catid=122:media-advisories&Itemid=55.

February 18, 2010, Henry A. Waxman, Chairman of the Subcommittee on Energy and Environment, and Subcommittee Chairman Edward Markey sent letters to eight oil and gas companies that use hydraulic fracturing "requesting information on the chemicals used in fracturing fluids and the potential impact of the practice on the environment and human health."[225]

On July 19, 2010, Congressmen Waxman and Markey sent another letter requesting additional information from companies involved in hydraulic fracturing, including a list of the total volume of flowback and produced water recovered from wells, how the water was disposed of and a variety of other well-specific data to determine the chemical content of flowback and produced water.[226] The companies did not thoroughly respond, and said "they were not able to provide data on the proximity of specific wells to underground sources of drinking water, or on the recovery and disposal of fluids and water that flowback to the surface of wells."[227]

10.11 THE EPA HYDRAULIC FRACTURING STUDY

In December 2009, six months after the introduction of the FRAC Act 2009, the US House of Representatives Appropriation Conference Committee concluded a focused study analyzing the relationship between hydraulic fracturing and drinking water.[228] The committee believed the EPA should conduct this study.[229] The EPA agreed with Congress that a study was warranted by the serious concerns from citizens about the potential impact on drinking water resources and public health, and

---

225 Ibid.

226 Letter from Rep. Henry A. Waxman, Chairman, Comm. on Energy and Commerce, to 10 Oil and Gas Companies, July 19, 2010, available at http://energycommerce.house.gov/documents/20100719/Letters.Hydraulic.Fracturing.07.19.2010.pdf; see also Press Release, Comm. on Energy and Commerce, *Committee Requests More Details on Hydraulic Fracturing Practices*, July 19, 2010, *available at* http://energycommerce.house.gov/index.php?option=com_content&view=article&id=2079:committee-requests-more-details-on-hydraulic-fracturing-practices&catid=154:correspondence&Itemid=55 (hereinafter *Committee Requests More Details*).

227 *Committee Requests More Details.*

228 Department of the Interior, Environment, and Related Agencies Appropriations Act, H. Rep. 111-316, at 109 (2010); Envtl Prot. Agency, *Hydraulic Fracturing*, http://water.epa.gov/type/groundwater/uic/class2/hydraulicfracturing/index.cfm (hereinafter *Hydraulic Fracturing Overview*).

229 Ibid.

environmental impacts in the vicinity of shale gas production areas employing hydraulic fracturing technology.[230]

In addition to examining the potential relationships between hydraulic fracturing and drinking water, a key goal of the EPA study is to generate data and information that can be used to assess risks and ultimately inform decision makers.[231] In conducting its study, the EPA intends to follow a case study approach, which is often used in in-depth investigations of complex issues like hydraulic fracturing. The EPA admits that "developing a single, national perspective on [hydraulic fracturing] is complex due to geographical variations in water resources, geologic formations, and hydrology."[232] Nonetheless, the EPA's intention is that "the types of data and information that are collected through case studies should provide enough detail to determine the extent to which conclusions can be generalized at local, regional, and national scales."[233] An initial set of research questions proposed by the EPA includes:[234]

1. What sampling strategies and analytical methods could be used to identify potential impacts on sources of drinking water, water supply wells, and receiving streams?
2. Are there vulnerable hydrogeologic settings where HF may impact the quality and availability of water supplies?
3. How does the proximity of HF to abandoned and/or poorly constructed wells, faults, and fractures alter expected impacts on drinking water resources and human health?
4. Is there evidence that pressurized methane or other gases, HF fluids, radionuclides, or other HF-associated contaminants can migrate into underground sources of drinking water? Under what conditions do these processes occur?

10.11.1 The EPA Study Plan and Case Sites

In April 2011, the EPA announced the experts, primarily composed of members of academia, chosen for the Science Advisory Board (SAB)'s

---

230 Ibid.

231 Envtl Prot. Agency, *Opportunity for Stakeholder Input on EPA's Hydraulic Fracturing Research Study: Criteria for Selecting Case Studies*, 1 (July 15, 2010), http://www.epa.gov/safewater/uic/pdfs/hydrofrac_casestudies.pdf (hereinafter *Opportunity for Stakeholder Input*).

232 Ibid. at 2.

233 Ibid.

234 Ibid.

study review panel[235] and the EPA subsequently submitted its draft study plan to the SAB for review.[236] On November 2, 2011, the EPA released details of its hydraulic fracturing study plan.[237] As set forth in the study, the EPA will focus on the entire hydraulic fracturing water life-cycle, from water acquisition to wastewater treatment and disposal.

The EPA will use a case study approach and has selected seven case studies that it believes will provide the most useful information about the potential impacts of hydraulic fracturing on drinking water resources under a variety of circumstances. Two sites are prospective case studies where the EPA will monitor key aspects of the hydraulic fracturing process at future hydraulic fracturing sites. Five sites are retrospective case studies, which will investigate reported drinking water contamination due to hydraulic fracturing operations at existing sites. The EPA is expected to issue its first report of findings in 2012 and its final report in 2014.

10.11.2 Other EPA Actions

While the EPA study is ongoing, there are a number of other activities underway by the EPA that could impact shale gas development going forward.[238]

Effluent guidelines for shale gas extraction

In October 2011, the EPA initiated a rulemaking to set discharge standards for wastewater from shale gas extraction.[239] In terms of background, according to the EPA, and based on information provided by industry, up to 1 million gallons of shale gas wastewater, or "flowback" or "produced

---

235 Envtl Prot. Agency Sci. Advisory Bd, *Members of the Hydraulic Fracturing Study Plan Review Panel*, http://yosemite.epa.gov/sab/sabpeople.nsf/WebExternalSubCommitteeRosters?OpenView&committee=BOARD&subcommittee=Hydraulic%20Fracturing%20Study%20Plan%20Review%20Panel.

236 *Hydraulic Fracturing Overview*.

237 US EPA, Plan to Study the Potential Impacts of Hydraulic Fracturing on Drinking Water Resources, available at http://water.epa.gov/type/groundwater/uic/class2/hydraulicfracturing/upload/hf_study_plan_110211_final_508.pdf.

238 There are numerous actions underway at EPA with respect to shale gas development and a detailed discussion of all actions underway is beyond the scope of this book. For illustrative purposes, two EPA actions pertaining to effluent guidelines and diesel use are discussed here.

239 EPA Fact Sheet, *EPA Initiates a Rulemaking to Set Discharge Standards for Wastewater From Shale Gas Extraction*, http://water.epa.gov/scitech/wastetech/guide/upload/shalereporterfactsheet.pdf.

water," may be produced from a single well within the first 30 days following fracturing. These produced waters generally contain elevated salt content (often expressed as total dissolved solids, or TDS), many times higher than that contained in sea water, conventional pollutants, organics, metals, and NORM (naturally occurring radioactive material). Additional data show that flowback waters contain concentrations of some of the fracturing fluid additives.

While some of the shale gas wastewater is re-used or re-injected, a significant amount still requires disposal. Some shale gas wastewater is transported to public and private treatment plants, many of which are not properly equipped to treat this type of wastewater. As a result, pollutants are discharged into surface waters such as rivers, lakes or streams where they can directly impact aquatic life and drinking water sources.

As part of the rulemaking process, the EPA plans to reach out to affected stakeholders to collect relevant data and information. The EPA also plans to collect financial data on the shale gas industry to determine the affordability of treatment options for produced water.

Guidance for diesel fuels

A key element of the SDWA UIC program is setting requirements for proper well siting, construction, and operation to minimize risks to underground sources of drinking water. The Energy Policy Act of 2005 excluded hydraulic fracturing, *except* when diesel fuels are used, for oil and gas production from permitting under the UIC program. This was because of concern about the risks to drinking water from diesel fuels.

Over the past few years, there has been some confusion over whether the industry must disclose the use of diesel fuel in hydraulic fracturing activities and, if so, how and when. In the summer of 2010, the EPA added a statement to its website that "Any service company that performs hydraulic fracturing using diesel fuel must receive prior authorization from the UIC program. Injection wells receiving diesel fuel as a hydraulic fracturing additive will be considered Class II wells by the UIC program."[240] Industry groups filed a lawsuit against the EPA in the US Court of Appeals for the DC Circuit contending that the website posting constituted a "final agency action" requiring certain procedural actions by the EPA prior to posting such as notice and public comment.[241] More

---

240 Envtl Prot. Agency, *Regulation of Hydraulic Fracturing by the Office of Water*, http://water.epa.gov/type/groundwater/uic/class2/hydraulicfracturing/wells_hydroreg.cfm#safehyfr.

241 Tom Zeller, *Gas Drilling Technique is Labeled Violation*, N.Y. Times, Feb. 1, 2011, at B1.

recently, the EPA announced it is formulating guidelines for the use of diesel fuel in hydraulic fracturing.[242]

10.12 STATE REGULATIONS AND ACTIONS PENDING POTENTIAL FEDERAL ACTION

The EPA study and any legislative action taken by Congress may ultimately take several years to resolve. In the meantime, and in response to the continued public scrutiny of shale gas drilling, some state governments have begun to amend or enact state laws and regulations in an effort to pre-empt the need for any eventual federal regulation of shale gas drilling operations. For example, New York is currently in the process of completing a Supplemental Generic Environmental Impact Statement (SGEIS) for horizontal drilling and hydraulic fracturing[243] with a revised draft SGEIS expected in June 2011.[244] In the interim, and pursuant to an order issued by New York Governor David Patterson, no permits for shale gas drilling may be issued.[245]

The suspension of drilling activity in New York may give that state time to learn lessons about hydraulic fracturing from its neighboring state, Pennsylvania, where more than 1,000 wells have been drilled in the Marcellus Shale since 2005.[246] Those lessons may be difficult for the industry to learn. A recent report from the Pennsylvania Land Trust Association indicates that drillers in Pennsylvania have been cited for 1,435 violations since 2008, 952 of which may affect the environment.[247] The article notes that "[i]ssues listed in the report include improper construction of waste-water compounds used to store [fracking] fluids and violations of the state's clean stream law."[248]

On a more positive note, some companies have begun to voluntarily disclose the chemicals they are using in hydraulic fracturing.[249] For example,

---

242 US EPA, *Underground Injection Control Guidance for Permitting Oil and Natural Gas Hydraulic Fracturing Activities Using Diesel Fuels*, available at http://water.epa.gov/type/groundwater/uic/class2/hydraulicfracturing/wells_hydroout.cfm.

243 N.Y. Dept. of Envtl Conservation, *Marcellus Shale*, http://www.dec.ny.gov/energy/46288.html.

244 Ibid.

245 Ibid.

246 Ibid.

247 Ibid.

248 Ibid.

249 See Range Resources, Press Release, *Range Resources Announces Voluntary*

on July 14, 2010, Range Resources announced a voluntary disclosure initiative for its Marcellus Shale operations whereby it will voluntarily submit to the Pennsylvania Department of Environmental Protection additional information about additives used in the hydraulic fracturing process.[250] The company's press release notes that the "disclosure initiative will provide regulators, landowners and citizens of the Commonwealth an accounting of the highly diluted additives used at each well site, along with their classifications, volumes, dilution factors, and specific and common purposes."[251]

More recently, in April 2011, the shale gas industry launched FracFocus.org, which is an online portal designed to serve as a repository for voluntary disclosure of chemicals used in hydraulic fracturing as well as provide a wide array of reference materials from federal, state, and independent sources. A number of US states, including Texas, Louisiana, and Montana, have incorporated the use of FracFocus in recent legislative and regulatory initiatives requiring disclosure.[252]

10.13 THE DOE SHALE GAS SUBCOMMITTEE[253]

On March 31, 2011, President Barack Obama stated that "recent innovations have given us [the US] the opportunity to tap large reserves – perhaps a century's worth" of shale gas.[254] In order to facilitate this development and ensure adequate environmental protections were in place, President Obama tasked the US Secretary of Energy, Steven Chu, to create a subcommittee of the Secretary of Energy Advisory Board (SEAB) to make recommendations to address the environmental and safety issues that had been raised pertaining to shale gas development in the US.

Disclosure of Marcellus Shale Hydraulic Fracturing, July 14, 2010, available at http://www.rangeresources.com/rangeresources/files/4a/4ad3b135-4c37-4ebb-8923-dec985a70bea.pdf.

250 Ibid.

251 Ibid.

252 Paula Dittrick, *Shale Gas Subcommittee Reviews Industry, Government Progress*, Oil & Gas Journal, Nov. 21, 2011, available at http://www.ogj.com/articles/print/volume-109/issue-47/general-interest/shale-gas-subcommittee-reviews-industry.html.

253 US Dept. of Energy, Secretary of Energy Advisory Board (SEAB), Shale Gas Production Subcommittee, http://www.shalegas.energy.gov/.

254 US Dept. of Energy, SEAB, Shale Gas Production Subcommittee, *90-Day Report*, Aug. 18, 2011, at p. 5, http://www.shalegas.energy.gov/resources/081811_90_day_report_final.pdf.

On August 18, 2011, the Shale Gas Production Subcommittee issued its *90-Day Report* presenting a number of recommendations "that if implemented will reduce the environmental impacts from shale gas production."[255] At the outset, the Subcommittee recognized that shale gas "has enormous potential to provide economic and environmental benefits for the country" including creating jobs across the country and climate change advantages due to the low carbon content of natural gas compared to coal."[256] However, the Subcommittee also recognized that, "[a]s will all energy use, shale gas must be produced in a manner that prevents, minimizes and mitigates environmental damage and the risk of accidents and protects public health and safety. *Public concern and debate about the production of shale gas has grown as shale gas output has expanded.*"[257]

Absent effective environmental policies and controls, the Subcommittee believed that public opposition to shale gas development would grow, thus "putting continued production at risk." The Subcommittee also believed that, absent effective action in terms of strong regulations and industry support, environmental consequences would grow.[258]

Nonetheless, the Subcommittee was optimistic that the industry would continue to pursue "more efficient operations that include minimizing waste, greater gas recovery, less water usage, and a reduced operating footprint."[259] The Subcommittee also believed that "a more systematic commitment to a process of continuous improvement to identify and implement best practices is needed, and should be embraced by all companies in the shale gas industry."[260]

In the Report, the Subcommittee identified four primary areas of concern: "(1) Possible pollution of drinking water from methane and chemicals used in fracturing fluids; (2) Air pollution; (3) Community disruption during shale gas production; and (4) Cumulative adverse impacts that intensive shale production can have on communities and ecosystems."[261]

In an effort to address these environmental concerns, the Subcommittee made a number of recommendations including: (1) improve public information about shale gas operations; (2) improve communication among state and federal regulators; (3) improve air quality by reducing emissions

---

255 Ibid. at p. 1.
256 Ibid. at p. 5.
257 Ibid. at pp. 8–9 (emphasis in the original).
258 Ibid. at p. 8.
259 Ibid. at 8–9.
260 Ibid. at 9–10.
261 Ibid. at 8.

of air pollutants, ozone precursers, and methane; (4) protection of water quality by adoption of a systems approach to water management; (5) disclosure of fracturing fluid composition; (6) reduction in the use of diesel fuel; (7) managing short-term and cumulative impacts on communities, land use, wildlife and ecologies; (8) organizing for best practice; and (9) increasing funding for research and development.[262]

The Subcommittee's 180-day final report was issued on November 18, 2011, and offered a review of the progress made on the 20 recommendations the subcommittee outlined in its August 18 initial report. While the subcommittee said in a November 10 news release that it was "gratified by the actions taken to date," it added that "the progress to date is less than what the subcommittee hoped."[263]

As this book goes to print, the global shale gas industry is still in its formative years, with major questions still to be answered about the extent of the resource base as well as the ability of the industry to develop it economically. There are also numerous legal, policy and environmental challenges that must be addressed, including how the social and environmental debate will play out in different parts of the world. As stated in the IEA's recent *Golden Rules for a Golden Age of Gas* report, "a continuous drive from governments and industry to improve performance is required if public confidence is to be maintained or earned" and "the industry needs to commit to apply the highest practicable environmental and social standards at all stages of the [shale gas] development process."[264] Although it is too soon to predict the ultimate outcome of global shale gas development, the production of US shale gas alone has already impacted global gas markets by giving rise to the prospects of North American LNG exports, a development which is discussed in detail in the next two chapters.

---

[262] Ibid. at 3–5. This is just a summary of the recommendations. More detail regarding the specifics of the recommendations can be found in the report.

[263] Paula Dittrick, *Shale Gas Subcommittee Reviews Industry, Government Progress*, Oil&Gas Journal, Nov. 21, 2011, available at http://www.ogj.com/articles/print/volume-109/issue-47/general-interest/shale-gas-subcommittee-reviews-industry.html.

[264] IEA Golden Rules Report, *supra* note 1, at p. 9.

11. The impact of shale gas on global gas markets and the prospects for US and Canadian LNG exports

11.1 OVERVIEW

As discussed in the preceding chapter, the tremendous boom in US shale gas has been a "game changer" for the US with numerous benefits deriving from shale gas development including economic growth, energy security, and the potential for emissions reductions as coal-fired power plants are replaced with gas-fired power plants. As a result, governments around the world are in the process of assessing their own shale gas reserves to determine whether they can replicate the success of US shale gas development in their own country.

Over time, and given the vast global shale gas resource base, increased global shale gas production could have numerous implications in terms of geopolitics and energy security. For example, European shale gas production could result in a reduced dependence on Russian natural gas imports if production is sufficient to offset the continuing decline in reserves. Whether or not enough European countries will develop their shale gas resources remains to be seen. At least one major study has suggested that while shale gas is not likely to be a game changer for European gas markets overall, it could have a significant impact on individual countries, such as Poland, that are committed to shale gas development.[1]

That study found that much more stringent European environmental standards and the difficulties of access to land and fresh water, as well as the lack of incentives for landowners to allow companies to drill, will require a completely different business model for unconventional gas development in Europe as compared to that in the US. The study also noted that although the impact of shale gas development could be greater

[1] Florence Gény, Can Unconventional Gas be a Game Changer in European Gas Markets? Oxford Institute for Energy Studies, 2010, http://www.oxfordenergy.org/2010/12/can-unconventional-gas-be-a-game-changer-in-european-gas-markets/.

in Poland and Germany, overall it would be surprising if unconventional gas provided more than 5 percent of European gas demand before the early 2020s.[2]

In terms of the impact of shale gas on LNG markets, it remains to be seen whether the global production of shale gas could ultimately displace some LNG production. At this point, this seems a distant possibility since LNG terminals have been and continue to be built around the world at enormous expense. Over time, however, if shale gas production ramps up, it is possible that some LNG production could be displaced by shale gas, depending on cost. It is also possible that, over the long run, increased production of global shale gas will actually strengthen the global LNG markets as increased shale gas supplies will lead to a further shifting of demand to natural gas. Moreover, increased shale gas production around the world could further strengthen LNG markets as more suppliers and buyers enter the market and the LNG market becomes more liquid.

While over the longer term successful shale gas development in countries outside of North America could have a significant impact on global gas markets, it remains to be seen how far the shale gas revolution will spread.[3] In the meantime, however, the sizable increase in shale gas production in the US alone has already impacted global gas markets in ways that were unthinkable just a few years ago.

11.2 THE US REPLACES RUSSIA AS THE WORLD'S TOP PRODUCER OF NATURAL GAS

One of the most dramatic impacts of the US shale gas revolution is the surprising shift of the world's top gas producing countries. For over a decade, the Russian Federation held the top spot as the number one producer of natural gas in the world. According to the *BP Statistical Review of World Energy 2012*, in 2011 the US surpassed Russia as the world's top natural gas producer for the third consecutive year.[4] (See Table 11.1.)

---

2 Ibid.

3 IEA WEO-2010.

4 BP Statistical Review of World Energy, June 2012, www.bp/com/statisicalreview. BP issues its statistical review of world energy every year and it is an excellent source of statistical data for all energy sources, including natural gas and LNG. It is worth noting that, according to BP, the US is also the world's largest consumer of natural gas with 2011 consumption of 690.1 bcm or 21.5% of the world's total. Ibid.

Table 11.1 World's top ten natural gas producers, 2005–11 (in billion cubic meters)

Rank (2011)	Country	2011	2011 % of total	2010	2009	2008	2007	2006	2005
1	United States	651.3	20.0	611.0	582.8	570.8	545.6	524.0	511.1
2	Russia	607.0	18.5	588.9	527.7	601.7	592.0	595.2	580.1
3	Canada	160.5	4.9	159.8	163.9	176.4	182.5	188.4	187.1
4	Iran	151.8	4.6	138.5	131.2	116.3	111.9	108.6	103.5
5	Qatar	146.8	4.5	116.7	89.3	77.0	63.2	50.7	45.8
6	China	102.5	3.1	96.8	85.3	80.3	69.2	58.6	49.3
7	Norway	101.4	3.1	106.4	103.7	99.3	89.7	87.6	85.0
8	Saudi Arabia	99.2	3.0	83.9	78.5	80.4	74.4	73.5	71.2
9	Algeria	78.0	2.4	80.4	79.6	85.8	84.8	84.5	88.2
10	Indonesia	75.6	2.3	82.0	71.9	69.7	67.6	70.3	71.2

Source: *BP Statistical Review of World Energy*, June 2012, www.bp/com/statisicalreview.

11.3 THE IMPACT ON THE US LNG IMPORT MARKET AND KNOCK-ON EFFECTS

The surge in production of shale gas in the US has essentially eliminated the need for the US to import LNG and most LNG originally destined for the US has had to be diverted to Europe or Asia. The dramatic shift in the US's LNG import needs has also resulted in certain "knock-on effects" for those countries that had planned LNG export projects with the US as the major destination. One such project is the $9.9 billion Angola LNG liquefaction project sponsored by state-owned Sonangol. This is the largest investment ever made in Angola and was expected to provide a huge economic boost for the country. When the project was first envisioned years ago, Angola LNG expected the US to take the majority of its LNG production, with the LNG being imported through the recently opened Gulf LNG Energy import terminal located in Pascagoula, Mississippi. Half of the terminal capacity of Gulf LNG Energy is contracted out to US-based Chevron, UK-based BP, France's Total, and Italy's Eni.[5] Angola LNG was expected to come online in early 2012 but many analysts question where the LNG will be exported since the increase in shale gas

[5] Gulf LNG Energy is a partnership between El Paso Corp (50%), Crest Group (30%) and Sonangol USA (20%).

production in the US makes it unlikely that it will go to the US. If the contracts have diversion clauses, then LNG cargoes could be diverted to the best markets.[6]

At an energy conference in late 2011, representatives from Sonangol indicated they still expected the US to import Angolan LNG but more recent reports indicate that Sonangol has been considering other options such as exporting to the Asia-Pacific region.[7] While the Angolan LNG will eventually find a home, it is unclear where and at what price. One analyst has suggested that Angolan LNG will most likely be diverted to Europe.[8] Most recently it was reported that Angola LNG's first cargo will ship in the second quarter of 2012 and will be on spot transactions in Europe and Asia. Chevron has indicated that, once some contractual issues are resolved, it may revisit prospects for selling at least some of the Angola LNG production under long-term agreements but for the moment there is a robust spot market for LNG. Moreover, the increase in the number of LNG import terminals being built around the world "signal[s] continued strength in demand" for LNG going forward.[9]

11.4 THE "SHALE SPREADS" AND THE MARKET OPPORTUNITY FOR NORTH AMERICAN LNG EXPORTS

The surge in US shale gas production over the past several years has led numerous companies in the US and Canada to seek authorization to export LNG in order to take advantage of the arbitrage opportunity resulting from the current supply overhang of shale gas and the price differentials between global gas markets. At least one expert has referred to the price differential due to shale gas production as the "shale spread" and has defined the various shale spreads as shown in Table 11.2.

By early 2010, it was increasingly apparent that the "shale spreads" were creating a market opportunity for US LNG exports. At the time, LNG sold to European markets at the UK National Balancing Point (NBP)

---

[6] Jessica Hatcher, *Questions Remain over Angola LNG Export Plans to the US*, Interfax, Natural Gas Review, Nov. 23, 2011, citing, CWC World LNG Conference, Rome, Italy.

[7] Ibid.

[8] Ibid.

[9] *Chevron's Angola LNG Cargoes to Rely on Spot Markets*, Argus Media, March 13, 2012, http://www.argusmedia.com/pages/NewsBody.aspx?id=78977&menu=yes.

Table 11.2 Shale spreads

Shale spread	Definition
Shale spread	The difference between low US prices due to shale gas production and high international prices that are indexed or benchmarked to oil.
Pacific shale spread	The difference between low US prices and Pacific LNG prices indexed to oil.
Atlantic shale spread	The difference between low US prices and European market prices that oscillate between oil and gas.

Source: Christopher Goncalves, Vice President, Charles River Associates, *Chasing the Shale Spread: Potential New Markets for LNG*, Presentation to the CWC 12th Annual World LNG Summit, November 14–17, 2011, Rome, Italy.

was trading at about a \$5.00 MMBtu premium over gas traded at the US Henry Hub price. (See Figure 11.1.)

The first US company to seize the opportunity created by the shale spreads was Cheniere Energy, which announced in June 2010 that it was initiating a project to add liquefaction facilities to its existing Sabine Pass LNG import terminal.[10] The addition of liquefaction would make Sabine Pass LNG the world's first bi-directional facility capable of *both* importing and exporting LNG. At the time, Cheniere expected to offer customers the bi-directional services for a capacity fee of approximately \$1.40 to \$1.75/MMBtu, which would provide customers the option to either import or export natural gas and would offer buyers an attractive opportunity to buy US sourced natural gas at prices indexed to Henry Hub (Figure 11.2).

---

[10] Cheniere Energy Partners, L.P., Press Release dated June 3, 2010, *Cheniere Energy Partners Initiating Project to Add Liquefaction Capabilities at the Sabine Pass LNG Terminal*, http://phx.corporate-ir.net/phoenix.zhtml?c=207560&p=irol-newsArticle&ID=1434470&highlight=. At the time, Cheniere's Chairman and CEO, Charif Souki summarized the market opportunity for US LNG exports as follows: 'We believe current market fundamentals have created an opportunity for the U.S. to offer natural gas to global markets at competitive prices. The U.S. is experiencing an increase in natural gas production, primarily driven by unconventional gas plays, while natural gas demand in the U.S. continues to lag behind market projections. Due to the depth of the markets in South Louisiana with an abundance of supply and existing pipeline infrastructure, we can provide an additional outlet for U.S. natural gas production while offering a low cost source of supply for global buyers seeking alternatives to oil-indexed contracts. The ability to buy or sell natural gas in one of the world's most liquid natural gas markets provides industry players with a very powerful tool to manage their portfolios.'

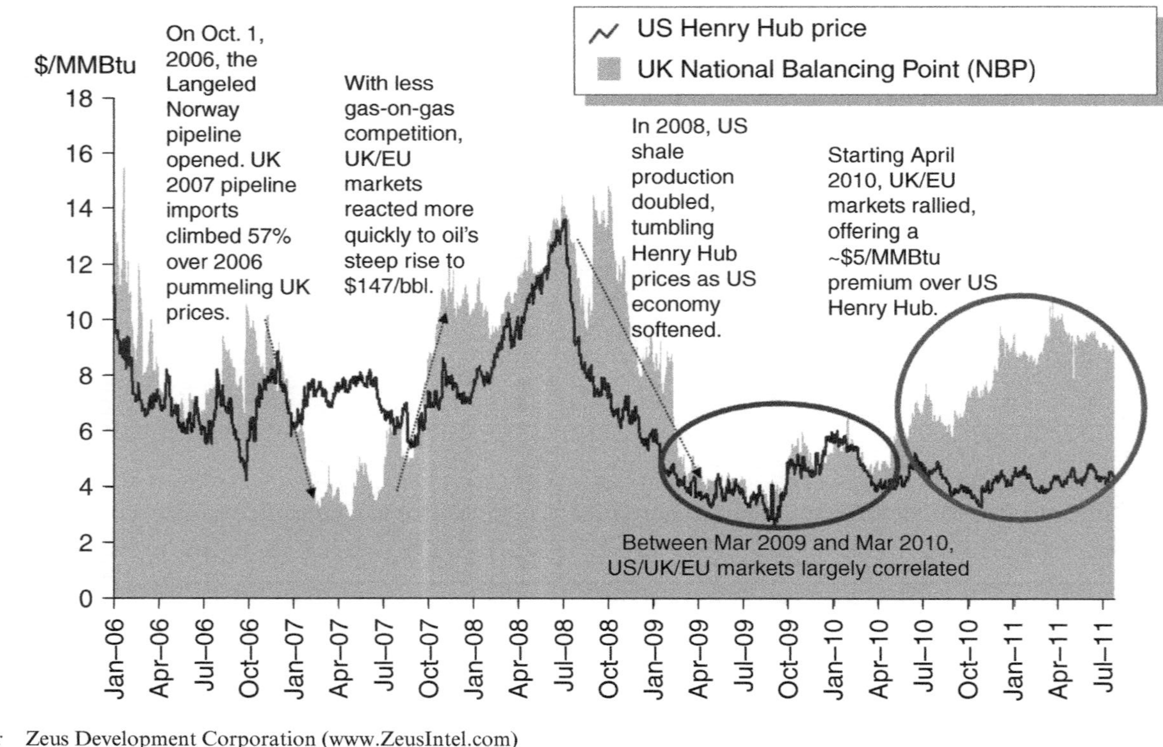

Source: Zeus Development Corporation (www.ZeusIntel.com)

Figure 11.1 UK/US price spread volatility, Jan. 2006–July 2011

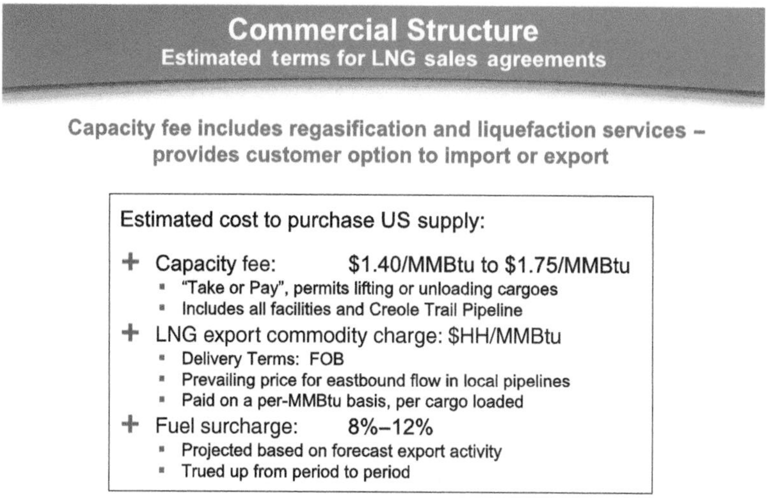

Commercial Structure
Estimated terms for LNG sales agreements

Capacity fee includes regasification and liquefaction services – provides customer option to import or export

Estimated cost to purchase US supply:

+ Capacity fee: $1.40/MMBtu to $1.75/MMBtu
  - "Take or Pay", permits lifting or unloading cargoes
  - Includes all facilities and Creole Trail Pipeline
+ LNG export commodity charge: $HH/MMBtu
  - Delivery Terms: FOB
  - Prevailing price for eastbound flow in local pipelines
  - Paid on a per-MMBtu basis, per cargo loaded
+ Fuel surcharge: 8%–12%
  - Projected based on forecast export activity
  - Trued up from period to period

Source: Cheniere Energy Partners, L.P. Cheniere Energy, Inc., "Proposed Sabine Pass LNG Facility Expansion – Adding Liquefaction Capabilities," Presentation dated June 4, 2010.

Figure 11.2 Cheniere commercial structure for LNG exports

With the existing market dynamics, Cheniere expected to be able to liquefy and ship LNG to Europe cheaper than oil-indexed pipeline gas in Europe on the margin. In its analysis, Cheniere assumed that the continued increase in US natural gas production driven by shale gas effectively capped Henry Hub prices at a mid-range of $6.50/MMBtu. Cheniere also claimed that if oil remains above $65/Bbl, Sabine Pass LNG is cheaper than oil-indexed pipeline gas in Europe on the margin, while oil prices above $77/Bbl justify it on an all-in basis. (See Figure 11.3.)

11.5 THE CASE FOR US LNG EXPORTS – CHENIERE ENERGY CASE STUDY

The actual and potential increase in shale gas production has led to a dramatic shift in natural gas markets in North America, particularly in the United States. Whereas just a few years ago it was predicted that the United States would become a large importer of LNG, the rise of shale gas has largely eliminated the need for the US to import LNG and LNG appears to come to the US only as a place of last resort.

Delivered Costs Compare Favorably to European Price Estimates

- Assuming continued increase in U.S. natural gas production, unconventional gas economics effectively cap Henry Hub at mid-range of $6.50/MMBtu.
- If oil remains above $65/Bbl, Sabine Pass LNG is cheaper than oil-indexed pipeline gas in Europe on the margin, while forecast prices above $77/Bbl justify it on an all-in basis.

$/MMBtu	Low	Mid	High
Henry Hub Price	$ 4.50	$ 6.50	$ 8.50
Terminal Fuel	0.45	0.65	0.85
Liquefaction Charge	1.50	1.50	1.50
Shipping Cost	1.00	1.00	1.00
Delivery Charges	$ 2.95	$ 3.15	$ 3.35
DES Price (Europe)	$ 7.45	$ 9.65	$ 11.85
Brent Crude @ 12.5%	$ 59.60	$ 77.20	$ 94.80
Brent Crude @ 15%	$ 49.67	$ 64.33	$ 79.00

Source: Cheniere Energy Partners, L.P. Cheniere Energy, Inc., "Proposed Sabine Pass LNG Facility Expansion – Adding Liquefaction Capabilities," Presentation dated June 4, 2010.

Figure 11.3 Cheniere US LNG exports – delivered costs to Europe

As a result, LNG importers in the US have had to adjust strategies to attempt to monetize their costly LNG infrastructure investments. In a significant turn of events, in 2010, several US companies began pursuing the option to export LNG from the US to other markets under a variety of scenarios including re-exporting LNG previously imported or exporting to countries with whom the US has a free trade agreement (FTA).[11]

Cheniere Energy, Inc. ("Cheniere"), through its wholly owned subsidiary Sabine Pass Liquefaction, LLC ("Sabine Pass"), was the first company to seek long-term, multi-contract authorization from DOE to export LNG to any country that has the capacity to import LNG and with which trade is not prohibited by US law or policy.[12]

---

11 US Dept. of Energy, 2010, – *Pending LNG Export and Long Term Natural Gas Applications*, available at http://www.fossil.energy.gov/programs/gas regulation/authorizations/2010_Long_Term_Applications.html. In addition to Sabine Pass Liquefaction, other companies seeking to export LNG at the time Cheniere initially filed included Freeport LNG, Lake Charles Exports LLC, Carib Energy (USA) LLC, and Dominion Cove Point LNG.

12 *Application of Sabine Pass Liquefaction, LLC for Long-Term Authorization to Export Liquefied Natural Gas*, FE Docket No. 10-111-LNG (Sept. 7, 2010),

Cheniere's early lead to export LNG is reminiscent of its attempt in 1999 to gain an "early-mover advantage" by being at the forefront of US LNG terminal development and construction by identifying potential terminal sites early on along the US Gulf Coast.[13] As a result of its quick action, Cheniere was the first company to build a new *import* terminal in the United States in anticipation of the US becoming a large LNG importer.[14]

11.5.1 Background on Cheniere's Import/Regasification Terminal

Cheniere's original plans called for the development of three 100% owned onshore LNG import terminals along the US Gulf Coast with an aggregate send-out capacity of 9.9 billion cubic feet of natural gas per day as well as associated pipeline development and natural gas marketing activities through various subsidiaries.[15]

Cheniere's 2.6 Bcf/d Sabine Pass LNG import terminal became Cheniere's "flagship" asset at an approximate cost of construction of $900–950 million, before financing costs.[16] Cheniere later sought FERC (Federal Energy Regulatory Commission) approval to expand Sabine Pass from its initial capacity of 2.6 Bcf/d to 4.0 Bcf/d with the cost of the planned expansion to be approximately $500 million, before financing costs. By 2005, Cheniere had successfully raised over $1.7 billion in debt financing.[17]

Ultimately, FERC authorized Sabine Pass to site, construct, and operate a liquefied natural gas (LNG) import, storage and vaporization terminal with a total send-out capacity of 4.0 billion cubic feet (Bcf) per day (4.0 Bcf/d).[18] The Sabine Pass terminal was designed to accommodate

---

available at http://www.fossil.energy.gov/programs/gasregulation/authorizations/Orders_Issued_2010/10_111sabine.pdf.

[13] Cheniere Annual Report 2005, available at http://library.corporate-ir.net/library/10/101/101667/items/213342/Cheniere_2005_Annual_Report.pdf.

[14] Clifford Krauss, *U.S. Company, in Reversal, Wants to Export Natural Gas*, NY Times, Jan. 27, 2011, available at http://www.nytimes.com/2011/01/28/business/economy/28gas.html?_r=1&emc=eta1.

[15] Ibid.

[16] "The Sabine Pass LNG terminal is located on 853 acres of land along the Sabine Pass River on the border between Texas and Louisiana, in Cameron Parish, Louisiana. It is located at the widest point on the Sabine River Navigation Channel, only 3.7 nautical miles from the open water and 23 nautical miles from the outer buoy. The channel is maintained at a depth of 40 feet and is not subject to tidal limitations. The terminal has two docks that are recessed far enough so that no part of the LNG vessel will protrude into the open waterway while docked."

[17] Cheniere Annual Report 2005, available at http://library.corporate-ir.net/library/10/101/101667/items/213342/Cheniere_2005_Annual_Report.pdf.

[18] U.S. Dept. of Energy, DOE Docket No. FE-08-77-LNG, *Sabine Pass LNG*

up to 400 LNG vessels per year ranging in size from 125,000 cubic meters (m^3) to up to 266,000 m^3, including the Q-max class vessels which are authorized to transit through the Sabine Pass Channel to the terminal.[19]

Phase 1 of Sabine Pass LNG commenced service in April 2008 with an initial 2.6 Bcf/d of send-out capacity and 10 Bcf of storage capacity.[20] Phase 2 was completed in mid-2009 and gave Sabine Pass a total send-out capacity of 4.0 Bcf/d and 16.8 Bcf of storage capacity, thus making the Sabine Pass terminal "the largest receiving terminal, by regasification capacity, in the world."[21]

According to Cheniere's corporate filings, the entire 4.0 Bcf/d of regasification capacity at Sabine Pass was originally reserved under three 20-year, firm commitment terminal use agreements (TUAs) that require payment regardless of whether or not the contracting parties use the terminal.[22] The initial three contracts are as follows:

- Total Gas and Power North America, Inc. ("Total") has reserved approximately 1.0 Bcf/d of regasification capacity and has agreed to make monthly capacity payments to Sabine Pass LNG of approximately $125 million per year for 20 years commencing April 1, 2009.
- Chevron U.S.A., Inc. ("Chevron") has reserved approximately 1.0 Bcf/d of regasification capacity and has agreed to make monthly capacity payments to Sabine Pass LNG of approximately $125 million per year for 20 years commencing July 1, 2009.
- Cheniere's wholly-owned subsidiary, Cheniere Marketing, reserved the remaining 2.0 Bcf/d of regasification capacity, and is entitled to use any capacity not utilized by Total and Chevron. Cheniere Marketing began making its TUA capacity reservation fee payments in the fourth quarter of 2008 and is required to make capacity payments totaling approximately $250 million per year from January 1, 2009, through at least September 30, 2028. Cheniere has guaranteed Cheniere Marketing's obligations under its TUA.

Export Project, available at http://www.fossil.energy.gov/programs/gasregulation/authorizations/Cheniere_Marketing_08-77-LNG.html.

19 Ibid.

20 *Cheniere Announces Opening of Sabine Pass Terminal*, Business Wire, April 21, 2008, available at http://phx.corporate-ir.net/phoenix.zhtml?c=101667&p=irol-newsArticle&ID=1132680&highlight=.

21 Cheniere, *Sabine Pass LNG Terminal*, available at http://www.cheniere.com/LNG_terminals/sabine_pass_lng.shtml.

22 Cheniere Energy, Inc., *2009 Annual Report*, Form 10-K, available at http://www.cheniere.com/corporate/2009_Cheniere_Annual_Report.pdf.

Under each of these TUAs, Sabine Pass LNG is entitled to retain 2 percent of the LNG delivered for the customer's account for its own use as fuel for revaporization and self-generated power at the Sabine Pass LNG receiving terminal.

11.5.2 Cheniere Marketing – Re-Export Application 2008

Although Cheniere ultimately won the race to construct the first new LNG import terminal in the US,[23] the dynamic LNG market had already shifted by the time Sabine Pass was completed and Cheniere found that very little LNG was coming to the US as suppliers sought higher paying Asian and European LNG market destinations.[24] In light of changing market dynamics, by late 2008 it was evident that one of Cheniere's key assets, the 2.0 Bcf/d of regasification capacity reserved by Cheniere Marketing, was essentially unused or underutilized capacity that needed to be monetized in some way.[25]

In August 2008, Cheniere Marketing, Inc., applied to the Office of Fossil Energy of the Department of Energy (DOE/FE) "requesting blanket authorization to export LNG that previously had been imported from foreign sources."[26] The application sought approval to export up to 64 Bcf over a two-year period from the Sabine Pass LNG terminal to the United Kingdom, France, Portugal, Spain, Belgium, Turkey, Italy, Brazil,

---

23 Clifford Krauss, *U.S. Company, In Reversal, Wants to Export Natural Gas*, NY Times, Jan. 27, 2011, available at http://www.nytimes.com/2011/01/28/business/economy/28gas.html?_r=1&emc=eta1, Cheniere Annual Report 2005.

24 Cheniere Marketing, Inc., *Application for Blanket Authorization to Export Imported Liquefied Natural Gas*, Docket No. 08-77-LNG (Aug. 8, 2008). In support of its application, Cheniere stated that "increasing world-wide demand for LNG in European and Asian markets, and relatively high prices, has resulted in a decrease in deliveries of LNG to the U.S. recently." Application, p. 4.

25 Cheniere's Annual Letter to Shareholders found in its 2009 Annual Report states that Cheniere's "key assets" include "the 2 Bcf/d of regasification capacity at the Sabine Pass LNG receiving terminal" and that Cheniere's "strategy continues to be to maximize the value of [its] assests" by "monetizing [its] capacity by entering into long-term TUAs, [and] developing a portfolio of long-term, short-term and spot LNG purchase agreements."

26 Cheniere Marketing, Inc., *Application for Blanket Authorization to Export Imported Liquefied Natural Gas*, Docket No. 08-77-LNG (Aug. 8, 2008). Cheniere's application was pursuant to Section 3 of the Natural Gas Act, 15 U.S.C. §717b, and Part 590 of the Department of Energy's regulations, 10 C.F.R. Part 590 (2008). All of the filings are available at http://www.fossil.energy.gov/programs/gasregulation/authorizations/Cheniere_Marketing_08-77-LNG.html.

Argentina, Mexico, Japan, Korea, Taiwan, China, India, Dominican Republic, Chile and possibly the Commonwealth of Puerto Rico.[27]

In its application, Cheniere stated that it was seeking to export LNG that had been imported into the US but for which there was no US demand – thus creating a market for "re-exports." Cheniere also indicated at the time that very little LNG was coming to the US since other markets were paying higher prices. Authorization to re-export would thus allow Cheniere to purchase LNG at current prices with the intent that such LNG would subsequently be exported to a foreign market paying higher prices at a later date.[28] For example, on November 8, 2010, the "month-ahead premium of UK ICE futures over US prices for [LNG] stood at $3.357/MMBtu [indicating] a significant arbitrage even after shipping is factored in."[29]

Indeed, this arbitrage opportunity ultimately resulted in the first LNG cargo from the US to the UK in 50 years. On November 6, 2010, the *Maersk Meridian* left Cheniere's Sabine Pass terminal with a cargo of LNG headed to the UK,[30] and it arrived at the UK's Isle of Grain import terminal some ten days later.[31] According to traders, this was the first US–UK LNG shipment since the world's first ever shipment of LNG was made by the *Methane Pioneer*'s maiden voyage on February 20, 1959, from an LNG storage facility near Lake Charles, Louisiana, to Canvey Island, UK.[32] The *Methane Pioneer* subsequently delivered seven additional cargoes, thereby demonstrating that "large quantities of liquefied natural gas could be transported safely across the ocean."[33] Fifty years after the *Methane Pioneer*'s historic voyage, is the US poised to become a major LNG exporter with Cheniere leading the way?

---

27 Cheniere Marketing, Inc., *Application for Blanket Authorization to Export Imported Liquefied Natural Gas*, Docket No. 08-77-LNG (Aug. 8, 2008).

28 Ibid. at pp.4–5.

29 Oleg Vukmanovic, *First LNG Cargo from US to UK in 50 Years Arrives at Grain LNG*, Platts LNG Daily, Nov. 19, 2010, available at http://www.platts.com/RSSFeedDetailedNews/RSSFeed/NaturalGas/8201994.

30 US Dept. of Energy, *LNG Imports – Nov. 2010, Monthly Report* (revised 1/20/11), available at http://fossil.energy.gov/programs/gasregulation/publications/Nov10LNG_rev_01-20-2011.pdf.

31 Oleg Vukmanovic, *First LNG Cargo from US to UK in 50 Years Arrives at Grain LNG*, Platts LNG Daily, Nov. 19, 2010, available at http://www.platts.com/RSSFeedDetailedNews/RSSFeed/NaturalGas/8201994.

32 Ibid.

33 Center of Energy Economics (CEE), *Brief History of LNG*, available at http://www.beg.utexas.edu/energyecon/lng/LNG_introduction _06.php.

11.5.3 Sabine Pass Liquefaction – Phase One – FTA Countries

On August 11, 2010, Sabine Pass filed the first part of a two-phased export application with the Office of Fossil Energy (FE) of the US Department of Energy (DOE) under section 3 of the Natural Gas Act (NGA)[34] for the long-term, multi-contract authorization to export up to 16 mtpa of LNG (803 billion cubic feet (Bcf) per year) for a 30-year term to any nation that the US has a free trade area agreement (FTA) with currently or in the future requiring the national treatment for trade in natural gas and LNG.[35]

Sabine Pass submitted its application under section 3(c) of the NGA which, as amended by the US Energy Policy Act of 1992,[36] requires DOE/FE to grant, on an expedited basis, an application for the exportation of natural gas if there is an FTA in effect requiring national treatment for trade in natural gas.[37] If an FTA is in force, then such exports are deemed to be in the public interest and must be granted without modification or delay.[38] Applications for export authorization absent an FTA require DOE/FE to review and analyze whether such exports are consistent with the public interest.[39] Since Sabine Pass's application fell under NGA

---

34 15 U.S.C. § 717b (2000), http://www.fossil.energy.gov/programs/gasregulation/authorizations/sect.pdf.

35 US Dept. of Energy, *Application of Sabine Pass Liquefaction, LLC to Request Long-term Authorization to Export Liquefied Natural Gas to Free Trade Agreement Countries*, Docket No. 10-85-LNG, http://www.cheniereenergypartners.com/liquefaction_project/DOE_filings.html.

36 Section 201, Energy Policy Act of 1992 (Pub. L. 102-486).

37 15 U.S.C. § 717b provides as follows:

(b) Free trade agreements

With respect to natural gas which is imported into the United States from a nation with which there is in effect a free trade agreement requiring national treatment for trade in natural gas, and with respect to liquefied natural gas –

(1) the importation of such natural gas shall be treated as a "first sale" within the meaning of section 3301(21) of this title; and

(2) the Commission shall not, on the basis of national origin, treat any such imported natural gas on an unjust, unreasonable, unduly discriminatory, or preferential basis.

38 15 U.S.C § 717b provides as follows:

(c) Expedited application and approval process

For purposes of subsection (a) of this section, the importation of the natural gas referred to in subsection (b) of this section, or the exportation of natural gas to a nation with which there is in effect a free trade agreement requiring national treatment for trade in natural gas, shall be deemed to be consistent with the public interest, and applications for such importation or exportation shall be granted without modification or delay.

39 15 U.S.C § 717b provides as follows:

section 3(c), as amended, DOE/FE summarily granted the application on September 7, 2010, and authorized Sabine Pass to export domestically produced LNG to any FTA country.[40]

11.5.4 Sabine Pass Liquefaction – Phase Two – WTO Nations

On September 7, 2010, Sabine Pass filed the second part of its two-phased export application seeking long-term, multi-contract authorization to export up to 16 million tonnes per annum (mtpa) of domestically produced LNG for a 20-year period to countries other than those that have negotiated FTAs with the US that are members of the World Trade Organization ("WTO Countries") and those countries that do not hold membership in the WTO ("non-WTO countries") and with which trade is not prohibited by US law or policy.[41]

In its application, Sabine Pass requested that DOE/FE review its request to export LNG to WTO countries under the same standard of review applicable to FTA countries and specifically requested that DOE/FE conduct its review under the standards set forth in section 3(c) of the NGA, 15 U.S.C. 717b(c) instead of section 3(a) of the NGA, 15 U.S.C. 717b(a).[42]

(a) Mandatory authorization order

After six months from June 21, 1938, no person shall export any natural gas from the United States to a foreign country or import any natural gas from a foreign country without first having secured an order of the Commission authorizing it to do so. The Commission shall issue such order upon application, unless, after opportunity for hearing, it finds that the proposed exportation or importation will not be consistent with the public interest. The Commission may by its order grant such application, in whole or in part, with such modification and upon such terms and conditions as the Commission may find necessary or appropriate, and may from time to time, after opportunity for hearing, and for good cause shown, make such supplemental order in the premises as it may find necessary or appropriate.

40 Sabine Pass Liquefaction, LLC, FE Docket No. 10-85-LNG, Order No. 2833 (Sept. 7, 2010). The DOE/FE order authorizes LNG exports to the following FTA countries: "Australia, Bahrain, Singapore, Dominican Republic, El Salvador, Guatemala, Honduras, Nicaragua, Chile, Morocco, Canada, Mexico, Oman, Peru, Singapore (sic) and Jordan, and to any nation which DOE subsequently identifies publicly (currently at http://www.fossil.energy.gov/programs/gas regulation/authorizations/How_to_Obtain_Authorizations.html) as having entered into a free trade agreement providing for national treatment for trade in natural gas, provided that the destination nation has the capacity to import LNG."

41 Sabine Pass Liquefaction, LLC, FE Docket No. 10-111-LNG, *Application of Sabine Pass Liquefaction, LLC for Long-Term Authorization to Export Liquefied Natural Gas* (Sept. 7, 2010) available at http://www.fossil.energy.gov/programs/gasregulation/authorizations/Orders_Issued_2010/10_111sabine. pdf.

42 Application at 2-3.

In making this novel request, Sabine Pass contended that US trade policy, as well as US obligations under the WTO, required the "automatic export authorization process" applicable for export of LNG to FTA countries and therefore sought DOE/FE's immediate approval to export LNG to WTO countries.[43] In support of its argument, Sabine Pass submitted Annex 1A to the WTO Agreement and "a compendious legal memorandum" entitled "A Review of International Trade-Related Legal Obligations and Policy Considerations Governing U.S. Export Licenses for Liquefied Natural Gas" (Aug. 23, 2010).[44]

Despite Sabine Pass's extensive briefing of the trade issues, DOE/FE found that Sabine Pass's request for review under section 3(c) was "not supported by law or policy."[45] With very little discussion of WTO issues, DOE/FE stated that section 3(a) of the NGA, not section 3(C), was the appropriate legal provision to an application to export LNG to any nation other than FTA countries. Since Sabine Pass had not pointed to any legislation authorizing or requiring a different result, DOE has no authority to grant Sabine Pass's request for section 3(c) review.[46] On October 21, 2010, DOE/FE issued an opinion and order denying Sabine Pass's request that its export application be reviewed under section 3(c) of the NGA and ordering that the application be reviewed under section 3(a) of the NGA.[47]

Under section 3(a) of the NGA, there is a rebuttable presumption that proposed exports of natural gas are in the "public interest."[48] In order to overcome this presumption in favor of exports, opponents of an export license must "make an affirmative showing of inconsistency with the public interest."[49] In evaluating whether a proposed export is within the public interest, DOE/FE applies the principles established by Policy Guidelines issues in 1984.[50] Under the Policy Guidelines, DOE's public interest analysis of export applications under section 3(a) is focused on the following:

---

[43] Application at 23-29.

[44] Sabine Pass Liquefaction, LLC, FE Docket No. 10-111-LNG, *Opinion and Order Denying Request for Review under Section 3(c) of the Natural Gas Act*, Oct. 21, 2010, at 3, available at http://www.fossil.energy.gov/programs/gasregulation/authorizations/Orders_Issued_2010/Sabine10_111dkt.html.

[45] Opinion and Order at 6.

[46] Opinion and Order at 7.

[47] Opinion and Order at 8.

[48] Opinion and Order at 4.

[49] Opinion and Order at 4-5, Order No. 1473, note 42 at 13, citing *Panhandle Producers and Royalty Owners Assoc. v. ERA*, 822 F.2d 1105, 111 (DC Cir. 1987).

[50] Policy Guidelines and Delegation Orders Relating to the Regulation of Imported Natural Gas, 49 Fed. Reg. 6,684 (Feb. 22, 1984).

1. The domestic need for the natural gas proposed to be exported;
2. Whether there is a threat to the domestic security of supply; and
3. Other factors to the extent they are shown to be relevant to a public interest determination.[51]

As an alternative to its WTO argument, Sabine Pass also addressed the "public interest" test in its application. At the outset, Sabine Pass stated that its proposed liquefaction plans are in response to the improved outlook for domestic natural gas production, and in particular increased shale gas production, which has lessened the uncertainties associated with future US natural gas production. In support of its application, Sabine Pass commissioned several reports to assess the domestic need for the natural gas proposed to be exported and, based on these reports, contended that the US has "significant natural gas resources available at modest prices to meet protected domestic demand and 16 mtpa of exports over the 20-year period as requested in its Application."[52]

In its application, Sabine Pass contended that exports of LNG would provide the following benefits, which are consistent with the pubic interest:[53]

1. The project will stimulate the local, regional, and national economies by creating jobs, increased economic activity and tax revenues.
2. The project will play an influential role in contributing to the growth of natural gas production in the US and a reduced reliance on foreign sources of oil.
3. LNG exports will further President Obama's National Export Initiative by improving the balance of payments with the rest of the world, thereby reducing the overall trade deficit.
4. Exports of LNG will raise domestic natural gas productive capacity and promote stability in domestic natural gas pricing.
5. Exports of LNG will promote liberalization of the global gas market by fostering increased liquidity and trade at prices established by market forces.
6. Exports of LNG will advance national security and the security of US allies through diversification of global natural gas supplies.

---

[51] Opinion and Order at 7.

[52] 75 Fed. Reg. 62512 (Oct. 12, 2010) available at http://www.fossil.energy.gov/programs/gasregulation/authorizations/Orders_Issued_2010/Sabine_Pass_10-111_FP1.pdf.

[53] Summary from the Federal Register Notice, 75 Fed. Reg. 62512 (Oct. 12, 2010).

7. Exports of LNG will advance initiatives underway to promote investment in energy infrastructure and to increase trade with Caribbean and Central/South American nations.

Sabine Pass's export application received significant support from local, state and national politicians[54] as well as support from natural gas companies such as Chesapeake.[55] In addition, Cheniere/Sabine Pass Liquefaction entered into eight non-binding MOUs with potential customers for the proposed bi-directional facility, representing a total of up to 9.8 mtpa of capacity.[56] While the MOUs were non-binding, they were indicative of at least some market interest in the project.

11.5.5 Opposition to Cheniere's Sabine Pass Export Application

Cheniere's Sabine Pass export application received opposition from two industry groups representing industrial users of natural gas – the Industrial Energy Consumers of America (IECA)[57] and the American Public Gas Association (APGA).[58] IECA filed a timely Motion to Intervene on December 13, 2010.[59] In its motion, IECA contended that exports of LNG

---

[54] See, for example, "Joint Letter to President Obama in Support of Application," filed December 13, 2010, by Senators James Inhofe and Mary Landrieu and Representatives Dan Boren and Charles Boustany along with other Members of Congress, available at http://www.fossil.energy.gov/programs/gasregulation/authorizations/Orders_Issued_2010/congressional_delagates_12_13_10.pdf. Other letters of support are listed on the Sabine Pass Docket page at http://www.fossil.energy.gov/programs/gasregulation/authorizations/Orders_Issued_2010/Sabine10_111dkt.html#Sabine%20Pass.

[55] See "Letter from James C. Johnson, Senior Vice President – Marketing, Chesapeake Energy Corporation in Support of Application", available at http://www.fossil.energy.gov/programs/gasregulation/authorizations/Orders_Issued_2010/chesapeake_12_13_10.pdf.

[56] Cheniere Energy Partners, L.P., Press Release Dated May 6, 2011, *Cheniere Energy Partners Reports First Quarter 2011 Results*, http://phx.corporate-ir.net/phoenix.zhtml?c=207560&p=irol-newsArticle&ID=1560809&highlight=.

[57] "IECA is a 501 (C) (6) nonprofit member-led organization created to promote the interests of manufacturing companies for which the availability, use and cost of energy, power or feedstock play a significant role in their ability to compete in domestic and world markets." *About IECA*, available at http://www.ieca-us.com/about.html.

[58] APGA is the national, non-profit association of publicly owned natural gas distribution systems and has over 700 members in 36 states. About APGA, available at http://www.apga.org.

[59] IECA Motion to Intervene: FE Docket No. 10-111-LNG, Sabine Pass Liquefaction, LLC, *Application for Long-Term Authorization to Export Liquefied*

are not in the public interest since such exports have the potential to increase demand and thereby the price of natural gas for the manufacturing sector and the public. IECA further claimed that increased natural gas prices erode US manufacturing competitiveness and could lead to further job losses in the manufacturing sector. While IECA acknowledged that recent estimates of US natural gas reserves had increased, IECA maintained that there were still many uncertainties that could reduce natural gas supplies, especially over the 20-year approval period that Sabine Pass is seeking in its export application.[60] IECA also argued that Sabine Pass is seeking export approval for a significant amount of natural gas – "as much as .803 trillion cubic feet per year for a 20 year period [which is] equivalent to 3.54 percent of 2009 US demand; 13.4 percent of industrial demand; 17.1 percent of residential demand and 11.6 percent of electric sector demand."[61]

In response, Sabine Pass filed an Answer to IECA's Motion contending that IECA's motion was unsupported by any empirical market studies or other data and therefore failed to overcome the statutory presumption in favor of granting the application.[62]

On March 4, 2011 the American Public Gas Association (APGA) filed a late Motion to Intervene in the Sabine Pass proceeding.[63] Although DOE/FE ultimately issued a procedural order denying APGA's late filed motion, resulting in APGA not being a party to the proceeding, DOE/FE nonetheless reviewed APGA's submission so that a complete evaluation of all relevant arguments was conducted in light of the "precedential nature of [the] proceeding."[64]

Natural Gas, available at http://www.fossil.energy.gov/programs/gasregulation/authorizations/Orders_Issued_2010/industrial12_13_10.pdf.

60 Ibid.

61 Ibid.

62 Answer of Sabine Pass Liquefaction, LLC to Motion to Intervene of Industrial Energy Consumers of America, available at http://www.fossil.energy.gov/programs/gasregulation/authorizations/Orders_Issued_2010/Answer10-111-lng_12_20_10.pdf.

63 Motion to Intervene Out-Of-Time and Protest on Behalf of the American Public Gas Association, FE Docket No. 10-111-LNG, Sabine Pass Liquefaction, LLC, *Application for Long-Term Authorization to Export Liquefied Natural Gas*, available at http://www.fossil.energy.gov/programs/gasregulation/authorizations/Orders_Issued_2011/american_public_gas_assc_03_04_11.pdf. In addition to its Motion to Intervene, the APGA Board of Directors will be considering a policy resolution on the issue of the exportation of domestically produced natural gas at its May meeting in Washington, DC. *APGA Files Motion to Intervene in Sabine Pass LNG Export Facility Application*, available at http://www.apga.org/i4a/headlines/headlinedetails.cfm?id=807&pageid=3284.

64 *Opinion and Order Conditionally Granting Long-Term Authorization to*

In its motion, APGA contended that "the export of large quantities of natural gas may have significant adverse implications for domestic consumers of natural gas, for U.S. energy supply, and national security." The motion also argued that the export of natural gas is inconsistent with a policy of energy independence, especially as APGA continues to push for domestic natural gas to play a larger role as a transportation fuel. APGA's motion contended that the replacement of existing gasoline-powered fleets and passenger vehicles with natural gas vehicles (and support infrastructure) would significantly reduce US dependence on foreign oil, and thereby enhance US security and strategic interests. Thus, instead of exporting domestic natural gas, the United States should maximize its use domestically in order to increase energy efficiency, reduce greenhouse gas emissions and increase energy independence.

11.5.6 DOE/FE Opinion and Order Conditionally Granting Cheniere's Sabine Pass Export Application

On May 20, 2011, DOE/FE issued its Opinion and Order in the Sabine Pass case.[65] In its summary of findings and conclusions, DOE/FE found that Sabine Pass had "submitted substantial evidence showing an existing and a projected future supply of domestic natural gas sufficient to simultaneously support the proposed export and domestic natural gas demand both currently and over the 20-year term of the requested authorization."[66]

The studies submitted by Sabine Pass indicated a modest increase in the domestic market price for natural gas through 2035, which reflected, according to DOE/FE, the costs of additional domestic production for LNG exports as opposed to an increase in price due to an alleged convergence of domestic natural gas prices with international gas prices linked

---

Export Liquefied Natural Gas from Sabine Pass LNG Terminal to Non-Free Trade Agreement Nations, DOE/FE Order No. 2961, May 20, 2011, FE Docket No. 10-111-LNG, Sabine Pass Liquefaction, LLC, *Application for Long-Term Authorization to Export Liquefied Natural Gas*, available at http://www.fossil.energy.gov/programs/gasregulation/authorizations/Orders_Issued_2011/ord2961.pdf.

65 *Opinion and Order Conditionally Granting Long-Term Authorization to Export Liquefied Natural Gas from Sabine Pass LNG Terminal to Non-Free Trade Agreement Nations*, DOE/FE Order No. 2961, May 20, 2011, FE Docket No. 10-111-LNG, Sabine Pass Liquefaction, LLC, *Application for Long-Term Authorization to Export Liquefied Natural Gas*, available at http://www.fossil.energy.gov/programs/gasregulation/authorizations/Orders_Issued_2011/ord2961.pdf.

66 Ibid. at p. 29.

to oil.[67] Significantly, DOE/FE noted that the opposing parties had *not* submitted any contrary studies.[68]

DOE/FE also found that the approval of the requested authorization would *not* threaten US energy security and that Sabine Pass had pointed to a number of significant economic and public benefits that would ensue from the granting of the application.[69] These benefits included potentially significant job creation[70] due to the construction of the project as well as improving the US balance of payments through the exportation of approximately 2 Bcf/d of natural gas, valued by the applicant at approximately $5 billion, and the displacement of $1.7 billion in natural gas liquids imports.[71] While the opponents had alleged a number of negative consequences to the public, they did not refute the alleged benefits of the project.[72]

11.5.7 DOE/FE's Continuing Duty to Protect the Public Interest

While DOE/FE found that Sabine Pass had provided substantial evidence supporting its export application, it also recognized that "no person can unqualifiedly warrant either that the present conditions that have yielded

---

[67] Ibid.

[68] Ibid.

[69] Ibid. at p. 30.

[70] Cheniere's Sabine Pass project appeared to have considerable support from both federal and local governments. For example, according to US Energy Secretary, Steven Chu, "Our long term economic strength depends on safely and responsibly harnessing America's domestic energy resources while developing new and innovative clean energy technologies . . . [The Sabine Pass] project reflects a broad, 'all of the above' approach that will put Americans to work producing the energy the world needs." US Senator Mary Landrieu said, "Cheniere Energy's plan to transform its existing terminal into a facility that can both import and export liquefied natural gas is a precedent-setting breakthrough that will bring substantial economic benefits to southwest Louisiana." DOE Press Release, *Energy Department Approves Gulf Coast Exports of Liquefied Natural Gas: Conditional Authorization for Sabine Pass LNG Terminal Could Bring Thousands of Jobs*, May 20, 2011, http://www.fossil.energy.gov/news/techlines/2011/11023-DOE_Approves_LNG_Export_Applicatio.html.

[71] *Opinion and Order Conditionally Granting Long-Term Authorization to Export Liquefied Natural Gas from Sabine Pass LNG Terminal to Non-Free Trade Agreement Nations*, DOE/FE Order No. 2961, May 20, 2011, FE Docket No. 10-111-LNG, Sabine Pass Liquefaction, LLC, *Application for Long-Term Authorization to Export Liquefied Natural Gas*, available at http://www.fossil.energy.gov/programs/gasregulation/authorizations/Orders_Issued_2011/ord2961.pdf, at p. 30.

[72] Ibid.

substantial new gas reserves [from shale] or the resulting projections of continuing supply increases contained in the studies presented by the applicant, will prove completely accurate over the entire 20-year projected term of the requested authorization."[73]

Some of the potential variables that could alter projections included the possibility that other nations might successfully develop their own shale gas resources, "as well as environmental, regulatory, and safety considerations."[74] In terms of the potential environmental impact of shale gas development, DOE/FE took administrative notice of the EPA study of the possible impacts from hydraulic fracturing discussed above in Chapter 10.[75] DOE/FE also noted the possibility that the results of the EPA's review could ultimately reduce the supplies of natural gas below the projections contained in the applicant's submission.

DOE/FE also recognized that projections regarding the demand for natural gas are also subject to change based on future demand for natural gas in the electricity sector and/or as a transportation fuel, both of which are difficult to predict. In short, DOE/FE indicated that it intended to monitor the conditions impacting natural gas supply/demand in the future to ensure that the exports of LNG authorized ultimately do not "lead to a reduction in the supply of natural gas needed to meet essential domestic needs."[76] In particular, DOE/FE noted it would "evaluate the cumulative impact of the instant authorization and any future authorizations for export authority when considering any subsequent application for such authority" to ensure that there was no threat to the public interest.[77]

After conducting its analysis, DOE/FE ordered that "Sabine Pass is authorized to export domestically produced LNG by vessel from the Sabine Pass LNG Terminal up to the equivalent of 803 Bcf per year[78] of natural gas for a term of 20 years to commence on the earlier of the date of first export or five years from the date of the issuance of this authorization on its own behalf or as agent for others pursuant to one or more long-term contracts (a contract with a term greater than two years) that do not exceed the term of this authorization."[79]

DOE/FE also ordered that "Sabine Pass must commence export

---

73 Ibid. at 31.

74 Ibid.

75 Ibid.

76 Ibid. at 32.

77 Ibid.

78 Sabine Pass applied for authorization to export up to the equivalent of 16 mtpa of LNG, which is approximately 2.2 Bcf/d or 803 Bcf per year.

79 Ibid. at 42.

operations using the planned liquefaction facilities no later than seven years from the date of issuance of this order" and that the "LNG may be exported to any country with which the United States does not have a FTA requiring the national treatment for trade in natural gas and LNG, which currently has or in the future develops the capacity to import LNG, and with which trade is not prohibited by United States law or policy."

The DOE/FE authorization was also "conditioned on the satisfactory completion of that environmental review process in FERC Docket No. PF10-24-000 and on issuance by DOE/FE of a finding of no significant impact or a record of decision pursuant to NEPA."

11.5.8 NEPA and FERC Review and Project Financing

As of November 2011, Cheniere's Sabine Pass Liquefaction project was still awaiting final environmental review pursuant to the National Environmental Policy Act (NEPA)[80] and ultimate approval by the Federal Energy Regulatory Commission (FERC), which has the exclusive authority to approve or deny an application for the siting, construction, expansion, or operation of an LNG terminal. As the lead agency, FERC will ultimately determine whether Cheniere will be issued an Order Granting Authority under Section 3 of the Natural Gas Act for the construction and operation of the Sabine Pass Liquefaction Project. FERC will also monitor all construction and restoration activities to insure that Cheniere complies with all federal, state, and local permits, plans, and regulations.[81]

On July 26, 2010, Cheniere requested authorization from FERC to engage in the Pre-Filing Environmental Review Process. The pre-filing process provides opportunities for federal and state cooperating agencies as well as other public stakeholders to comment on the project's impacts

---

80 To fulfill the requirements of NEPA, FERC must prepare an Environmental Assessment (EA) or an Environmental Impact Statement (EIS). The purpose of the EA/EIS is to inform the public and the permitting agencies about the potential adverse and/or beneficial environmental and safety impacts of proposed projects and their alternatives. To assist FERC, Cheniere prepared Resource Reports covering the following: General project description; Water use and quality; Fish, wildlife, and vegetation; Cultural resources; Socioeconomics; Geological resources; Soils; Land use, recreation, and aesthetics; Air and noise quality; Alternatives; and Reliability and safety. Cheniere Energy Partners, L.P., FERC Process, http://www.chenierenergypartners.com/liquefaction_project/ferc_process.html.

81 Cheniere Energy Partners, L.P., FERC Process, http://www.cheniereenergypartners.com/liquefaction_project/ferc_process.html.

prior to an application to FERC being submitted. Permission was granted to begin the pre-filing process on August 4, 2010.[82]

In terms of the project schedule, Cheniere estimated that the Environmental Assessment/Environmental Impact Statement would be issued in the second half of 2011 along with the Issuance of Authorization and the filing of the Initial Implementation Plan. Cheniere estimates that it will commence construction in the first quarter of 2012 and the liquefaction facility will be placed in service in 2015.[83] However, some analysts have cautioned that Cheniere's project timeline may be too optimistic and that 2016 is a more realistic date.[84]

At the time, analysts also cautioned that even assuming FERC approval, Cheniere still needed to secure project financing to construct the $6.5 billion liquefaction facility.[85] In order to underpin the financing, Cheniere would most likely need to enter into binding agreements with customers reserving bi-directional capacity at Sabine Pass LNG, and at the time Cheniere only had in place non-binding MOUs. At least one analyst commented that obtaining financing might be difficult and that

> [t]he goal for Cheniere will be to lock in a Henry Hub/Japanese Crude Cocktail price spread of future cash flows in order to get the right backing. Securing supplies at a low enough price will be key in carrying the project forward. Any terminal looking to export US gas would face a major hurdle: finding producers willing to sell their gas below the $6/MMBtu breakeven price necessary to keep such projects economically viable in 2015 and beyond.[86]

Timing might also be of the essence since "any export terminal faces a rapidly-closing window to meet Asian – particularly Chinese – demand,

---

[82] Ibid.

[83] Ibid.

[84] Samantha Santa Maria, *US LNG Export Proposals May Be Too Optimistic: SocGen Analyst*, Platts, May 25, 2011, http://www.platts.com/RSSFeedDetailedNews/RSSFeed/NaturalGas/6135424, quoting Societe Generale's analyst Laurent Key as stating that even though there is some existing infrastructure in place, "late 2016 would seem to be a more reasonable start-up date."

[85] "Cheniere's anticipated investment of $6.5 billion will expand its Sabine Pass facilities with one of the largest capital investments in Louisiana history." Louisiana Economic Development, News Release July 19, 2011, *Cheniere Energy Announces More Than $6 Billion Natural Gas Facility in Louisiana*, http://www.louisianaeconomicdevelopment.com/led-news/news-releases/cheniere-energy-announces-more-than-$6-billion-natural-gas-facility-in-louisiana.aspx.

[86] Samantha Santa Maria, *US LNG Export Proposals May Be Too Optimistic: SocGen Analyst*, Platts, May 25, 2011, http://www.platts.com/RSSFeedDetailedNews/RSSFeed/NaturalGas/6135424, quoting Société Génerale's analyst Laurent Key.

as China ramps up its shale production. Acting fast is thus a necessity for future US exporters to build a new value-creating, successful LNG export facility."[87] Suffice it to say, back in 2010 and 2011, many experts were skeptical about Cheniere's export plans and the feasibility of US LNG exports in general.[88] As discussed in detail in Chapter12, Cheniere's export project started to gain traction in 2012.

11.6 THE CASE FOR CANADIAN LNG EXPORTS

Canadian gas production is connected to the North American gas market through a network of thousands of kilometers of pipelines that allows buyers to purchase and transport natural gas from a number of supply sources across the continent. While many gas market transactions in North America are based on the pricing at Henry Hub, the main Canadian pricing point is the AECO-C hub in southeast Alberta with another pricing point being the Dawn hub in Ontario.[89]

Since the Canadian and US natural gas markets operate as one large integrated market where events in any region such as changes in transportation costs, infrastructure constraints or weather will have effects on the other regions, the dramatic increase in US shale gas production has also had a significant impact on Canada.[90] Most significantly, US shale gas production has effectively "destroyed export demand for Canadian gas" from the US, resulting in Canada now looking to other markets to export its gas.[91]

While Canada itself has significant shale gas resources that could be developed, the price of Canada's AECO-C hub has tracked the downward trajectory of the US Henry Hub, with the AECO-C hub price actually

---

[87] Ibid.

[88] Edward McAllister, *Pricing and Politics to Scupper US LNG Export Plan*, Reuters, June 16, 2010, quoting analysts calling into question the economics of Cheniere's export plan, available at http://www.forexpros.com/news/general-news/analysis-pricing-and-politics-to-scupper-us-lng-export-plan-142980. See also Ryan Dezember, *Cheniere Doubles Down on Its LNG Bet*, Wall Street Journal, May 23, 2011, noting that Cheniere was "wagering big again" and that Cheniere's export project faced challenges even with DOE's approval.

[89] Canada National Energy Board (NEB), *Natural Gas – How Canadian Markets Work*, http://www.neb-one.gc.ca/clf-nsi/rnrgynfmtn/prcng/ntrlgs/cndnmrk-eng.html.

[90] Ibid.

[91] Darrell Stonehouse, *Fenced In: Canada Needs New Oil and Gas Markets for Sustained Growth*, Profiler, 2011 Gas & Oil Expo Publication, June 2011.

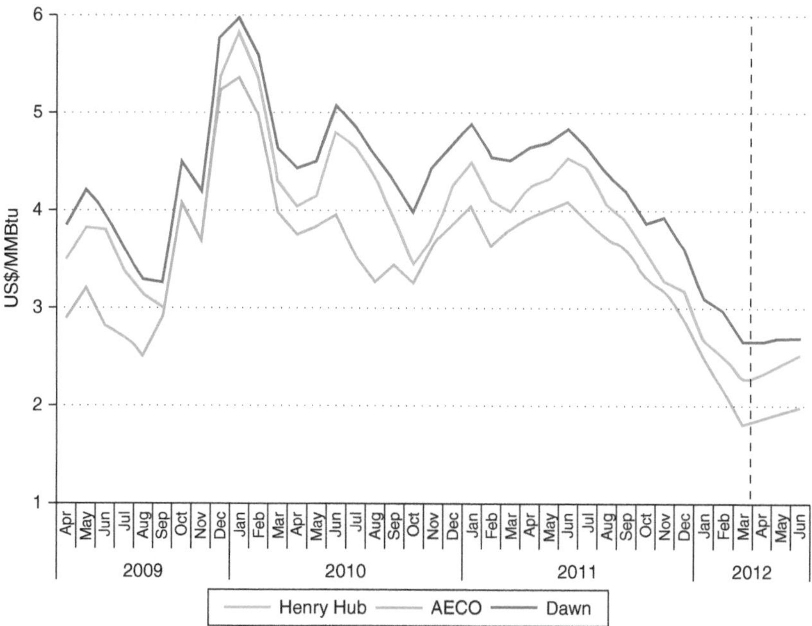

Source: Canada National Energy Board (NEB)

Figure 11.4 North American natural gas prices (monthly averages)

lower than Henry Hub since Canada's natural gas is now desperately in need of a market, having lost the US as its primary market.[92] According to Canada's National Energy Board (NEB), prices for natural gas have declined over the first quarter of 2012, averaging US$2.96/MMBtu at Dawn in Ontario (top line), US$2.53/MMBtu at Henry Hub (middle line) and $2.12/GJ at AECO in Alberta (bottom line) from January to March 2012[93] (Figure 11.4). According to the NEB, prices for the second quarter of 2012 for these hub points could move lower by about $0.50 due to an excess volume of gas in storage and reduced demand due to mild weather.[94]

---

[92] Rebecca Penty, *Canada Must Invest $50 Billion to Stay in LNG Race: Report*, Calgary Herald, May 10, 2012, http://www.calgaryherald.com/story_print.html?id=6598438&sponsor=curriebarracks.

[93] Canada National Energy Board (NEB), *Natural Gas – Current Market Conditions April–June 2012*, http://www.neb-one.gc.ca/clf-nsi/rnrgynfmtn/prcng/ntrlgs/crrntmrktcndtn-eng.html.

[94] Ibid. As explained by the NEB, spring is considered a shoulder season

The loss of the US as Canada's main market, coupled with growing supplies of shale gas and low prices, has led to Canada seeking to export its natural gas as LNG, with Asia being the prime market. One of the most significant hurdles for Canada is the need to build the Pacific Trail Pipelines, which will carry gas from the gas fields to the port of Kitimat where, as discussed in detail below, several LNG export projects are planned.[95]

According to some experts, Canada does not really have any choice but to export its gas as LNG and must move quickly before other competitors move in.[96] At the same time, Canada itself is expected to produce large amounts of shale gas that will need a market. These dynamics mean that Canada will need to look to new customers to monetize its natural gas reserves, with the Asia-Pacific market being the prime target.[97]

In recent years, Asia-Pacific markets, led by emerging economies like China and India, have been driving global economic growth. Despite the incredible growth in these countries and their high demand for all goods, particularly energy resources, the Asia-Pacific region currently imports very little from Canada. "For example, Canada accounted for just 1.1% of China's total imports in 2008 and its share of the Chinese markets has fallen since 1988. Western Canada's share of Canadian exports to China has also fallen over the same period (from 72.5% to 66.3%)."[98] With

---

in the natural gas markets. Demand for natural gas tends to be lower as milder temperatures reduce the need for space heating or cooling. Natural gas prices tend to move in response to the need to withdraw gas from storage to avoid penalties and the pace of the storage refill. Mild winter temperatures in 2012 combined with strong production of natural gas are resulting in record storage inventories in Canada and the US and are causing prices to decline. The price of natural gas at the Henry Hub pricing point throughout the April to June 2012 outlook period is expected to be US$1.75 to 2.50/MMBtu but could fluctuate in response to cold spells or an early onset of hot temperatures.

[95] Justine Hunter, *Kitimat Weighs Up the Risks of Oil and Gas*, The Globe and Mail, Jan. 12, 2012, http://www.theglobalandmail.com, noting that Canadian Premier Christy Clark is very much in favor of building the $1 billion Pacific Trail Pipelines project to facilitate LNG exports.

[96] Rebecca Penty, *Canada Must Invest $50 Billion to Stay in LNG Race: Report*, Calgary Herald, May 10, 2012, http://www.calgaryherald.com/story_print.html?id=6598438&sponsor=curriebarracks, citing a recent report by Ernst & Young finding that Canada must speed up development of its LNG export projects to compete with other countries, such as Australia, Qatar, Russia and Malaysia, seeking to export to Asia.

[97] Darrell Stonehouse, *Fenced In: Canada Needs New Oil and Gas Markets for Sustained Growth*, Profiler, 2011 Gas & Oil Expo Publication, June 2011.

[98] Canada West Foundation Publication, *Window on the West: A Look at Life and Policy in Western Canada*, March 2011, at p. 5. Canada West Foundation website, www.cwf.ca.

abundant natural resources, western Canada seems well positioned to increase trade with the Asia-Pacific region.[99]

11.6.1 Canada's LNG Export Projects – Kitimat Case Study

At present, there is only one LNG terminal operating in Canada – Canaport LNG's receiving and regasification terminal at Saint John, New Brunswick.[100] At one point, there were several proposed LNG regasification projects in Canada but all of these have been canceled or suspended due to difficulties securing long-term supply commitments, concerns over existing excess regasification capacity in North America, and the prospects for domestic shale gas as a new long-term source of natural gas.[101]

In order to monetize the abundant natural gas and shale gas resources found in Canada, a number of companies have been exploring opportunities to build an LNG export facility. As of the time of this writing, there are currently two LNG export applications before Canada's National Energy Board (NEB) – Kitimat LNG and Douglas Channel LNG.[102] It is also expected that additional LNG export applications will be filed in Canada.[103]

---

[99] Canada West Foundation Publication, *Window on the West: A Look at Life and Policy in Western Canada*, March 2011, www.cwf.ca, "Forging closer economic ties with Asia-Pacific – and with China and India in particular – could ensure the long-term prosperity of western Canada for years to come."

[100] Natural Resources Canada, http://www.nrcan.gc.ca/eneene/sources/natnat/imppro-eng.php.

[101] Ibid.

[102] The Proposed Douglas Channel LNG project is a barge-based liquefaction plant which will be fabricated and assembled in a shipyard and then towed to its permanent location in British Columbia on the west bank of the Douglas Channel. This project is much smaller than the proposed Kitimat LNG project and calls for up to 125 million cubic feet of gas per day, or approximately 900,000 tonnes per annum of LNG. Douglas Channel LNG, http://douglaschannelenergy.com/?page_id=42. The Applicant for the Douglas Channel project is BC LNG Export Co-operative LLC (BC LNG), which was formed to purchase LNG at the outlet of the LNG facility. The LNG facility will be developed and owned by Douglas Channel LNG and operated by Douglas Channel Energy Partnership. National Energy Board (NEB), *Application of BC LNG Export Co-operative LLC*, March 8, 2011, https://www.neb-one.gc.ca/ll-eng/livelink.exe/fetch/2000/90466/94153/552726/674445/674203/704633/674343/B1-2-_BC_LNG_Export_Co-operative_LLC_Application_-_A1Y0J3_.pdf?nodeid=674344&vernum=0.

[103] For example, in October 2011, Shell announced it had acquired the Kitimat Marine Terminal, in Kitimat, British Columbia, from Cenovus Energy Inc and "is now exploring the potential for an LNG export terminal on the site." LNG World News, *Shell Buys Site for LNG Export Project*, Oct. 21, 2011, http://www.

The Kitimat LNG export project is particularly noteworthy since it was originally slated to be an LNG import facility but when the market for imports shifted in 2008, Kitimat reversed its proposal and sought approval for an export facility instead. This move reflected the increased optimism over new shale gas developments in Canada, and North America more broadly, as well as the expectation that natural gas prices in Asia would continue to exceed those in North America, thereby creating an export opportunity for Canadian companies.[104]

11.6.2 Kitimat LNG – Project Background

Kitimat LNG is 40 percent owned by KM LNG Operating General Partnership (KM LNG) through its managing partner Apache Canada Ltd.,[105] 30 percent owned by EOG Resources Canada Inc.,[106] and 30 percent owned by EnCana Corporation.[107] KM LNG is the operator.[108]

The Kitimat LNG facility will be located at Bish Cove near Kitimat, British Columbia, about 650 kilometers/400 miles north of Vancouver. The site seems ideal since it is located in a designated industrial area with a natural deepwater port that will not require dredging, thereby minimizing the impact on the marine environment.[109] The Kitimat LNG facility will be built on First Nations land under a unique partnership with the Haisla First Nation.[110]

---

lngworldnews.com/canada-shell-buys-site-for-lng-export-project/. As discussed in Chapter 12, Shell has recently announced plans for a proposed LNG export terminal.

104 Natural Resources Canada, http://www.nrcan.gc.ca/eneene/sources/natnat/imppro-eng.php.

105 Apache Canada Ltd. is a subsidiary of Apache Corporation and is one of Canada's top oil and gas producers with operations in Alberta, British Columbia, and Saskatchewan. www.apachecorp.com/Canada.

106 EOG Resources Canada Inc. is a wholly owned subsidiary of EOG Resources, Inc., which is one of the largest independent (non-integrated) oil and natural gas companies in the United States. www.eogresources.com.

107 Encana is a leading North American natural gas producer focused on growing its portfolio of natural gas resource plays in key basins from northeast British Columbia to east Texas and Louisiana. www.encana.com.

108 Kitimat LNG, *Ownership*, http://kitimatlngfacility.com/Project/project_ownership.aspx.

109 Kitimat LNG, Project Site, http://kitimatlngfacility.com/Project/project_site.aspx. Other measures designed to mitigate environmental impact include a small facility footprint to minimize land use and the use of hydroelectric power to reduce overall emissions.

110 Under the partnership, the Haisla Nation received a $56 million payment

Kitimat LNG will include natural gas liquefaction, LNG storage and marine on-loading facilities. Natural gas will be delivered via a pipeline lateral of approximately 14 kilometers from the Pacific Trail Pipelines,[111] which will connect to the existing Spectra Energy Westcoast Pipeline system. The proximity of Kitimat LNG to existing pipeline infrastructure is one of the key advantages of the project and ensures a ready supply of natural gas.[112]

The feed gas for Kitimat LNG will come from Apache's and EOG Resource's fields in British Columbia and Alberta, which contain approximately 19 trillion cubic feet (Tcf) of combined marketable/technically recoverable natural gas resources. The initial LNG plant capacity will be 5 million metric tons per annum (mmtpa) of LNG output with potential capacity expansion to 10 mmtpa or more. The projected number of shipments is five to seven per month.[113]

According to a project update, Kitimat LNG is currently carrying out a front-end engineering and design (FEED) study,[114] which is expected to provide certainty around project design, construction timelines, costs and labor force requirements. The study is scheduled for completion in late 2011 or early 2012. Kitimat LNG expects to be in a position to make a final investment decision (FID) on the facility near the end of 2011 or

---

for the sale of their equity in Kitimat LNG, and also will receive "long-term, regular lease and property tax payments combine[d] with the employment and business opportunities associated with the project to provide a greater measure of economic stability," Kitimat LNG, Press Release, *Canada, BC, Join Haisla Nation and Kitimat LNG Partners in Marking Project Go-Ahead; "A Very Big Day for Our People" Says Chief Councilor Pollard*, March 9, 2011, http://media-center.kitimatlngfacility.com/Mediacenter/view_press_release.aspx?PressRelease.ItemID=2807.

[111] In February 2011, Kitimat LNG Partners entered into an agreement to purchase the remaining 50 percent interest in Pacific Trail Pipelines Limited Partnership, thus securing full ownership in the infrastructure to transport natural gas from production areas to the Kitimat LNG facility. PennEnergy, *KBR Wins Kitimat LNG FEED Contract for Liquefaction, Export off Canada's West Coast*, March 4, 2011, http://www.pennenergy.com/index/petroleum/display/0763847826/articles/pennenergy/petroleum/refining/2011/03/kbr-wins_kitimat_lng.html.

[112] Kitimat LNG, *Project Description*, http://kitimatlngfacility.com/Project/project_description.aspx.

[113] Kitimat LNG, *Project Summary*, http://kitimatlngfacility.com/index.aspx.

[114] Kitimat LNG awarded the FEED contract to KBR, noting that "KBR is a recognized leader in LNG developments and brings extensive experience to [the] project." PennEnergy, *KBR Wins Kitimat LNG FEED Contract for Liquefaction, Export off Canada's West Coast*, March 4, 2011, http://www.pennenergy.com/index/petroleum/display/0763847826/articles/pennenergy/petroleum/refining/2011/03/kbr-wins_kitimat_lng.html.

early in 2012. Once the FID is granted, construction of the Kitimat LNG facility is expected to begin in 2012, with commercial operations expected to start in late 2015.[115]

The Kitimat LNG Terminal's initial processing capacity in 2015 will be approximately 19,800 $10^3m^3/d$ (700 MMcf/d) with a send-out capacity of up to 5 million tonnes per year (5MMt/y) of LNG (natural gas equivalent of approximately 6,600,000 $10^3m^3/y$ (234 Bcf/y)) ("Phase 1"). The approximate cost of Phase 1 is $3 billion. An additional train is currently proposed for the 2017–18 timeframe at a cost of an incremental $1.5 billion, although this timing could be accelerated based on market conditions and engineering studies ("Phase 2"). This expansion will double the processing capacity of the Kitimat LNG Terminal to 39,600 $10^3m^3/d$ (1.4 Bcf/d) with a send-out capacity of 10 MMt/y of LNG (natural gas equivalent of approximately 13,300,000 $10^3m^3/y$ (468 Bcf/y)).[116]

11.6.3 Kitimat LNG Export Application

On December 9, 2010, KM LNG filed its application for a license authorizing the export of LNG with the National Energy Board (NEB)[117] of Canada pursuant to Section 117 of the National Energy Board Act and the National Energy Board Part VI (Oil and Gas) Regulations. In its application, KM LNG sought approval for a 20-year license to export up to 10 million tonnes of LNG per year (10 MMt/y), the natural gas equivalent of approximately 13,300,000 $10^3m^3/y$ or 468 Bcf/y.[118]

KM LNG's application noted that "this is the first time that an export of LNG has been applied-for under the present National Energy Board

[115] Kitimat LNG, *Community Update*, Fall 2011, http://www.kitimatlngfacility.com/Resources/Upload/ProjectUpdate/Files/2811/Kitimat%20LNG%20Nwsltr-%20Sept%2013.pdf.

[116] National Energy Board, *Application by KM LNG Operating General Partnership, for a License Authorizing the Export of Liquefied Natural Gas*, at p.4, https://www.neb-one.gc.ca/ll-eng/livelink.exe/fetch/2000/90466/94153/552726/657379/657474/670503/657060/B1-3_-_KM_LNG_Application_-A1W6S5.pdf?nodeid=657064&vernum=0.

[117] The NEB is an independent federal regulator that regulates several aspects of Canada's energy industry. The NEB's primary purpose is to regulate pipelines, energy development and trade in a manner consistent with the Canadian public interest.

[118] National Energy Board, *Application by KM LNG Operating General Partnership, for a License Authorizing the Export of Liquefied Natural Gas*, https://www.neb-one.gc.ca/ll-eng/livelink.exe/fetch/2000/90466/94153/552726/657379/657474/670503/657060/B1-3_-_KM_LNG_Application_-A1W6S5.pdf?nodeid=657064&vernum=0.

Part VI (Oil and Gas) Regulations, for the purpose of accessing offshore markets." The application also noted that since the NEB's Filing Manual does *not* expressly contemplate offshore LNG exports, KM LNG was therefore applying the Filing Manual by taking into account the unique circumstances presented as compared to a conventional onshore pipeline-transmitted export to the US.[119]

In its application, KM LNG acknowledged that the application was in response to a rapidly changing North American gas market and the previously unforeseen abundance of shale gas driven by recent technological advances. The majority of the gas sought to be exported under KM LNG's license would be sourced from Northeast British Columbia, a region widely considered to hold significant gas resources, although it is in a relatively early stage of development.[120]

KM LNG's export application noted that gas production will be liquefied at the Kitimat LNG Terminal and directly transported by LNG carriers, primarily to Asia-Pacific markets. According to KM LNG, Asia-Pacific presently represents a new, long-term and stable market for Canadian natural gas supply, with the Kitimat LNG Terminal well positioned to take advantage of the growing Asian demand for LNG.

KM LNG also stated that, unlike continental North American natural gas markets served by onshore pipelines, Asia-Pacific LNG buyers are seeking long-term secure gas supply arrangements with regulatory certainty before committing to long-term contractual commitments. Therefore, KM LNG was seeking a long-term gas export license prior to completing gas export sales contracts. This contrasts with the National Energy Board Part VI (Oil and Gas) Regulations requirement for gas export sales contract information as part of the gas export.

Lastly, KM LNG indicated that it had retained energy consultants Poten & Partners Inc. ("Poten")[121] to provide an independent assessment ("Kitimat LNG Market Assessment") of the international demand for LNG relevant to the potential for the Kitimat LNG Terminal to meet some of this demand, especially in the Asia-Pacific market.[122]

---

119 Ibid. at p. 4.

120 Ibid. at pp. 3–4 (quoted directly from the application).

121 Poten & Partners is a global broker and commercial advisor for the energy and ocean transportation industries and they are recognized leaders in LNG, crude and petroleum products, liquefied petroleum gas (LPG), fuel oil, naphtha and asphalt market sectors. Poten & Partners, http://www.poten.com.

122 National Energy Board, *Application by KM LNG Operating General Partnership, for a License Authorizing the Export of Liquefied Natural Gas*, at p. 6, https://www.neb-one.gc.ca/ll-eng/livelink.exe/fetch/2000/90466/94153/552726

The key conclusions of the Kitimat LNG Market Assessment were:[123]

(a) The Asia Pacific region is the natural market for the Kitimat LNG Terminal due to its location. Strong LNG demand growth is expected in the Asia Pacific region due primarily to growth in gas for power generation in China, Japan, South Korea and Taiwan. LNG demand is expected to grow in Asia Pacific an average of 3.25%/year during the 2014–2035 period.
(b) The largest supply shortfall is projected for the period 2014–2018 and it is anticipated that Japan, South Korea, Taiwan and China will need additional LNG supplies to fill this supply gap in the time period when the Kitimat LNG Terminal is scheduled to start operation. These countries are currently seeking secure long-term supply contracts with proponents whose projects are most likely to commence commercial operations during the 2014–2018 period.
(c) The Kitimat LNG Terminal is well positioned to capture long-term market share given that it meets the qualities valued by Asia Pacific buyers including: (1) security of a long-term contract backed by long-term supply commitments from Apache and EOG; (2) a diversification of supply previously relied on by the market; (3) high likelihood of the project being built and operational to meet the forecasted demand/supply gap; (4) political stability and regulatory certainty in British Columbia, and in Canada generally; and (5) strong equity players in the project, namely Apache and EOG.

KM LNG also retained additional energy consultants to assess the potential impact that LNG exports might have on the ability of Canadians to meet their expected energy requirements over the proposed 20-year license term.[124] According to these consultants' findings, the proposed exports were not expected to have a significant effect on future natural

/657379/657474/670503/657060/B1-3_-_KM_LNG_Application_-A1W6S5.pdf?nodeid=657064&vernum=0.

[123] Ibid. at pp. 6–7.

[124] KM LNG retained Ziff Energy Group, a premier consulting practice, providing "leading edge" long-term forecasts for the gas industry, benchmarking reports, and oil and gas production consulting, http://www.ziffenergy.com, and Mr Roland Priddle, a well-established energy expert and inductee of the Canadian Petroleum Hall of Fame, http://www.canadianpetroleumhalloffame.ca/roland-priddle.html.

gas prices and the Canadian gas-producing sector would be able to meet Canadian energy demand even with the proposed exports.[125]

11.6.4 The NEB Grants KM LNG's Application

On October 13, 2011, the NEB granted KM LNG's export application and authorized KM LNG to export 200 million tonnes of LNG (equivalent to approximately 265 million 10^3m^3 or 9,360 Bcf of natural gas) over a 20-year period. The maximum annual quantity allowed for export will be 10 million tonnes of LNG (equivalent to approximately 13 million 10^3m^3 or 468 Bcf of natural gas).[126]

The NEB took note of the fact that this was the first application for an LNG export license that the Board had considered since the deregulation of the natural gas market in 1985. The NEB also found that the exported LNG will not only open new markets for Canadian gas production, but will also serve to promote the ongoing development of shale gas resources which will ultimately further increase the availability of natural gas for Canadians. In light of Canada's significant gas resources, the Board was satisfied that the quantity of gas to be exported does not exceed the amount required to meet foreseeable Canadian demand.[127]

Prior to approving the license, the NEB also considered the environmental and related socio-economic effects of KM LNG's application. These effects included matters related to marine shipping, and the proposed LNG terminal and Pacific Trail Pipelines. The NEB also acknowledged the potential economic benefits associated with KM LNG's project. These benefits include employment opportunities due to the development of the LNG terminal and the Pacific Trail Pipelines.

---

125 National Energy Board, *Application by KM LNG Operating General Partnership, for a License Authorizing the Export of Liquefied Natural Gas*, at p.9, https://www.neb-one.gc.ca/ll-eng/livelink.exe/fetch/2000/90466/94153/552726/657379/657474/670503/657060/B1-3_-_KM_LNG_Application_-A1W6S5.pdf?nodeid=657064&vernum=0.

126 NEB News Release, *NEB Grants 20-year Export License to KM LNG*, Oct. 13, 2011, http://www.neb-one.gc.ca/clf-nsi/rthnb/nwsrls/2011/nwsrls31-eng.html. See also Kitimat LNG Press Release, *Kitimat LNG Partners Announce Export License Approval by National Energy Board*, Oct. 14, 2011, http://media-center.kitimatlngfacility.com/Mediacenter/view_press_release.aspx?PressRelease.ItemID=2816.

127 NEB News Release, *NEB Grants 20-year Export License to KM LNG*, Oct. 13, 2011, http://www.neb-one.gc.ca/clf-nsi/rthnb/nwsrls/2011/nwsrls31-eng.html.

11.6.5 The Expected Market for Canadian LNG is Asia

As part of its application, KM LNG was required to provide a description of the market to be served by the proposed exportation. In the NEB's Reasons for Decision, the NEB highlighted the relevant market conditions and noted that no parties took an opposing view regarding the information submitted by KM LNG about its proposed markets for LNG that would be exported from the terminal.[128]

In the Reasons for Decision, the NEB noted that KM LNG submitted that it will follow a traditional LNG business model that includes the marketing and selling of LNG directly to Asia-Pacific buyers who seek large volumes and long-term firm SPAs. This type of alignment provides the security of demand for the seller which is required to underpin the large-scale financial investment, as well as security of supply for the LNG buyer.

Long-term supply reliability was identified to be a cornerstone requirement of Asia-Pacific LNG buyers. KM LNG stated that LNG buyers see KM LNG's access to the highly liquid North American natural gas market as an opportunity because it contributes to project longevity, which distinguishes KM LNG from other LNG projects. According to KM LNG's consultants, LNG sold under new long-term supply contracts to the Asia-Pacific market is currently priced around 90 percent of oil on an equivalent heat-value basis. In Asia, oil is considered to be a substitute fuel to natural gas and oil prices are considered a regional energy price benchmark.

In terms of international demand for LNG, KM LNG expects demand for LNG in Asia-Pacific markets to grow on average by 2.7 percent per year between 2014 and 2035, driven primarily by gas-fired power generation growth in China, Japan, South Korea and Taiwan. Starting around 2015, when KM LNG's terminal is projected to come on-stream, a supply deficit emerges between long-term contracts and forecast demand. The capacity of KM LNG's terminal represents approximately 8 percent of the global LNG supply deficit currently forecast for 2035.

The NEB also acknowledged the significance of the Asia-Pacific region to the Kitimat LNG export project by stating:

---

[128] National Energy Board (NEB), *Reasons for Decision in the Matter of KM LNG Operating General Partnership Application Dated 9 December 2010 for a Long-term Export License to Export Liquefied Natural Gas under Section 117 of the National Energy Board Act (NEB Act)*, GH-1-2011, October 2011, https://www.neb-one.gc.ca/ll-eng/livelink.exe/fetch/2000/130635/728336/A2F3D2_-_RfD_GH-1-2011_KM_LNG.pdf?nodeid=728337&vernum=0.

The Board recognizes that forecast demand growth for LNG in the Asia Pacific region provides a new opportunity for Canadian producers to diversify their export markets. The Board also recognizes that long-term oil-indexed sales contracts could provide for higher netbacks to Canadian producers. While the Board noted the existence of competing sources of global LNG, given the size of Canada's natural gas resource, proximity to markets in Asia and Canada's stable political and regulatory environment, the Board was of the view that KM LNG has the opportunity to compete in the global LNG market. Having regard to the size and potential growth of the Asia Pacific market, the Board concluded that the proposed export volume is likely to flow.[129]

129 Ibid.

12. Emerging issues in the LNG industry

12.1 OVERVIEW

As this book goes to print, there are a number of emerging issues that highlight some of the opportunities and challenges the industry will encounter in the coming decades, although there will no doubt be many more than can be envisioned at the time of this writing, with new potential issues emerging on an almost daily basis.[1] Perhaps one of the most dramatic opportunities is the possibility of North American LNG exports, which was discussed in detail in Chapter 11 but, because of its potential significance, is highlighted again in this chapter with a discussion of the most recent developments (Section 12.2). A related emerging issue that may make US LNG exports more viable is the expansion of the Panama Canal, which will allow larger ships to transit the canal upon its expected completion in 2014 (Section 12.3).

On the technical side, one of the key emerging issues is the development of the world's first floating LNG (FLNG) liquefaction terminal, the Shell Prelude project. If successful, the Prelude project could pave the way for other floating LNG projects, which are expected to be a growing segment in the global LNG industry (Section 12.4).

In terms of market developments, the role of the Gas Exporting

---

[1] For example, as this book was going to print, Argentina's president Cristina Fernandez announced the planned seizure and nationalization of a 51 percent stake in Argentina's energy company, YPF SA, from majority stakeholder Repsol YPF SA, contending that Repsol's lack of investment had led to a decline in Argentina's oil and gas production. Spanish based Repsol has sued Argentina (Carlos Ruano and Jonathan Stempel, *Repsol Sue Argentina over Giant YPF Seizure*, Reuters, May 15, 2012, http://www.reuters.com/article/2012/05/15/repsol-ypf-idUSL5E8GF9RN20120515) and has also canceled LNG cargoes to Argentina. Alejandro Lifschitz, *Exclusive: Repsol Cancels LNG Cargoes to Argentina*, Reuters, May 18, 2012, http://www.reuters.com/article/2012/05/18/us-repsol-argentina-idUSBRE84H0SI20120518 (noting that Argentina relies on LNG imports to meet 20–30 percent of domestic natural gas consumption and that Repsol had agreed to supply Argentina with 10 LNG cargoes in 2012).

Countries Forum (GECF) emerging as a potential "Gas OPEC" has been discussed in various publications but merits mention here since the GECF only recently became a full-fledged organization that is starting to participate in various forums throughout the world (Section 12.5).

Lastly, as the world continues to focus on reducing emissions and on environmental sustainability, the emergence of LNG as a potential shipping fuel and vehicle fuel could aid in emission reduction and environmental sustainability efforts around the world (Section 12.6).

12.2 WILL NORTH AMERICAN LNG EXPORTS BECOME A REALITY?

12.2.1 The Arbitrage Opportunity for US LNG Exports Increases

Initially there was much skepticism about the prospects of US LNG exports with analysts predicting that pricing and politics would hinder US export plans.[2] Analysts doubted that the "economics [would] work out in Cheniere's favor," calculating that LNG prices would have to be about $8–10/MMBtu in Europe to make the project economically feasible. More skepticism was raised about the project costs (Cheniere's original estimate was $2–3 billion) and whether Cheniere's proposed capacity fee would have to be increased to reflect higher construction costs.[3] Doubts were also raised whether DOE would authorize US LNG exports since there was little precedent and likely to be opposition.[4]

More recent articles have indicated that the prospects for US LNG

---

[2] Edward McAllister, *Pricing and Politics to Scupper US LNG Export Plan*, Reuters, June 16, 2010, http://www.forexpros.com/news/general-news/analysis-pricing-and-politics-to-scupper-us-lng-export-plan-142980 (quoting analysts calling into question the economics of Cheniere's export plan).

[3] At least one analyst cautioned that "Sabine Pass' projected in-service date of 2015 is unrealistic, that its $4 billion price tag for liquefaction facilities was 'to be taken with a grain of salt,' and that LNG buyers would be hard-pressed to find US producers willing to sell gas forward at prices below $6/MMBtu for 2015 or 2016." Samantha Santa Maria, *US LNG Export Proposals May Be Too Optimistic: SocGen Analyst*, Platts, May 25, 2011, http://www.platts.com/RSSFeedDetailedNews/RSSFeed/NaturalGas/6135424 (quoting Société Générale analyst Laurent Key).

[4] Edward McAllister, *Pricing and Politics to Scupper US LNG Export Plan*, Reuters, June 16, 2010, http://www.forexpros.com/news/general-news/analysis-pricing-and-politics-to-scupper-us-lng-export-plan-142980.

exports are now much brighter due to higher than expected Japanese demand coupled with a widening of the "shale spreads." For example, in April 2012, Japan alone consumed a record amount of 4.56 million tonnes of LNG equivalent, primarily due to increased use of natural gas for electricity generation after Japan's loss of nuclear power due to the Fukushima disaster in 2011.[5] Going forward, Japan's demand for LNG is expected to remain strong since the Fukushima crisis has eroded public confidence in nuclear power and prevented the restart of Japan's nuclear reactors.[6]

At the same time, US shale gas production has continued to grow, keeping Henry Hub prices at ten-year lows of about $2/MMBtu as of May 2012, while Asian LNG prices are at four-year highs of around $18/MMBtu.[7] These spreads make US LNG exports much more attractive with at least one major consulting firm opining that "a spread of just $4 for Europe and $6 for Asia would justify the infrastructure investment" for US LNG exporters.[8] As indicated in Figure 12.1, Cheniere expects to be able to deliver LNG from the US to the Americas, Europe or Asia for $7–9/MMBtu. Since most LNG prices around the world are linked to an oil-based index, global LNG prices are expected to continue to trade in a range of $11–23/MMBtu, assuming oil remains at $100/bbl or higher. If Cheniere's estimates are correct, then US LNG exports would be competitive at prices in the range of $7–9/MMBtu to most destinations.

12.2.2 The Commercial Structure for US LNG Exports

As the markets started to warm up to the prospect of US LNG exports in 2011, more details began emerging about the possible commercial structure for such exports. Under Cheniere's proposed plans, customers will pay a take-or-pay capacity fee plus a fuel surcharge to Cheniere, which gives the customer the option to either import or export LNG. Customers

---

5 Osamu Tsukimori, *Japan Utilities Burn Record LNG Equivalent for April*, Reuters, May 16, 2012, http://www.reuters.com/article/2012/05/16/utilities-japan-electricity-idUSL4E8GG5RM20120516.

6 Ibid.

7 Edward McAllister, *Excelerate to Build First Floating US LNG Export Plant*, Reuters, May 15, 2012, http://uk.reuters.com/article/2012/05/15/excelerateenergy-idUKL4E8GF74920120515.

8 Christopher Swann and Neil Unmack, *Outlook Brightens for U.S. Gas Exports*, New York Times, Oct. 13, 2011, http://www.nytimes.com/2011/10/14/business/outlook-brightens-for-us-gas-exports.html (quoting consultancy firm IHS).

Sabine Pass LNG exports will provide global LNG buyers with an attractive, long-term, alternative source of supply

Worldwide LNG prices predominantly based on oil prices = $11–23/MMBtu

LNG Contract price		
Indexation %	11%	15%
at $100/bbl	$ 11.00	$ 15.00
at $150/bbl	$ 16.50	$ 22.50

Cost of delivered gas from Sabine Pass to Americas/Europe/Asia = $7–9/MMBtu

($/MMBtu)	Americas	Europe	Asia
Henry Hub	$ 3.00	$ 3.00	$ 3.00
Capacity charge	3.00	3.00	3.00
Shipping	0.75	1.25	3.00
Fuel/basis	0.35	0.35	0.35
Delivered cost	$ 7.10	$ 7.60	$ 9.35

Current LNG market	30–40 Bcf/d	LNG contracts indexed to oil prices – rule of thumb 11% to 15% of crude oil prices
Growth market	100 Bcf/d	Power generators switching from oil to gas – paying $13 to $19/MMBtu for fuel oil and diesel

Source: Cheniere Energy Partners, L.P., Corporate Presentation dated June 2012.

Figure 12.1 Cheniere/Sabine Pass Liquefaction arbitrage opportunity

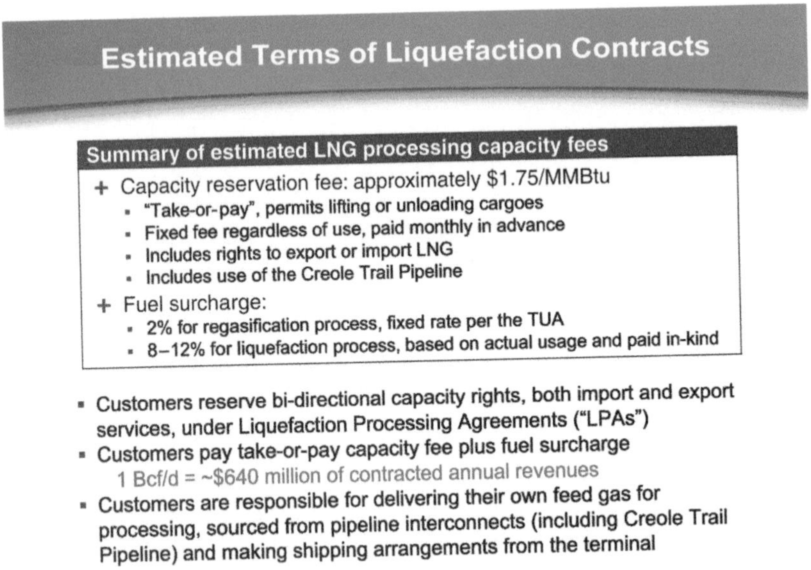

Estimated Terms of Liquefaction Contracts

Summary of estimated LNG processing capacity fees

+ Capacity reservation fee: approximately $1.75/MMBtu
  - "Take-or-pay", permits lifting or unloading cargoes
  - Fixed fee regardless of use, paid monthly in advance
  - Includes rights to export or import LNG
  - Includes use of the Creole Trail Pipeline
+ Fuel surcharge:
  - 2% for regasification process, fixed rate per the TUA
  - 8–12% for liquefaction process, based on actual usage and paid in-kind

- Customers reserve bi-directional capacity rights, both import and export services, under Liquefaction Processing Agreements ("LPAs")
- Customers pay take-or-pay capacity fee plus fuel surcharge
  1 Bcf/d = ~$640 million of contracted annual revenues
- Customers are responsible for delivering their own feed gas for processing, sourced from pipeline interconnects (including Creole Trail Pipeline) and making shipping arrangements from the terminal

Source: Cheniere Energy Partners, L.P., Corporate Presentation dated June 2011

Figure 12.2 Cheniere/Sabine Pass Liquefaction commercial structure

are responsible for securing the feed gas at prevailing Henry Hub prices and delivering the feed gas to Sabine Pass for liquefaction processing and shipping (FOB) from Sabine Pass to any destination permitted under export licenses. (See Figure 12.2.)

12.2.3 Cheniere's Sabine Pass Liquefaction Project Secures the First Contracts for US LNG Exports

On October 26, 2011, Cheniere's export project received a significant boost when Cheniere announced that it had signed an LNG Sale and Purchase Agreement (SPA) with BG Gulf Coast LNG ("BG"), an affiliate of BG Energy Holdings Limited ("BG Parent"), a major player in the LNG world.[9] A summary of the key terms of the SPA is as follows:[10]

---

9 Cheniere Energy Partners, Press Release, Oct. 26, 2011, *Cheniere and BG Sign 20-Year LNG Sale and Purchase Agreement*, http://www.prnewswire.com/news-releases/cheniere-and-bg-sign-20-year-lng-sale-and-purchase-agreement-132606318.html.

10 Cheniere's contract with BG Group was submitted to the Securities and

- BG agreed to purchase 3.5 million tonnes per annum (mtpa) of LNG and pay Sabine Liquefaction a fixed sales charge for the full annual contract quantity of 182,500,000 MMBtu (equivalent to approximately 3.5 mtpa).
- BG will pay Sabine Liquefaction a fixed sales charge of $2.25 per MMBtu for the full 182,500,000 MMBtu annual contract quantity regardless of whether BG purchases any cargoes of LNG. The fixed sales charge will be paid ratably on a monthly basis, and 15 percent of the fixed sales charge will be subject to annual adjustment for inflation.
- BG will also pay Sabine Liquefaction a contract sales price for each MMBtu of LNG delivered under the SPA. The contract sales price will be equal to 115 percent of the final settlement price for the New York Mercantile Exchange Henry Hub natural gas futures contract for the month in which the relevant cargo is scheduled.
- BG will have the right to cancel all or any part of a scheduled cargo of LNG by a timely advance notice, in which case BG will continue to be obligated to pay the full monthly fixed sales charge but will forfeit its right to receive the canceled quantity and will not be obligated to pay the contract sales price for the forfeited quantity.
- The LNG delivery, payment and related provisions of the SPA will have a 20-year term, commencing on the date designated for the first commercial delivery of LNG. BG will have the right to extend the 20-year term for an additional period of up to 10 years.

The SPA is subject to certain conditions precedent, including but not limited to Sabine Liquefaction's receiving regulatory approvals, securing necessary financing arrangements and making a final investment decision to construct the liquefaction facilities.

In November 2011, Cheniere's project received another boost when it secured its second foundation customer, Gas Natural Aprovisionamientos, a subsidiary of the Spanish Gas Natural Fenosa. The 3.5 mtpa SPA with Gas Natural Fenosa represented another milestone for Cheniere, allowing the project to reach its contract capacity target of 7 mtpa. With the two foundation contracts in place, Cheniere planned to proceed towards

---

Exchange Commission (SEC) with Cheniere's SEC 8-K filing on October 26, 2011, and is available at www.cheniereenergypartners.com.

making a final investment decision in order to start construction[11] on the first two liquefaction trains in early 2012.[12]

12.2.4 Cheniere's Sabine Pass Liquefaction Project Gains Traction with Additional Contracts, FERC Approval and Financing

As a further sign of increased interest and conviction in the feasibility of US LNG exports, in January 2012 BG Group announced that it would purchase an additional 2 mpta of LNG over a 20-year period from the Sabine Pass Liquefaction terminal, building on the prior "ground-breaking" agreement entered into in 2011. In the press release announcing the revised deal, BG Group stated that the agreement to add additional volumes was due to "the material increases in US gas reserves as well as a favourable long-term outlook for global LNG demand."[13] The additional capacity contracted for by BG Group was under the same commercial terms as the earlier deal except that the fixed liquefaction fee paid by BG to Cheniere was increased from \$2.25/MMBtu to \$3.00/MMBtu, putting the weighted average cost of BG's secured capacity at \$2.52/MMBtu. As per the original agreement, BG will pay an additional 115 percent of the NYMEX Henry Hub contract to source volumes from the US pipeline system before lifting them free on board (FOB) at the Sabine Pass Liquefaction facility.[14] According to media reports, since suppliers are not comfortable in signing long-term supply deals, customers for Sabine Pass will secure feed gas based on Henry Hub pricing, which should pose

---

[11] In November 2011, Cheniere, through its subsidiary, Sabine Pass Liquefaction, also entered into a lump sum turnkey contract with Bechtel Oil, Gas and Chemicals, Inc. ("Bechtel") for the engineering, procurement and construction (EPC) of the first two liquefaction trains at the Sabine Pass LNG terminal. Cheniere Energy Partners, PR Newswire, Press Release, Nov. 14, 2011, *Cheniere Partners Enters into Lump Sum Turnkey Contract with Bechtel*, http://www.prnewswire.com/news-releases/cheniere-partners-enters-into-lump-sum-turnkey-contract-with-bechtel-133799293.html.

[12] Cheniere Energy Partners, PR Newswire, Press Release, Nov. 21, 2011, *Cheniere and Gas Natural Fenosa Sign 20-Year LNG Sale and Purchase Agreement*, http://www.prnewswire.com/news-releases/cheniere-and-gas-natural-fenosa-sign-20-year-lng-sale-and-purchase-agreement-134267813.html.

[13] BG Group Press Release, *BG Group to Increase LNG Volumes from Sabine Pass*, Jan. 26, 2012, http://www.bg-group.com/MediaCentre?Press/Pages/26Jan2012.aspx.

[14] *BG Group Adds 2.0mpta [sic] of Export Capacity at Sabine Pass*, ICIS Heren, Jan. 27, 2012, http://www.icis.com/heren/articles/2012/01/26/9527261/lng/glm/bg-group-adds-2.0mpta-of-export-capacity-at-sabine-pass.html.

no security of supply issues since the US gas market is well supplied and highly liquid.[15]

In addition to the additional volumes contracted for by BG, Cheniere has most recently entered into an SPA with KOGAS for 3.5 mtpa, commencing with the start of train three operations. As the largest LNG importer in the world, KOGAS is viewed as a strong addition to Cheniere's portfolio of customers for its liquefaction project.[16] Cheniere has also entered into a contract with GAIL (India) Ltd. that commences with the start of train four operations,[17] bringing the total long-term "take-or pay" style contracts for Sabine Pass exports to approximately 16 mtpa. (See Figure 12.3.)

In April 2012, Cheniere's Sabine Pass Liquefaction project received regulatory approval from the Federal Energy Regulatory Commission (FERC), thus becoming the first LNG export project to receive FERC approval.[18] In its order, FERC found that since the proposed liquefaction project was located entirely within the footprint of the previously approved and operating Sabine Pass LNG terminal site, the environmental impacts were limited and well defined. Nonetheless, FERC's order was subject to Sabine Pass adhering to 55 mitigation measures to limit the environmental impact of the new facility and also required the project to be complete and made available for service within five years of the date of issuance of the order.[19]

As this book goes to print, financing the approximately $4.5 to 5.0 billion project appears to be the final milestone for Cheniere's Sabine Pass Liquefaction project (Table 12.1). Cheniere has indicated that it intends to finance the project with a combination of debt and equity. In

---

15 *Cheniere Primed to Sanction Sabine Pass Project*, ICIS Heren Global LNG Markets Report, April 20, 2012, www.icis.com/heren.

16 Cheniere Energy Partners, L.P., News Release, *Cheniere and KOGAS Sign 20-Year LNG Sale and Purchase Agreement*, Jan. 30, 2012, http://phx.corporate-ir.net/phoenix.zhtml?c=207560&p=irol-newsArticle&ID=1653972&highlight=.

17 Cheniere Energy Partners, L.P., News Release, *Cheniere and GAIL India Sign 20-Year LNG Sale and Purchase Agreement*, Dec. 22, 2011, http://phx.corporate-ir.net/phoenix.zhtml?c=207560&p=irol-newsArticle&ID=1638423&highlight=. Commencement of construction for the third (KOGAS) and fourth (GAIL) trains is subject to, but not limited to, entering into an EPC contract, obtaining financing and Cheniere Partners making a final investment decision.

18 Federal Energy Regulatory Commission (FERC), News Release, *FERC Approves LNG Export Project*, April 16, 2012, http://www.ferc.gov/media/news-releases/2012/2012-2/04-16-12-sabine.asp#skipnav.

19 *Sabine Pass LNG, L.P.*, 139 FERC ¶ 61,039 (2012). Order Granting Section 3 Authorization at p. 40, http://www.ferc.gov/EventCalendar/Files/20120416164846-CP11-72-000.pdf.

Sale and Purchase Agreements (SPAs)

Long-term, "take-or-pay" style commercial contracts equating to ~16 mtpa

	BG GROUP BG Gulf Coast LNG (1)	gasNatural fenosa Gas Natural Fenosa (1)	KOGAS Korea Gas Corporation (1)	GAIL GAIL (India) Limited (1)
Annual Contract Quantity (MMBtu)	286,500,000	182,500,000	182,500,000	182,500,000
Annual Fixed Fees (5)	~$723 MM	~$454 MM	~$548 MM	~$548 MM
Fixed Fees $/MMBtu (2)	$2.25 - $3.00	$2.49	$3.00	$3.00
Term (4)	20 years	20 years	20 years	20 years
Guarantor	BG Energy Holdings Ltd.	Gas Natural SDG S.A.	N/A	N/A
Guarantor Credit Rating (3)	A2/A	Baa2/BBB	A / A1	Baa2/NR/BBB-
Fee During Force Majeure	Up to 24 months	Up to 24 months	N/A	N/A
Contract Start Date	Train 1 + additional volumes with Trains 2,3,4	Train 2	Train 3	Train 4

Notes:

(1) Conditions precedent must be satisfied by December 31, 2012 for BG Group and Gas Natural Fenosa and by June 30, 2013 for KOGAS and GAIL (India) Ltd. or either party can terminate. CPs include financing, regulatory approvals, positive final investment decision, issuance of notice to preceed and entering into common facilities agreements (other than KOGAS and GAIL (India) Ltd.).

(2) A portion of the fee is subject to inflation, approximately 15% for BG Group, 13.6% for Gas Natural Fenosa and 15% for KOGAS and GAIL (India) Ltd.

(3) Ratings may be changed, suspended or withdrawn at any time and are not a recommendation to buy, hold or sell any security.

(4) SPAs have a 20 year term with the right to extend up to an additional 10 years. Gas Natural Fenosa has an extension right up to an additional 12 years in certain circumstances

(5) BG will provide annual fixed fees of approximately $520 million for trains 1–2 and $203 million for trains 3–4.

Source: Cheniere Energy Partners, Corporate Presentation, June 2012, www.cheniereenergypartners.com

Figure 12.3 Contracted capacity at Cheniere's Sabine Pass Liquefaction

Table 12.1 Cheniere's Sabine Pass Liquefaction project milestones

Milestone	Target Date	
	Trains 1 and 2	Trains 3 and 4
DOE export authorization	Received	Received
Definitive commercial agreements	Completed 7.7 mtpa	Completed 8.3 mtpa
• BG Gulf Coast LNG, LLC	4.2 mtpa	1.3 mtpa
• Gas Natural Fenosa	3.5 mtpa	
• KOGAS		3.5 mtpa
• GAIL (India) Ltd.		3.5 mtpa
EPC contract	Complete	4Q12
Financing commitments		1Q13
• Equity	Complete	
• Debt	1H12	
FERC authorization	Received	Received
• Certificate to commence construction	Received	2013
Commence construction	1H12	2013
Commence operations	2015/16	2017/18

Source: Cheniere Energy Partners, L.P.

February 2012, Cheniere entered into an arrangement with Blackstone for a $2 billion equity investment in the project[20] and in May 2012, Cheniere agreed to sell $2 billion of equity in the project to Blackstone and Cheniere Energy, Inc., thus completing the equity portion of the financing.[21]

In terms of the debt financing, in April 2012 Cheniere engaged eight financial institutions to act as Joint Lead Arrangers to assist in the structuring and arranging of up to $4 billion of debt facilities and it remains to be seen what terms will be agreed upon for this portion of the project's financing.[22] Assuming Cheniere takes a final investment decision on the

[20] Cheniere Energy Partners, L.P., News Release, *Blackstone Enters into an Arrangement for $2B Equity Investment for the Sabine Pass Liquefaction Project*, Feb. 27, 2012, http://phx.corporate-ir.net/phoenix.zhtml?c=207560&p=irol-newsArticle&ID=1665723&highlight=.

[21] Cheniere Energy Partners, L.P., News Release, *Cheniere Agrees to Sell $2B of Equity for Sabine Pass Liquefaction Project*, May 15, 2012, http://phx.corporate-ir.net/phoenix.zhtml?c=207560&p=irol-newsArticle&ID=1695942&highlight=.

[22] Cheniere Energy Partners, L.P., News Release, *Cheniere Engages Eight Joint Lead Arrangers to Arrange the Debt Financing for the Sabine Pass Liquefaction Project*, April 16, 2012, http://phx.corporate-ir.net/phoenix.zhtml?c=207560&p=irol-newsArticle&ID=1683242&highlight=.

project, it also remains to be seen whether Cheniere's export/liquefaction project will encounter a better, and more profitable, fate than Cheniere's import/regasification project, which has been to the brink of bankruptcy and back in the past year.[23]

It also remains to be seen whether Cheniere can get its liquefaction terminal built in time to take advantage of the arbitrage opportunity that now exists but may disappear before an export terminal can be built.[24] For now, it is a risk that Cheniere and a growing number of companies in the US and Canada are willing to take.[25] According to the US DOE, as of April 2012, there are 11 proposed/potential LNG export terminals in North America with more likely on the way[26] (Appendix E). As discussed in more detail below, all except Cheniere's Sabine Pass Liquefaction are awaiting DOE approval to export to WTO nations (Appendix F) and it

---

[23] In October 2011, Standard & Poors issued a report indicating that Cheniere's contract with BG Group would probably not avoid a cash crunch that threatened to trigger a default on debt maturing in May 2012. As of June 30, 2011, Cheniere had $162.6 million in cash and a $298 million debt payment due in May 2012. Cheniere has $3.14 billion in outstanding debt, most of it from the largely idle regasification terminal it built to import natural gas. Joe Carroll, *Cheniere's BG Agreement May Not Prevent Default, S&P Says*, Bloomberg, Oct. 31, 2011, http://www.bloomberg.com.

[24] According to Michelle Michot Foss, chief energy economist at the University of Texas Center for Energy Economics, while the arbitrage opportunity for exports is huge (shipping 18 mtpa of LNG is worth approximately $1.7 billion at today's prices) it is "tricky" because no one is positioned to take advantage of the opportunity right now. By the time an export terminal can be built, the opportunity might disappear as competition increases or market conditions change. Joe Carroll, *LNG Export Plant Verges on U.S. Approval Amid Shale Glut*, Bloomberg Businessweek, April 13, 2012, http://www.businessweek.com/printer/articles/43480?type=bloomberg.

[25] As Cheniere's project has advanced, the interest among buyers has also increased, with other projects also gaining traction. For example, in April 2012, Sempra Energy announced that it had signed a commercial development agreement with Mitsubishi Corporation and Mitsui & Co., Ltd. to develop and construct an LNG export facility at the site of Sempra's existing import terminal, Cameron LNG, in Hackberry, LA. Sempra Energy Media Release, *Sempra Energy Unit Signs Commercial Development Agreements with Mitsubishi Corporation, Mitsui & Co., Ltd. to Develop Louisiana Liquefaction Facility*, Apr 17, 2012, http://sempra.mediaroom.com/index.php?s=19080&item=126964.

[26] For example, in March 2012, the State of Alaska "reached a settlement with Exxon Mobil Corp and its partners to develop a huge, long-fallow oil and gas field, possibly paving the way for a $26 billion pipeline and an export plant for liquefied natural gas." Yereth Rosen, *Alaska, Exxon Deal Opens Way for LNG Exports*, Reuters, March 30, 2012, http://www.reuters.com/assets/print?aid=USBRE82T12I20120330.

remains to be seen how many export terminals will ultimately be built in North America.[27]

12.2.5 Will Political Pressure and the Potential Impact on Domestic Prices of Natural Gas Hinder US LNG Exports?

While the market opportunity for US LNG exports seems to have solidified, concerns about the effect of US LNG exports on the domestic price of natural gas began to mount in late 2011. While many factors can have a bearing on the price of domestic gas, exports are likely to have an effect since they represent an additional source of demand. At the same time, over the long run, an increase in demand is also likely to increase supply. The extent to which the price of natural gas interacts with its supply and demand has been a cause of much speculation in the US, which in turn led to a US Senate Subcommittee hearing in November 2011 to address the issues raised by the possibility of US LNG exports.[28]

At the outset of the hearing, it was noted that the last time the Senate held a hearing on LNG was in 2005, when the US was anticipating the need to *import* growing quantities of LNG, whereas the current hearing was to discuss the role that LNG *exports* might play in the US energy future. As stated in the opening statement, there were two main objectives of the hearing. The first was "to understand the laws and regulations that govern LNG exports. Those laws were put into place assuming the United States would be an importing country, not an exporting country."[29] Accordingly, US policy makers considered whether it made sense to take a new look at existing US laws in light of the new market situation.

The second objective was to understand how exports might affect the domestic natural gas market. While the implications of increased gas exports for US job creation and the balance of payments could be very positive, it was also noted that "U.S. energy security requires reliable and

---

[27] Richard Nemec, *LNG Exports: The Newest Economic Engine, or a Fad That Will Pass?*, Pipeline and Gas Journal, April 2012, http://pipelineandgasjournal.com/lng-exports-newest-economic-engine-or-fad-will-pass?page=show. See also Eric Martin, *Export Floodgates are Ready to Burst*, LNG Unlimited, May 14, 2012, www.lngunlimited.com (noting that according to PFC Analyst Nikos Tsafus, "the US has a great history of getting LNG completely wrong" and that only a few US export terminals will ultimately be built).

[28] United States Senate Committee on Energy and Natural Resources, *LNG Export Approvals, Market Impacts*, Nov. 8, 2011, http://energy.senate.gov/public/index.cfm?FuseAction=PressReleases.detail&PressRelease_id=9c9aea0c-74d4-4b2c-ad.

[29] Ibid.

affordable energy *prices*, not just reliable *supplies*. Therefore, understanding how exports might affect domestic prices is also critical."[30] Since US gas prices are considerably lower than prices in most of the rest of the world, Senators questioned how the US could "ensure that our export policy is consistent with our continued ability to reap the benefits of our newfound abundance of natural gas."[31]

12.2.6 The Impact of US LNG Exports on US Natural Gas Prices

At the hearing, several Senators questioned a Department of Energy (DOE) official about the impact LNG exports could have on domestic natural gas prices.[32] In addition to concerns about gas price, the Senators also questioned regulators about whether the current export licensing system can protect US interests, given that they were written when the US was assumed to be a gas-importing – not exporting – country.[33]

At the hearing, US Senator Ron Wyden, a Democrat from Oregon, noted that "it's very understandable why North American natural gas producers would want to build LNG export terminals so they can sell natural gas to Asia and other overseas markets at four or five times the prices here. What's less clear is how this is going to be beneficial for our businesses and our consumers who are going to have to compete with these prices."[34]

Senator Wyden also raised concerns that DOE appeared to be satisfied with Sabine Pass's projections that exports from the proposed export facility could raise gas prices by 10 percent by 2015. In response to questions about how much of a price increase DOE would find acceptable, DOE acknowledged that the analysis was complicated and that when DOE does the public interest determination, a range of factors are considered such as the impact on jobs, the balance of trade, and the impact on price. Since some of the factors are influenced by price itself, DOE explicitly recognized the importance that price holds.[35]

---

30 Ibid.

31 Ibid.

32 Kate Winston, *Senators Question Price Impact of LNG Exports*, Platts Energy Week TV, Nov. 9, 2011, http://plattsenergyweektv.com/story.aspx?storyid=174167&catid=293. Several witnesses presented testimony before the subcommittee and that testimony can be found at http://energy.senate.gov/public/index.cfm?FuseAction=Hearings.Hearing&Hearing_ID=45ea3a80-f5d7-bfba-b023-31a9da05de0a.

33 Ibid.

34 Ibid.

35 Statement of Christopher Smith, Deputy Assistant Secretary for Oil and Natural Gas, Office of Fossil Energy, US Department of Energy, Before the

At the hearing, DOE also noted that when it approved the Sabine Pass export application, it indicated that DOE has a continued duty to monitor the cumulative impact of such exports. DOE was mindful of the fact that there was a growing interest in exporting domestically produced LNG and that DOE had four other long-term applications pending before it to export US-produced LNG to countries with which the United States does not have a free trade agreement.[36]

The volumes of LNG that could be authorized for export in those non-free trade agreement applications, including the 2.2 Bcf/d authorized for export in Sabine Pass, would total 6.6 Bcf/d, which represents 10 percent of total current domestic natural gas daily consumption in the United States. Consistent with the Natural Gas Act, DOE has already granted authorization from these five facilities to export the same volume to free trade agreement countries but has not ruled on the non-free trade applications yet.[37]

In order to address the potential cumulative impact of granting the pending export applications, DOE indicated at the Senate hearing that it had commissioned two studies. One study will be conducted by DOE's Energy Information Administration (EIA) and will look at the price increase that will result from an incremental increase in demand for exports. A private contractor, who will use the EIA data to study the net impact of exports on the economy, including job creation and the US balance of trade, will do the second study. Taken together, the "studies will address the impacts of additional natural gas exports on domestic energy consumption, production, and prices, as well as the cumulative impact on the U.S. economy, including the effect on gross domestic product, jobs creation, and balance of trade, among other factors."[38]

12.2.7 Price Studies on the Impact of US LNG Exports on the Domestic Price of Natural Gas

At the US Senate hearing in November 2011, DOE indicated that it was mindful of the need to promptly decide the other export applications

---

Committee on Energy and Natural Resources, United States Senate, *The Department of Energy's Role in Liquefied Natural Gas Export Applications*, Nov. 8, 2011, available at http://energy.senate.gov/public/_files/SmithDOETestimony110811.pdf.

36 Ibid.

37 Ibid.

38 Kate Winston, *Senators Question Price Impact of LNG Exports*, Platts Energy Week TV, Nov. 9, 2011, http://plattsenergyweektv.com/story.aspx?storyid=174167&catid=293.

pending before it but noted that the two pricing studies were an important component of a sound evidentiary record on which DOE would ultimately base its decision on further US LNG exports.[39]

In January 2012, the US EIA issued the first study analyzing the impact of US LNG exports on the domestic energy market. As requested by DOE, the EIA's study looked at how specified scenarios of natural gas exports could affect the US energy markets, focusing on consumption, production, and prices. The study was not intended to give an estimate of what LNG exports would likely be in the future, but to assume that the levels of exports would be either 6 Bcf/d or 12 Bcf/d, discounting other possible scenarios. In summary, the US EIA concluded that "increased natural gas exports lead to higher domestic natural gas prices, increased domestic natural gas production, reduced domestic natural gas consumption, and increased natural gas imports from Canada via pipeline."[40]

In terms of the impact of exports on US natural gas prices, the EIA noted at the outset that US natural gas prices are expected to increase even before considering the possibility of additional exports. Nonetheless, increased natural gas exports are expected to lead to higher domestic natural gas prices, with the amount dependent on the ultimate level of exports and the rate at which increased exports are phased in. For example, under the low/slow scenario, it is assumed that 6 Bcf/d of exports are phased in at a rate of 1 Bcf/d per year over six years. Under this scenario, the wellhead price impacts peak at about 14 percent ($0.70/Mcf) in 2022 but the wellhead price differential falls below 10 percent by about 2026.[41] Although the impact of LNG exports varies depending on the assumptions about resource availability and economic growth, the basic assumption remains the same: "higher export levels would lead to higher prices, rapid increases in exports would lead to sharp price increases, and slower export increases would lead to slower but more lasting price increases."[42]

In contrast to the potentially severe impacts on price found in the EIA study, an independent assessment done by Deloitte MarketPoint LLC (DMP) found that any price increase resulting from US LNG exports would be quite minimal.[43] In the DMP assessment, Deloitte applied its

---

39 Ibid.

40 US EIA, *Effect of Increased Natural Gas Exports on Domestic Energy Markets*, (January 2012), available at http://www.eia.gov/analysis/requests/fe/pdf/fe_lng.pdf, p. 6.

41 Ibid., p. 8.

42 Ibid., p. 9.

43 *Made in America: the Economic Impact of LNG Exports from the United*

Table 12.2 Study-by-study comparison of the average price impact from US LNG exports, 2015–35

Study	Average price without exports ($/MMBtu)	Average price with exports ($/MMBtu)	Average price increase (%)
EIA*	5.28	5.78	9
Deloitte	7.09	7.21	2
Navigant (2010)** (2 bcf/day of exports)	4.75	5.10	7
Navigant (2012)***	5.67	6.01	6
ICF International***	5.81	6.45	11

Notes:
6 bcf/day of LNG exports (unless otherwise noted)
* Price impact figure for EIA study reflects the reference case, low-slow export scenario.
** The Navigant study did not analyze exports of 6 bcf/day.
*** Navigant (2010 and 2012) and ICF International studies are based on Henry Hub price.

Source: Dr Charles K. Ebinger, *Liquid Markets: Assessing the Case for U.S. Exports of LNG*, the Brookings Institution Energy Security Initiative, Presentation dated May 11, 2012

integrated "North American Power, Coal, and World Gas Model" to analyze the price and quantity impacts of LNG exports on the US gas market. Using the Model and assuming 6 Bcf/d of exports, the DMP study projected that the average impact on US prices from 2016 to 2035 would be $0.12/MMBtu. This represents a 1.7 percent increase in the projected average US citygate gas price of $7.09/MMBtu over the same time period. The projected impact on the Henry Hub price is $0.22/MMBtu and significantly higher due to the close proximity to the proposed export terminals. According to DMP, the price impacts diminish further away from the Gulf Coast export terminals. For example, the projected price impacts for markets in the New York and Chicago areas are less than $0.10/MMBtu.[44]

In May 2012, the Brookings Institution released a report that analyzed the various pricing studies that had been conducted on the impact of US LNG exports[45] (see Table 12.2). While there does not seem to be consensus on the exact price impact of US LNG exports, there does seem

States, Deloitte MarketPoint LLC and the Deloitte Center of Energy Solutions (Dec. 2011), available at http://www.deloittemarketpoint.com/Papers/Papers.aspx#3.

44 Ibid. at p. 2.

45 The Brookings Institution, *Liquid Markets: Assessing the Case for U.S.*

to be consensus that the US domestic price of natural gas will increase by some amount in the range of 2–11 percent due to US LNG exports. What impact this will have in terms of a US policy or regulatory response is unclear, with Brookings advocating for what could be called a "do nothing" approach:

> The study recommends that U.S. policy makers should refrain from introducing legislation or regulations that would either promote or limit additional exports of LNG from the United States. The nature of the LNG sector, both the costs associated with producing, processing, and shipping the gas, and the global market in which it will compete, will place upper bounds on the amount of LNG that will be economic to export. Incremental increases in the price of domestic gas (as a result of domestic demand or export) negatively impact the economics of each additional proposed export project, which even with government approval will still require private financing and interested buyers. Efforts to intervene in the market by policy makers are likely to result in subsidies to consumers at the expense of producers, and to lead to unintended consequences. They are also likely to weaken the position of the United States as a supporter of a global trading system characterized by the free flow of goods and capital.[46]

While the various studies are analyzed and additional studies are no doubt underway by a variety of groups, numerous export applications remain pending at DOE with the most recent news indicating that political opposition to US LNG exports may be growing. Adding to the uncertainty about the future of US LNG exports was the March 2012 announcement by DOE that the much-anticipated second study on the impact of LNG exports on the US gas market, originally expected in early 2012, is now delayed until sometime later in 2012.[47]

While the length of time DOE has taken in considering the remaining export applications has caused a level of market uncertainty for the projects awaiting approval, it has also allowed more time for other opponents to voice concerns about the impact of US LNG exports. For example, US Representative Ed Markey (Democrat–Massachusetts) has introduced two bills in the US Congress with the stated purpose of protecting US consumers from increased natural gas prices and ensuring that America's natural gas

---

Exports of Liquefied Natural Gas, May 2, 2012, http://www.brookings.edu/research/reports/2012/05/02-lng-exports-ebinger.

[46] The Brookings Institution, *Liquid Markets: Assessing the Case for U.S. Exports of Liquefied Natural Gas*, May 2, 2012, http://www.brookings.edu/research/reports/2012/05/02-lng-exports-ebinger.

[47] *US Government Further Delays LNG Export Decision*, ISIS Heren, March 30, 2012, available at http://www.icis.com/heren/articles/2012/03/30/9546199/us-government-further-delays-lng-export-decision.html.

stays in America.[48] The first bill, the North America Natural Gas Security and Consumer Protection Act,[49] would preclude FERC from approving new LNG export terminals. The second bill, the Keep American Natural Gas Here Act,[50] would require natural gas extracted from federal lands to be resold only to American consumers. Subsequent to FERC's approval of Cheniere's Sabine Pass, Congressman Markey issued a press release continuing to express his concern that US LNG exports would increase domestic prices for natural gas, which would harm individual as well as industrial users of natural gas such as the steel, plastics, and fertilizer industries.[51]

Congressman Markey is not alone in his view that "America should exploit her competitive advantage with lower natural gas prices to create jobs in the United States,"[52] with leaders from other industries calling on the US to use its cheap natural gas to convert to products for export, as opposed to exporting the natural resource itself. For example, the CEO of Dow Chemical has argued that US LNG exports should be limited since there is up to eight times more value in using America's abundant and cheap natural gas as the raw material to create high-value products that can be exported, as opposed to simply exporting the natural gas itself.[53]

12.2.8 Will Environmental Opposition to Shale Gas Development Hinder US LNG Exports?

In addition to political opposition to US LNG exports, there is some risk that environmental opposition to shale gas development will spill over

---

[48] Natural Resources Committee, Press Release, Feb. 14, 2012, *Markey Introduces Legislation to Keep American Natural Gas in America*, http://democrats.naturalresources.house.gov/press-release/markey-introduces-legislation-keep-american-natural-gas-america.

[49] H.R. 4024, 112th Cong., 2nd Sess. (2012), http://www.govtrack.us/congress/bills/112/hr4024.

[50] H.R. 4025, 112th Cong., 2nd Sess. (2012), http://www.govtrack.us/congress/bills/112/hr4025.

[51] Congressman Ed Markey, Press Release, April 17, 2012, *Markey Sabine LNG Export Facility Approval Would Help Export US Manufacturing Jobs*, http://markey.house.gov/press-release/markey-sabine-lng-export-facility-approval-would-help-export-us-manufacturing-jobs.

[52] Kate Winston, *Senators Question Price Impact of LNG Exports*, Platts Energy Week TV, Nov. 9, 2011, http://plattsenergyweektv.com/story.aspx?storyid=174167&catid=293.

[53] Christopher Helman, *Dow Chemical Chief Wants to Limit U.S. LNG Exports*, Forbes, March 8, 2012, http://www.forbes.com/sites/christopherhelman/2012/03/08/dow-chemical-chief-wants-to-limit-u-s-lng-exports/ (citing Dow CEO Andrew Liveris).

into opposition to US LNG exports. This has already occurred to some extent with the Sierra Club filing an opposition to Cheniere's Sabine Pass export project[54] as well as Dominion's Cove Point liquefaction project.[55] In its Opposition to Dominion's Cove Point export project, the Sierra Club argued that in addition to issues about whether LNG exports will increase the domestic price of natural gas, the environmental impacts associated with natural gas production must also be considered in determining whether US LNG exports are in the "public's interest."[56] While the Sierra Club contends that all environmental impacts from natural gas production need to be considered, their brief highlighted the particular environmental concerns that have been raised pertaining to hydraulic fracturing and shale gas development. In summary, the Sierra Club maintains that DOE's approval for LNG exports could have "major environmental impacts through the [United States], and especially in the Northeast, where [US LNG exports] will intensify Marcellus Shale extraction activities."[57]

---

[54] Although the Sierra Club filed an opposition to Cheniere's Sabine Pass export project, it appeared to have limited impact in that case.

[55] While Dominion's Cove Point export project received DOE authorization in October 2011 to export LNG to countries that have free trade agreements with the United States, it is one of the pending projects awaiting DOE approval to export to WTO nations. Dominion Cove LNG, http://www.dom.com/business/gas-transmission/cove-point/index.jsp.

[56] Sierra Club's Motion to Intervene, Protest, and Comments, *In the Matter of Dominion Cove Point LNG, LP.*, FE Docket No. 11-128-LNG (Feb. 6, 2012), http://www.fossil.energy.gov/programs/gasregulation/authorizations/2011_applications/Motion_to_Intervene_Sierra_Club_02_06_12.pdf. The Sierra Club and Dominion also appear to be in dispute over whether Dominion even has a right to construct its export facilities. The Sierra Club alleges that a 2005 agreement with Dominion precludes the export project in its entirety but Dominion claims otherwise and is seeking a declaratory judgment that the prior agreement does not preclude the export project. See *Sierra Club Letter to Paul E. Ruppert*, Senior Vice President–Transmission, Dominion, April 26, 2012, http://content.sierraclub.org/sites/default/files/documents/4_26_12_DCP_Cease_and_Desist_Letter.pdf (stating that a 2005 settlement between the Sierra Club and Dominion seeks to protect the area surrounding Cove Point and precludes the expansion of the facility for exports due to the environmental impacts), and Dominion Press Release, May 18, 2012, *Dominion Files Lawsuit to Confirm Right to Construct Cove Point Natural Gas Liquefaction Project*, http://www.prnewswire.com/news-releases/dominion-files-lawsuit-to-confirm-right-to-construct-cove-point-natural-gas-liquefaction-project-152047405.html (stating that Dominion would seek judicial declaratory judgment that the 2005 agreement with the Sierra Club does not preclude constructing an export facility).

[57] Sierra Club's Motion to Intervene, Protest, and Comments, *In the Matter of Dominion Cove Point LNG, LP.*, FE Docket No. 11-128-LNG (Feb. 6, 2012), at p. 47, http://www.fossil.energy.gov/programs/gasregulation/authorizations/2011_applications/Motion_to_Intervene_Sierra_Club_02_06_12.pdf.

The Sierra Club's opposition to Dominion's Cove Point export project also appears to coincide with an intensified effort on the part of the Sierra Club to ensure that as coal-fired power plants are retired, they are not replaced with natural gas power plants. To that end, the Sierra Club recently announced that it is launching a new "Beyond Gas" campaign that represents a significant expansion of the group's ongoing efforts against other major fossil fuels and is modeled after the decade-old "Beyond Coal" campaign that sought to phase out coal-fired power plants. According to the Sierra Club, it will seek to "prevent new gas plants from being built wherever we can."[58]

While it remains to be seen whether the environmental opposition to US LNG exports will grow more intense over time, some reports have acknowledged that since the case for US LNG exports depends on the continued development of shale gas, the public's concerns over the environmental impacts of shale gas development must be resolved.[59] In the meantime, as a result of the growing controversy over US LNG exports, it appears that DOE will not make a decision on any of the pending export licenses until after the US presidential election in November 2012.[60] This would once again give Cheniere a first-mover advantage, with Cheniere's Sabine Pass Liquefaction project being, to date, the only project to secure DOE approval to export to all WTO nations and the only export project to secure FERC approval. While some have expressed concern that DOE could ultimately revoke or amend even Cheniere's license to export, it is unclear whether DOE would have the legal authority to do so and, thus far, DOE has appeared reluctant to consider this possibility.[61]

Some policy makers and business leaders have urged the US to approve

---

58 Amy Harder, *War Over Natural Gas About to Escalate*, National Journal, May 3, 2012, http://www.nationaljournal.com/energy-report/war-over-natural-gas-about-to-escalate-20120503.

59 Charles Ebinger, Kevin Massy, and Govinda Avasarala, *Evaluating the Prospects for Increased Exports of Liquefied Natural Gas from the United States*, Energy Security Initiative, Brookings Institution, Jan. 2012, http://www.brookings.edu/~/media/research/files/papers/2012/1/natural%20gas%20ebinger/natural_gas_ebinger_2.pdf.

60 *US Government Further Delays LNG Export Decision*, ICIS Heren, March 30, 2012, http://www.icis.com/heren/articles/2012/03/30/9546199/us-government-further-delays-lng-export-decision.html; Tennille Tracy, *U.S. Gas Exports Put on Back Burner*, May 31, 2012, Wall Street Journal at B3 (indicating that "The Obama administration is telling Japan and other allied countries they will have to wait before moving forward on plans to buy American natural gas").

61 Ayesha Rascoe, *US Will Not Revoke LNG Exports to Control Prices–DOE*, Reuters, March 1, 2012, http://uk.reuters.com/article/2012/03/01/usa-lng-exports-idUKL2E8E15XA20120301.

export licenses and have expressed the view that the market should dictate whether US LNG exports happen or not.[62] For example, at the November 2011 Senate hearing, Senator Lisa Murkowski (Republican–Alaska) indicated she is inclined to let the market sort out the issue. "Our proper course won't be sweeping legislation or layers of new regulation. Instead, it will be to ensure a degree of comfort that our newfound energy security can be maintained under current export rules."[63]

12.2.9 Will Canada Win the Race to Export?

While the debate over LNG exports in the US continues, it appears that LNG export projects in Canada may be gaining an advantage with strong government support[64] amidst a growing recognition that Canada's LNG export projects should move forward with haste to exploit the market advantage that now exists.[65]

With the final investment decision (FID) still pending on Kitimat LNG as this book goes to print, it is unlikely that Kitimat LNG will become operational by 2015 as originally planned. There are, however, a number of encouraging developments that will help the project move forward. One positive development was the NEB's approval of Kitimat LNG, which was considered "momentous" since it was the first LNG export license considered by the Board and the first LNG project to move forward since several had collapsed in the 1980s.[66]

---

62 Tennille Tracy, *U.S. Gas Exports Put on Back Burner*, May 31, 2012, Wall Street Journal at B3 (noting that ExxonMobil is looking at exporting from the US and Canada and that Exxon's CEO, Rex Tillerson, "laid out the case for exports at Exxon's shareholder meeting [on May 30, 2012], saying they would create jobs and help the U.S. trade balance.") See also Eric Martin, *Domestic Price Impact Elephant in the Room*, LNG Unlimited, May 14, 2012, www.lngunlimited.com (noting that as a free trade nation, the US may have little choice but to approve LNG exports).

63 Kate Winston, *Senators Question Price Impact of LNG Exports*, Platts Energy Week TV, Nov. 9, 2011, http://plattsenergyweektv.com/story.aspx?storyid=174167&catid=293.

64 British Columbia News Release, *Premier Applauds Progress on Kitimat Project: LNG Canada*, May 15, 2012, http://www.newsroom.gov.bc.ca/2012/05/premier-applauds-progress-on-kitimat-project-lng-canada.html (noting Canada's Premier Christy Clark's support of several proposed Canadian LNG projects).

65 Nathan Vanderklippe, *Shell Urges Quick Action to Secure LNG Markets for Kitimat Terminal*, The Globe and Mail, May 15, 2012, http://www.theglobaandmail.com.

66 Vaughn Palmer, *Opinion: Kitimat LNG Export Project Gliding Along with Little Opposition. NEB Approval Grabs Attention of Would-be Operators of Oregon's*

Another encouraging development for the project is the interest of Korea Gas Corp (KOGAS), the world's largest importer of LNG. In June 2009, Kitimat LNG signed a Memorandum of Understanding (MOU) with KOGAS pursuant to which KOGAS will acquire up to 40 percent of Kitimat LNG's production, or 2 million tons per year for 20 years with an option to acquire an equity stake in the terminal. The total purchase value over that period would be more than US$20 billion. In addition, earlier in 2009, Kitimat LNG had entered a nonbinding agreement with Mitsubishi whereby Mitsubishi would buy 1.5 million tonnes a year of terminal capacity and acquire a minority equity interest in the terminal.[67] In 2010, KOGAS and EnCana entered into a three-year deal whereby KOGAS will invest $565 million to acquire a 50 percent working interest in expected production from three shale gas fields in the Canadian Horn River and Montney formations. The deal added a "lift" to Kitimat LNG's plans and "sets the stage for Canada to access Asian markets and end its total reliance on the United States as an export outlet."[68]

Despite the boost from the KOGAS deals, there is still some uncertainty whether FID will be taken on Kitimat LNG, whether the project could compete with Australia's LNG projects, and whether, when shipping costs are factored in, Kitimat LNG could be profitable.[69] In addition, Kitimat LNG is now facing growing competition from larger companies seeking to take advantage of the market opportunity for Canadian LNG exports. For example, in May 2012, Shell, KOGAS, Mitsubishi, and PetroChina announced they are developing a proposed LNG export facility (LNG Canada) in Western Canada, near Kitimat, British Columbia.[70] Although the proposed 12 mtpa (1.2 bcf/d) LNG Canada project will be a large, complex project, Shell has indicated it will start front-end engineering and

Coos Bay Terminal, Vancouver Sun, Oct. 16, 2011, http://www.vancouversun.com/Opinion+Kitimat+export+project+gliding+along+with+little+opposition/5559395/story.html#ixzz1b1fUivMQ

67 *Kitimat LNG Signs Deal with Korea Gas Corp. Worth $20 Billion*, Horn River News, June 9, 2009, http://hornrivernews.com/2009/06/09/kitimat-lng-signs-deal-with-korea-gas-corp-worth-20-billion/

68 Some analysts have suggested that shipping LNG from Canada to Asia might cost "C$5 per thousand cubic feet, which raises questions about the profitability of the Kitimat project." Kitimat believes shipping will cost C$3–4 per thousand cubic feet. Gary Park, *Kogas Cuts EnCana Deal, Farm-in Gives Lift to Kitimat Plans; Kogas Could Spend C$1.1B over 5 Years*, Petroleum News, March 7, 2010, http://www.petroleumnews.com/pntruncate/248230748.shtml.

69 Ibid.

70 Shell Press Release, *Shell and Partners Announce LNG Project in Canada*, May 15, 2012, http://www.shell.com.

design (FEED) work on the project in 2013 with FID on the estimated $9.6–13.2 billion project expected sometime in 2015.[71]

12.2.10 Can North American LNG exports compete?

One of the issues that has been raised regarding the proposed North American LNG export projects is whether they can compete with other LNG projects, especially the mega projects in Qatar and Australia. According to Cheniere's analysis, the range of liquefaction costs for many of the global LNG projects is $200–2,000 per ton, with several of the Australian projects at the higher end of the range and Cheniere's Sabine Pass Liquefaction project estimated to be $400 per ton (Figure 12.4).

In a research report analyzing the competitiveness of North American LNG exports, Barclays Capital (BarCap) concluded that US LNG exports from the Gulf Coast might be competitive with other global LNG projects, but it is likely to be at the higher end of the supply curve.[72] The BarCap report found while US Gulf Coast LNG exports might be competitive with the costly Australian projects, they would have a much harder time competing with Qatari projects. This is because the revenue the Qatari projects earn from selling the natural gas liquids (NGLs) associated with the feed gas in Qatar covers not only the upstream costs of development but also the majority of the capital costs of liquefaction. As such, the Qatari projects yield a breakeven LNG cost as low as the cost of operation.[73] Whether or not Qatar can maintain its competitive advantage in the longer term remains to be seen given the recent focus on liquids-rich US shale plays[74] as well as a moratorium Qatar has on developing additional LNG capacity pending a study on the sustainability of the North Field due to be completed in 2013.[75]

---

[71] Nathan Vanderklippe, *Shell Urges Quick Action to Secure LNG Markets for Kitimat Terminal*, The Globe and Mail, May 15, 2012, http://www.theglobaand-mail.com.

[72] Barclays Capital, Natural Gas Weekly Kaleidoscope, *Can North American LNG Exports Compete?*, April 12, 2011, www.barcap.com.

[73] Barclays Capital, Natural Gas Weekly Kaleidoscope, *Can North American LNG Exports Compete?*, April 12, 2011, www.barcap.com.

[74] Due to low natural gas prices, there has been a shift in US shale gas production from dry gas to liquids-rich plays that improve the plays' economics. Ben DuBose, *WPC '12: US Shale Gas Development Shifts from Dry to Liquids*, Hydrocarbon Processing, March 29, 2012, http://www.hydrocarbonprocessing.com/Article/3004044/WPC-12-US-shale-gas-development-shifts-from-dry-to-liquids.html?Print=true, noting that US shale plays are rapidly shifting away from dry gas towards areas with NGLs to take advantage of the extremely high spread between US oil and gas plays.

[75] *Qatar's LNG Threat*, Zawya.com, May 17, 2012, http://www.zawya.com/

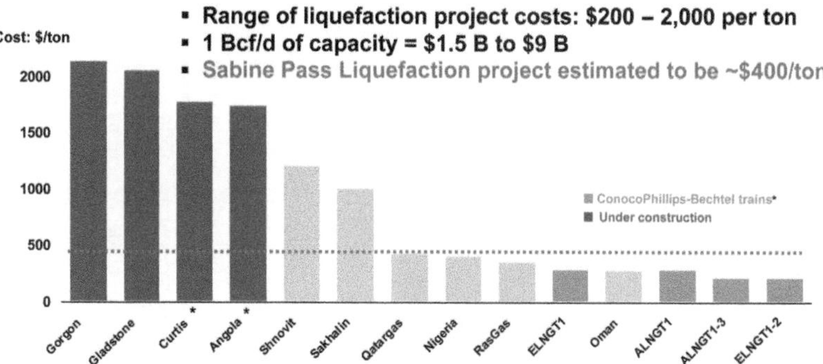

Notes: Projects costs per ton are total project costs divided by mtpa capacity of LGN trains. Figures do not attempt to isolate, where applicable, the cost of liquefaction facilities within a major LNG complex. Chart includes a representative sample of liquefaction facilities and does not include all liquefaction facilities existing or under construction.

Source: Cheniere Energy Partners, L.P., Corporate Presentation dated June 2011

Figure 12.4 Estimated liquefaction project costs for global LNG projects

Other experts have indicated that whether or not North American LNG exports can compete beyond 2015 is far less certain as more uncontracted LNG supply will come online at this time at prices that are yet to be determined (Figure 12.5).

In addition, a number of so-called "wildcard" factors could also weigh heavily on the shale spreads that have existed in recent years. These factors could push US prices up and global prices down, which could impact the ability of North American LNG exports to compete on a global basis. (See Figure 12.6.)

---

story/pdf/Qatars_LNG_threat-ZAWYA20120517041238/, noting that Qatar faces competition from other LNG players such as Australia but has moved to maintain its economic advantage with other projects such as Pearl GTL.

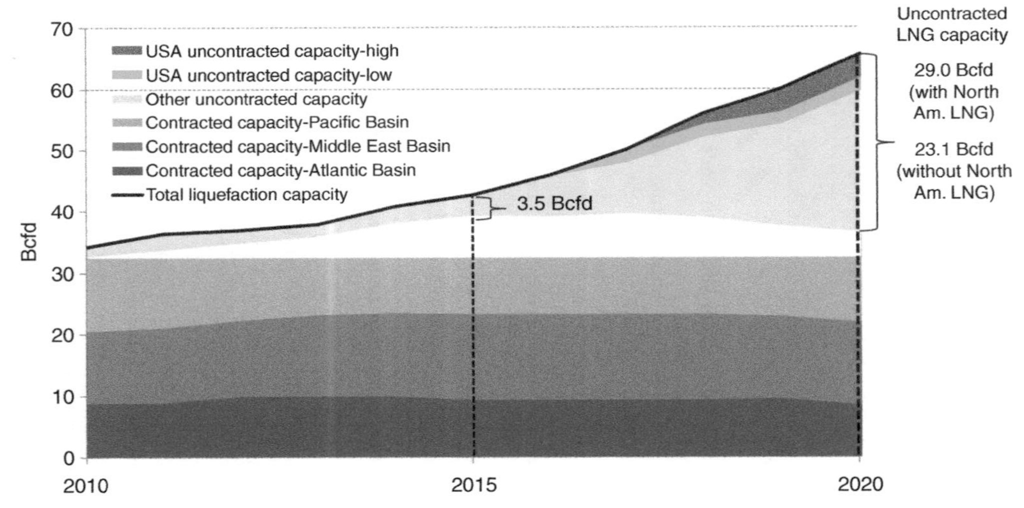

Notes:
Uncontracted capacity equals CRA projected liquefaction capacity minus contracted capacity including both LNG SPA and LNG MOU/HOA/LOO agreements in place. It does not reflect spot LNG trade. The liquefaction capacity reflects CRA research and judgement as to project status, timing, and delays.
Charles Rivers Associates (CRA) (www.crai.com) is a leading global consulting firm that offers economic, financial, and business management expertise to major law firms, corporations, accounting firms, and governments around the world. "These conclusions are based on independent research, and publicly available material, and do not reflect or represent the views of Charles River Associates or any of the organizations with which the author is affiliated. Neither the author nor Charles River Associates accept a duty of care or liability of any kind whatsoever to any party, or any responsibility for damages, if any, suffered by any party as a result of decisions made, or not made, or actions taken, or not taken, based on these materials."

Source: Christopher Concalves, Vice President, Charles River Associates (CRA), *Chasing the Shale Spread: Potential New Markets for LNG*, Presentation to the CWC 12th Annual World LNG Summit, November 14–17, 2011, Rome

Figure 12.5 Uncontracted LNG capacity 2015 and beyond

The leading fundamental drivers and "wildcards" could weigh heavily on shale spreads, pushing US prices up and global prices down – unless the wave of LNG regasification demand swells to tsunami status

Shale spread uncertainty	Natural gas price impacts		Shale spread	Incremental volume
Fundamental Price Drivers by 2020	North Am (HH)	Global LNG	Impacts	(Bcfd)
US shale net impact on production	↓	NA	↑	6.6
North American LNG exports	↑	↓	↓	3.5–7.3
US emissions regulation and coal retirements	↑	NA	↓	3.6
US NGV penetration	↑	NA	↓	0.4–1.6
China and ROW unconventional production growth	NA	↓	↓	7.4–16.8
Nuclear safety policy and retirements	NA	↑	↑	1.8–4.2
New regasification / LNG demand growth	NA	↑	↑	7.6–21.4
Factors pushing US prices down / shale spreads up	↓		↑	**6.6**
Factors pushing US prices up / shale spreads down	↑		↓	**7.5–12.5**
Factors pushing global prices down / shale spreads down		↓	↓	**7.4–16.8**
Factors pushing global prices up / shale spreads up		↑	↑	**9.4–25.6**

Source: Christopher Goncalves, Vice President, Charles River Associates, Presentation to the CWC 12th Annual World LNG Summit, November 14–17, 2011

Figure 12.6 "Wildcard" factors and shale spreads

12.2.11 Will North American LNG Exports Impact the Global LNG Market?

At the US Senate hearing regarding US LNG exports, the issue was raised whether US exports would definitively link US gas prices to higher international prices. Some of the related questions that have been raised are:

- How much LNG can be exported before it becomes self-limiting?
- What will limit volumes first – rising US gas prices that make exports uneconomic or the fear of price increases that drives vested interests to persuade regulators to restrict exports?
- Will the threat of North American LNG exports impose a future ceiling for European gas prices?

While these questions are difficult to answer, most experts seem to be of the opinion that the global inter-dependencies among gas markets will continue over the next decade and that this will influence global gas pricing more than North American LNG exports. While the North America region is expected to remain largely disconnected from global gas markets as oversupply continues to depress Henry Hub prices, some experts have noted that it is possible that the mere movements of North America toward becoming an LNG exporter could have an impact on price even before LNG exports become a reality.[76]

As LNG becomes a more widely traded commodity, it is expected that the mechanisms determining its price will evolve.[77] For the time being, however, it appears that the LNG trading system is not advanced enough to erase the boundaries between US and global prices.[78] As explained at the US Senate hearing, at the relatively low levels of exports expected from

---

[76] Warren R. True, *Gastech: WoodMac Expects Global Gas Dependencies to Continue*, Oil & Gas Journal, Mar. 28, 2011, http://www.ogj.com/articles/print/volume-109/issue-13/general-interest/gastech-woodmac-expects-global-gas.html (citing Noel Tomnay, Head of Global Gas Research, Wood Mackenzie, Gastech Conference, Amsterdam, Mar. 21, 2011).

[77] Some experts have argued that as gas becomes a more important fuel for countries around the world, it will become increasingly necessary for its pricing to reconnect with economic and market fundamentals, rather than continue to be determined by crude oil and oil product prices, or politically-driven subsidies. Jonathan Stern (ed.), THE PRICING OF INTERNATIONALLY TRADED GAS, The Oxford Institute for Energy Studies (2012), www.oxfordenergy.org.

[78] Kate Winston, *Senators Question Price Impact of LNG Exports*, Platts Energy Week TV, Nov. 9, 2011, http://plattsenergyweektv.com/story.aspx?storyid=174167&catid=293.

the US, natural gas will remain an "inherently local domestic commodity," especially since there is not the infrastructure in place at the moment to allow prices in the US to become coupled with prices in Asia.[79]

More recently, some experts have also opined that LNG exports from the US – if and when they happen – would have little if any effect on the European gas supply. While the immediate effect of shale gas development in the US was to divert LNG shipments away from the US, the markets have already absorbed that effect.[80] Moreover, according to some analysts, since natural gas is currently at a price disadvantage to coal in Europe, and with the likely demand-depressing effects of the ongoing financial crisis, Europe is unlikely to attract much US-produced LNG, no matter how much or how little liquefaction eventually gets built there.[81]

For now, with export terminals not even yet built, it remains to be seen what the prospects will be for North American LNG exports. Nonetheless, a few energy analysts are already forecasting that, while there may be a short-term window for North American LNG exports, after 2018 North American LNG exports, particularly from the US, will be limited by competition and demand.[82]

12.3 THE PANAMA CANAL EXPANSION AND POTENTIAL IMPACT ON LNG TRADE

An emerging issue that may open up new trade routes for LNG, including making US LNG exports more viable, is the expansion of the Panama Canal, which is targeted for completion in 2014. In 2006, the government of Panama announced plans to expand the Panama Canal to accommodate increased trade between China and the West and also to accommodate the advent of much larger ships that currently cannot

---

79 Ibid., quoting Christopher Smith, Deputy Assistant Secretary for Oil and Natural Gas Office of Fossil Energy, US Department of Energy.

80 Warren R. True, *WPC20: US LNG Exports to Have Little Effect on Europe*, Oil & Gas Journal, Dec. 8, 2011, http://www.ogj.com/articles/2011/12/wgc20-us-lng-exports-to-have-little-effect-on-europe.html, citing E.On Energy Trading Chairman Klaus Schafer in comments to Oil & Gas Journal made on the final day of the 20th World Petroleum Congress in Doha, Qatar.

81 Ibid.

82 *Long-Term LNG Projects in US May Face Stiffer Competition*, JPT Online, April 5, 2012, http://www.jptonline.org/index.php?id=1600, citing a Wood Mackenzie energy study that forecast that incremental US LNG exports into Asia after 2018 will be limited by competition and demand and that a positive scenario would be for the US to achieve exports of 5.1 Bcf/D by 2022.

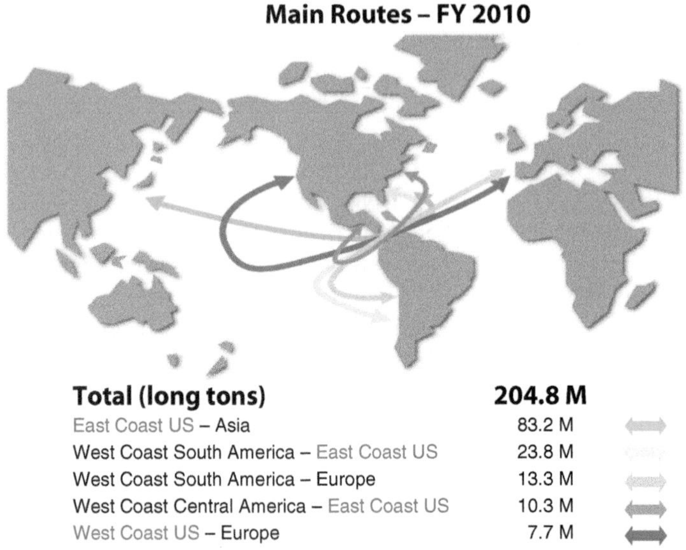

Source: Rodolfo Sabonge, Vice President, Market Research and Analysis, Panama Canal Authority, *The Panama Canal Expansion: Further Developing Trade Integration*, Presentation to the University of Texas Center for Transportation Research Annual Symposium, April 2011, available at http://www.utexas.edu/research/ctr/symposium/symp_2011/Panama_2011.pdf

Figure 12.7 Panama Canal main trade routes

transit the canal due to size.[83] As envisioned, the $5.25 billion expansion project will double the canal's capacity by creating a new lane of traffic along the canal through the construction of a new set of locks that will allow more ships to transit through the Panama Canal trade routes.[84] (See Figure 12.7.)

The Panama Canal expansion is also designed to open up new trades that are not coming through the canal today, such as LNG. Currently, only about 6 percent of the world's LNG fleet can use the canal. As shown in Figure 12.8, the expansion will increase the existing locks' maximum vessel size to accommodate typical LNG vessel sizes of 150,000–180,000 cubic meters. According to the Panama Canal Authority, the maximum

---

[83] John Lyons, *Panama Takes Step Toward Expanding the Canal*, Wall Street Journal at A8, April 24, 2006.

[84] Panama Canal Authority, Expansion Project, http://www.pancanal.com/eng/expansion/index.html.

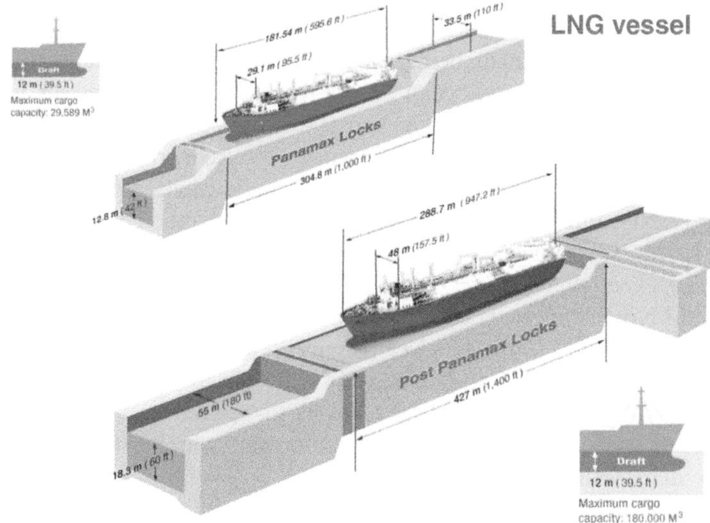

Source: Panama Canal Authority

Figure 12.8 Panama Canal lock and LNG vessel size

vessel size that will fit the new locks is 366 meters long (LOA), 49 meters wide (beam) and 15.2 meters draft. To the extent that an LNG vessel meets these dimensions, it will be able to transit the new locks.[85] Once the expansion is completed, it is expected that about 80 percent of the world's LNG fleet will be able to fit through the locks.[86] The only exceptions to the LNG carrier feet appear to be the mega ships such as the Q-Max (266,000 cubic meters) and the Q-Flex (210,000 cubic meters), which are larger than 180,000 cubic meters and also wider than 49 meters.[87]

Some analysts have also stated that the expansion of the Panama Canal will give US Gulf coast LNG exporters, such as Cheniere, direct access to

---

[85] Silvia de Marucci, Leader, Liquid Bulk Segment, Panama Canal Authority, May 8, 2012, email communication with author (on file with author).

[86] *Special Report: LNG Update: Panama Canal Expansion Will Loosen LNG Trade*, Oil & Gas Journal, March 15, 2010, available at http://www.ogj.com/articles/prnt/volume-108/issue-10/technology/special-report-lng-update-panam-canal-expansion-will-loosen-lng-trade.html.

[87] Zeus Development Corporation Conference, *LNG: Panama, Suez Canals' Expanding Influence in World Trade*, Feb. 17, 2011, available at http://www.zeuslibrary.com/Canals2011/index.asp.

the Asia-Pacific market, which may possibly threaten Australian LNG projects.[88] While the competitiveness of US LNG exports remains to be seen, another development being eyed by the Panama Canal Authority is the opening-up of cross-ocean access to new LNG markets in Latin America. According to the Panama Canal Authority, the "expanded Canal will be the first route choice for LNG trades between Trinidad–Chile and Peru–USG."[89] Other experts have noted that the canal expansion could open up LNG trade from the Caribbean directly into the Pacific, and vice versa, saving some 2,000 nautical miles through the Magellan Straits.[90] Still other experts have noted that if the shale spreads between the Pacific Basin and North America persist, this could put pricing pressure on LNG opportunity costs to Latin America (Figure 12.9).

Whether the Panama Canal expansion will offer any profound benefits to the global LNG tanker market remains to be seen and the final cost in terms of transit fees may ultimately determine whether the expansion project increases LNG trade or not.[91]

12.4 THE GROWING ROLE OF FLOATING LNG

12.4.1 FLNG: The Way Forward?

On the technical side, one of the key emerging issues is the growing role of floating LNG (FLNG) production facilities. The term FLNG includes both floating regasification and liquefaction facilities.[92] In terms of FLNG regasification, there are currently eight terminals in operation

---

[88] Kwok W. Wan, *Panama Opens Asia to US LNG Exports*, Petroleum Economist, Nov. 2, 2011, excerpt available at www.petroleumeconomist.com.

[89] University of Texas Center for Transportation Research Annual Symposium, April 2011, *Panama Canal Expansion and World Trade*, Presentation of Mr Rodolfo Sabonge, http://www.utexas.edu/research/ctr.symposium/symp_2011/Panama_2011.pdf.

[90] *Panama Canal Expansion to Boost LNG Trade*, The Oil Daily, April 15, 2010, and Zeus Development Corporation Conference, *LNG: Panama, Suez Canals' Expanding Influence in World Trade*, Feb. 17, 2011, http://www.zeuslibrary.com/Canals2011/index.asp.

[91] For example, at least one shipping expert has indicated that the transit fees for crude oil tanker traffic through the expanded Panama Canal might prove to be cost prohibitive despite the expansion's opportunity for more efficient trade. Gibson Consultancy and Research, *Panama – No Shortcut for Tankers*, July 22, 2011, available at www.gibson.co.uk.com.

[92] Other floating structures also exist such as floating production storage and offloading (FPSO) and floating storage and regasification unit (FSRU). There are

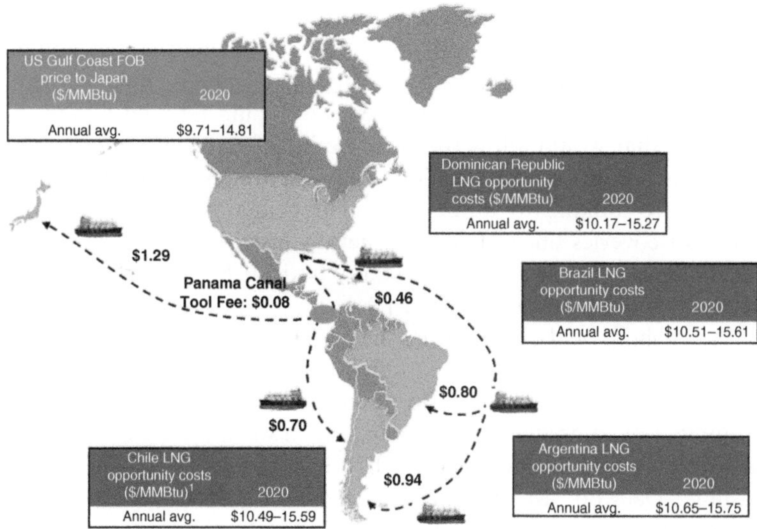

Note: LNG opportunity costs are calculated by US Gulf Costs FOB netback prices for Japan plus shipping costs to various markets. Panama Canal toll fees apply to Japan and Chile shipping costs.

Source: Christopher Goncalves, Vice President, Charles River Associates, Presentation to CWC LNG Americas Summit, San Antonio, TX (April 2012), www.cwclngamericas

Figure 12.9 US LNG exporters' opportunity cost of supply to Latin America and the Caribbean

throughout the world, in the US, Brazil, the UK, Kuwait and Argentina.[93] A key benefit of FLNG regasification is that the facility can be quickly constructed to meet seasonal, peak, or short-term demand. For example, Kuwait imports LNG through an FLNG terminal in order to satisfy peak summer demand for LNG.[94] Other countries including the United Arab Emirates (UAE), Pakistan, Bangladesh and Indonesia are expected to

also gravity based structures (GBS), which are not very common. A discussion of these other floating structures is beyond the scope of this book.

[93] *Floating LNG Terminals – Technological Innovation and Low Cost Monetization of Offshore Gas Reserves Will Play Key Role in Global LNG Industry Growth*, GlobalData Report, July 2011.

[94] Ibid., noting that in 2010 Kuwait National Petroleum Co. chartered Excelerate Energy's *Explorer* vessel to receive cargoes under an agreement with Shell International Trading Middle East Limited to meet peak summer demand during 2010–13.

establish FLNG regasification terminals, with some experts predicting that FLNG regasification terminals will play a major role in the global regasification market by 2015 when additional terminals come online.[95]

More recently, companies have been developing floating liquefaction technology that would allow the offshore liquefaction of natural gas. In recent years, as the number of offshore gas discoveries has increased, more countries are looking to utilize FLNG liquefaction to monetize those offshore discoveries since FLNG is generally faster to construct and more economical than capital-intensive land-based facilities, especially as the costs in remote areas have risen. FLNG places gas liquefaction directly over the offshore fields, eliminating the need for long-distance pipelines to shore as well as extensive onshore infrastructure. Thus, the use of FLNG liquefaction could result in significant cost savings.[96]

A number of companies are participating in small-scale FLNG projects around the world including Flex LNG, Hoegh LNG, Rift Oil Plc, Samsung, Daewoo, Mitsubishi, and Peak Petroleum. In May 2015, US-based Excelerate Energy announced it would develop the US's first FLNG export facility off the US Gulf Coast. Excelerate's Lavaca Bay LNG project will initially be able to ship 3–4 mtpa of LNG but could be expanded to 8 mtpa. Although Excelerate expects to start exporting by 2017, it should be noted that it is now the eighth company awaiting export approval from DOE.[97]

12.4.2 Shell Prelude – the World's Most Closely Watched FLNG Project

While Excelerate's Lavaca Bay LNG project appears to be the most recently announced FLNG project, the most closely watched FLNG liquefaction project is the Shell Prelude project, which took FID in May 2011.[98] Unlike most of the other planned FLNG projects, Shell's Prelude project is large-scale and it is expected to cost up to $12.6 billion. Once completed, it is likely to be the world's first FLNG liquefaction facility.[99]

---

95 *Floating LNG Terminals – Technological Innovation and Low Cost Monetization of Offshore Gas Reserves Will Play Key Role in Global LNG Industry Growth*, GlobalData Report, July 2011.

96 World Gas Conference 2012, *Floating LNG: The Way Forward*, New Sunday Times, June 3, 2012.

97 Edward McAllister, *Excelerate to Build First Floating US LNG Export Plant*, Reuters, May 15, 2015, http://uk.reuters.com/article/2012/05/15/excelerateenergy-idUKL4E8F74920120515.

98 Shell, *Prelude FLNG – An Overview*, available at http://www.shell.com/home/content/aboutshell/our_strategy/major_projects_2/prelude_flng/overview/.

99 Andrew Burrell, *$12bn Prelude Floating Plant has Shell Fired for LNG*, The

It is being closely watched because if Shell can prove that FLNG liquefaction is economically viable and operationally reliable, then it could open the door for more FLNG projects in areas where the natural gas reserves have been stranded due to the lack of appropriate onshore plant sites or the high costs of pipeline transportation from offshore.[100]

The Prelude FLNG facility will produce an estimated 3 Tcf of gas and equivalent resources from the Prelude field, which was discovered by Shell in 2007 and is approximately 125 miles off the coast of Western Australia.[101] The first production of approximately 3.6 million tonnes per year of LNG is expected by 2017.[102] When constructed, the Prelude FLNG facility will be the world's largest offshore FLNG facility ever built, measuring 1,062 ft (488 m) long by 243 ft (74 m) wide, and comparable in size to many of the world's most iconic structures[103] (see Figure 12.10).

The practical realization of FLNG requires the successful integration of many complex technological elements. Even though each element is based on existing technology, the ability to mix them optimally and integrate them wholly is required. Moreover, there are numerous technological issues unique to floating facilities, such as the impact on the capacity of liquefaction due to wave vibration (declining liquefaction efficiency, lowering plant operation rates, shortened life-cycles and potential accidents); an impact on storage (potential accidents due to sloshing liquid levels in tanks); and an impact on shipping (lowering operation rates and higher potential accidents).[104]

Construction of Shell's Prelude project will take place at Samsung Heavy Industries' Geoje Shipyard in South Korea, which is one of the few locations in the world with a dry dock large enough to accommodate a facility

---

Australian, May 21, 2011, http://www.theaustralian.com.au/archive/business-old/bn-prelude-floating-plant-has-shell-fired-for-lng/story-e6frg9ef-1226059923612.

[100] IEA MTOGM 2011, at p. 248.

[101] Joel Parshall, *Shell to Build World's First Floating LNG Facility*, Journal of Petroleum Technology, Sept. 2011, http://www.spe.org/jpt/print/archives/2011/09/11FLNG.pdf.

[102] Andrew Burrell, *$12bn Prelude Floating Plant has Shell Fired for LNG*, The Australian, May 21, 2011, http://www.theaustralian.com.au/archive/business-old/bn-prelude-floating-plant-has-shell-fired-for-lng/story-e6frg9ef-1226059923612.

[103] Joel Parshall, *Shell to Build World's First Floating LNG Facility*, Journal of Petroleum Technology, Sept. 2011, http://www.spe.org/jpt/print/archives/2011/09/11FLNG.pdf; Shell, *Prelude FLNG in Numbers*, http://www.shell.com/home/content/aboutshell/our_strategy/major_projects_2/prelude_flng/by_numbers/.

[104] Chiyoda Corp, http://www.chiyoda corp.com/technology/en/lng/offshore.html.

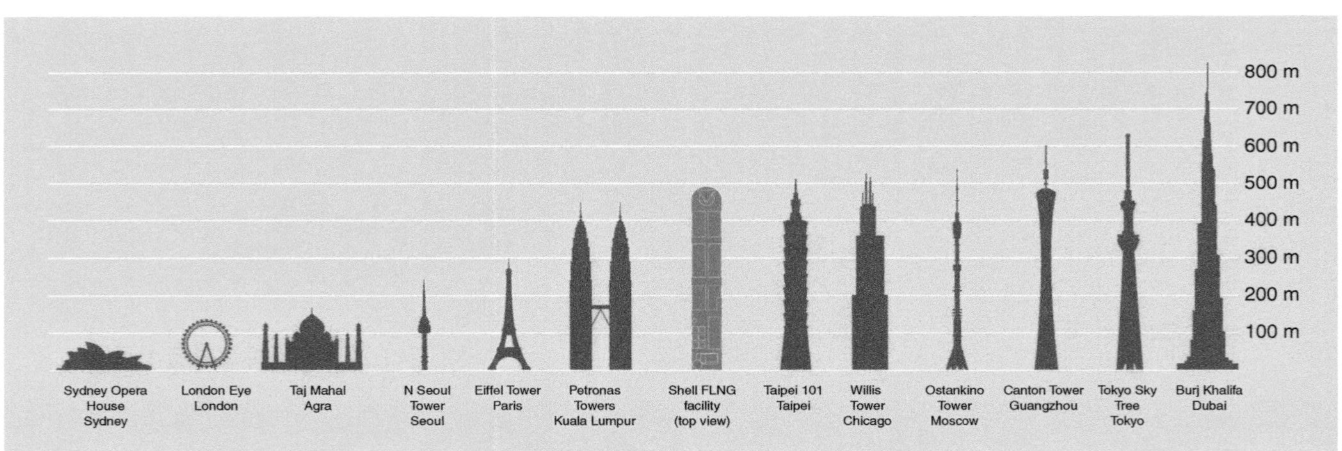

Source: Shell International Limited

Figure 12.10 Length of Shell Prelude compared with heights of major structures worldwide

of this size.[105] Once constructed, the facility will be towed to location – a six-week journey – where it will be fully moored and commissioning of the facility will begin. Once operational, Shell Prelude is expected to produce 4 million tons/yr of LNG, 1.4 million tons/yr of condensate, and 0.4 million tons/yr of LPG.[106]

The facility will be equipped with dual-row membrane for LNG/LPG storage and have storage capacities of 7.8 MMcf for LNG, 3.2 MMcf for LPG, and 4.5 MMcf for condensate. All reservoir, subsea control, processing, storage, and offloading systems are operated from the facility, which has an integrated system design combining upstream production with the FLNG process to optimize efficiency, safety, and reliability in operation.[107] One of the world's largest mooring systems will secure the facility and four groups of mooring lines will anchor it to the seabed. The system allows the facility to turn slowly to absorb the impact of strong weather conditions while remaining moored over the gas field. The sheer size of the facility should help it withstand severe weather and it is designed to stay safely moored at sea even during Category 5 cyclones.[108]

12.5 THE GAS EXPORTING COUNTRIES FORUM (GECF) – THE NEXT OPEC?

In terms of market developments, the potential of the Gas Exporting Countries Forum (GECF) to act as a "Gas OPEC" has been discussed in various publications in the past, but merits discussion here since the GECF only recently became a full-fledged organization with the potential impact of the organization going forward unknown.

12.5.1 Early Years of the GECF

In May 2001, the GECF[109] held its first ministerial meeting in Tehran, Iran, to enhance cooperation between the major gas producing countries, including Algeria, Brunei, Iran, Indonesia, Malaysia, Nigeria, Oman,

---

105 Samsung Heavy Industries, Geoje Shipyard, http://www.shi.samsung.co.kr/Eng/pr/shipyard01.aspx.

106 Joel Parshall, *Shell to Build World's First Floating LNG Facility*, Journal of Petroleum Technology, Sept. 2011, http://www.spe.org/jpt/print/archives/2011/09/11FLNG.pdf.

107 Ibid.

108 Ibid.

109 Gas Exporting Countries Forum, www.gecforum.org.

Qatar, Russia, Norway and Turkmenistan. Between 2001 and 2008, the GECF operated as a loose association with seemingly very limited coordination and without a clear mandate. Although ministerial meetings were held yearly, the countries participating in the meetings tended to vary, with little detail about the meetings made public. As an informational organization without a secretariat, the GECF also received very little international recognition although many speculated whether the GECF would operate as a "Gas OPEC".[110]

This speculation continued throughout the early to mid-2000s, with most analysts concluding that the GECF was unlikely to become an OPEC-like cartel, for a number of reasons. First, the GECF had too many members with competing interests such that the ability to constrain new supply projects in the near term was unlikely. Second, it was likely to take a decade or more before the GECF members could collectively assert sustainable monopoly power over the world's gas markets, although a "swing" gas producer such as Qatar or Russia could potentially influence markets by manipulating supplies. Lastly, since the world's natural gas markets were still primarily regional and lacking a single pricing structure, it would be difficult for any cartel to significantly influence price.[111] In addition to these factors, much of the early questions regarding the motives of the GECF appeared to be driven by the fact that since very little official information was made available by the GECF, its mission and motives were ripe for speculation.[112]

12.5.2 The GECF Becomes an International Organization

In December 2008, after seven years of operating as an informal organization, the GEFC became a full-fledged international organization with a secretariat and headquarters located in Doha, Qatar, upon the signing of two documents, which, according to the GECF, form the legal basis

---

110 Ronald Soligo and Amy Myers Jaffe, *Market Structure in the New Gas Economy: Is Cartelization Possible?* Prepared for the Geopolitics of Natural Gas Study, a joint project of the Program on Energy and Sustainable Development at Stanford University and the James A. Baker III Institute for Public Policy of Rice University (May 2004).

111 Ibid. and see Hadi Hallouche, *The Gas Exporting Countries Forum: Is it Really a Gas OPEC in the Making?*, Oxford Institute for Energy Studies (June 2006); Monika Ehrman, *Competition is a Sin: An Evaluation of the Formation and Effects of a Natural Gas OPEC*, 27 Energy L.J. 175 (2006).

112 Hadi Hallouche, *The Gas Exporting Countries Forum: Is it Really a Gas OPEC in the Making?*, Oxford Institute for Energy Studies (June 2006).

for the GECF – the Agreement on the functioning of the GECF and the Statute of the GECF.[113]

The GECF became operational on October 1, 2009, when five member countries deposited their ratification instruments: Russia, Qatar, Libya, the Republic of Trinidad and Tobago, and Algeria. In March 2010, two more countries joined the GECF – Egypt and the Republic of Equatorial Guinea, followed by Venezuela, Nigeria, Iran and Bolivia.[114] In addition to the 11 regular members, three other countries have observer status at the GECF – Kazakhstan, the Netherlands, and Norway. Together, the 11 members of the GECF control about 70 percent of the world's natural gas reserves.[115]

Since 2008, the GECF has continued to evolve as an organization and expand its influence in the world's natural gas markets. These efforts have been aided by the appointment of the Russian national Leonid Bokhanovsky as Secretary General in late 2009. In October 2010, the GECF was invited to participate in the JODI Gas Data Transparency Conference. JODI stands for the Joint Organisations Data Initiative, which was first launched for oil in 2001 with the goal of increasing transparency in world oil markets. The initiative was later extended to gas markets. The JODI Gas Data Transparency Conference invited the GECF in recognition of its work in establishing a data collection mechanism and its increased role in joining inter-organizational efforts to enhance gas data transparency.[116] The IEA has also recognized the cooperation of the GECF in the JODI Gas initiative.[117]

While not necessarily significant events, these two instances indicate that the GECF is stepping up to take a more active role in international dialogue concerning global gas markets. Moreover, these events also show that other international organizations are beginning to recognize

---

113 Presentation of HE Mr Leonid Bokhanovsky, Secretary General of the GECF, to the 12th IEF Ministerial & 4th IEBF, Cancun, Mexico, March 30–31, 2010, available at http://www.ief.org/Events/Documents/CancunPresentationBook.pdf.

114 Ibid.

115 Gas Exporting Countries Forum, www.gecforum.org.

116 Joint Organisations Data Initiative (JODI), Gas Data Transparency Conference, Oct. 26, 2010, Moscow, available at http://www.jodidata.org/events. More recently, the GECF participated in the second JODI Transparency Conference held in Doha, Qatar, in May 2012. *Gas Data Meeting Opens in Doha*, Gulf Times, May 23, 2012, http://www.gulf-times.com/site/topics/printArticle.asp?cu_no=2&item_no=507483&version=1&template_id=36&parent_id=16.

117 IEA Energy Statistics Cooperation, Presentation of Jean-Yves Garnier, Head, Energy Statistics Division, *The Role of the IEA in Strengthening International Cooperation*, IEA, Paris, April 13–14, 2011.

the GECF as a legitimate organization. As such, it is useful to once again review the GECF's mission and objectives with a view towards determining what influence the organization might seek to exert over global gas markets going forward.

12.5.3 The GECF's Mission and Objectives

The stated objectives of the GECF are to "support the sovereign rights of member countries over their natural gas resources and their abilities to independently plan and manage the sustainable, efficient and environmentally conscious development, use and conservation of natural gas resources for the benefit of their peoples."[118] These objectives are promoted through the exchange of experience, views and information focused on the following key goals of the GECF:

- to develop and implement necessary steps to guarantee member-countries derive the most value from their gas resources;
- to protect sovereign rights of member-countries over their natural gas reserves;
- to exploit and preserve natural gas reserves to benefit of their nations;
- to promote integration of gas markets and its stability.[119]

12.5.4 Is the GECF a Cartel Designed to Influence Global Gas Prices?

Despite the rather benign stated objectives, the GECF's actions and statements have always been scrutinized for indications of cartel-like behavior, with some members of the GECF more likely to be willing to try to influence global gas markets than others. For example, in March 2010, Russian's Prime Minister Vladimir Putin indicated that Russia expects the GECF to become the "gas equivalent" of the OPEC oil cartel and an "effective tool for coordinating the international gas industry."[120]

In April 2010, Algeria proposed that GECF member countries reduce

---

118 Gas Exporting Countries Forum, www.gecforum.org.

119 Presentation of HE Mr Leonid Bokhanovsky, Secretary General of the GECF, to the 12th IEF Ministerial and 4th IEBF, Cancun, Mexico, March 30–31, 2010, available at http://www.ief.org/Events/Documents/CancunPresentationBook.pdf.

120 Aleksey Nikoiskyi, *Russia Says GECF Gas Forum Should Become Effective Market Tool*, RIA Novosti, March 24, 2010, http://en.rian.ru/world/20100324/158301797.html.

gas exports to restore balance in global gas markets.[121] Other members, most notably Qatar, seemed to shy away from this suggestion, with the GECF ministers as a whole agreeing that the overall aim was to support the linking of gas to oil parity and also that there should not be gas-on-gas competition between pipeline gas and LNG.

In October 2010, it was reported that the GECF was studying OPEC's experience as the GECF attempted to build a model for the global natural gas market. The basis for the model would be the gathering of information and data swaps between the GECF countries, with the model itself being "based on a positive practice of creation of the oil market . . . worked out with participation of Opec and International Energy Agency (IEA)."[122]

In December 2010, Russia's Energy Minister Sergei Shmatko indicated that the GECF would become fully operational in 2010 and that it would play a key role in coordinating the market for LNG, especially since its membership consists of the world's major LNG-producing countries.[123]

12.5.5 The GECF Seeks to Maintain Oil-linked Prices

For the moment, the main goal of the GECF appears to be maintaining the traditional gas–oil parity. Prior to the global economic crisis of 2008, there was a growing spot trade of LNG – largely due to high LNG prices in Asia and Europe. Following the economic crisis in 2008–09, oil and gas demand dropped – causing prices to drop as well. For the most part, 2010–11 was characterized by low natural gas prices and rising oil prices. This led to a "de-linking" of the historical gas–oil parity of approximately 10–1 (for example, at $100 per barrel of oil, natural gas should trade at $10).

Many have questioned whether this de-linking is a trend that is likely to continue and, if so, whether the traditional oil-linked gas markets – Asia and Europe – will seek to renegotiate existing long-term oil-linked contracts or otherwise take advantage of lower spot market prices of natural gas. The IEA has questioned whether this de-linking is a "separation" or a "divorce."

---

[121] IEA WEO-2010, p. 196, "The GECF seeks oil price parity and ponders how to achieve it."

[122] Gulf Times, *Gas Exporters Study Opec for LNG Market Model*, Oct. 25, 2010, available at http://www.gulf-times.com/site/topics/article.asp?cu_no=2&item_no=394360&version=1&template_id=48&parent_id=28.

[123] Aleksey Nikoiskyi, *Russia Says GECF Gas Forum Should Become Effective Market Tool*, RIA Novosti, March 24, 2010, http://en.rian.ru/world/20100324/158301797.html.

Although it remains to be seen longer term, the global gas markets in 2011 appeared to be tightening as Asian demand reached record highs. For example, by June 2, 2011, when the GECF ministers met for their 12th Meeting of the GECF in Cairo, Egypt, the supply glut had largely dissipated with the global gas markets tightening due to Japan's nuclear crisis, Middle Eastern turmoil, and rising demand. Accordingly, the GECF stated that there was no need to consider cutting output of natural gas because any excess supply would soon be taken due to growing demand. This appeared to have played out as predicted by the GECF in 2011.

12.5.6 Will the GECF Evolve into a Cartel Like OPEC?

There is no doubt that the GECF has "come of age" in the past two years although it is still not as transparent an organization as many would like. This may change over time as the organization continues to take shape. In the meantime, the GECF is increasing its participation as an international organization in a number of forums and this trend is likely to continue in the future.

For its part, the US has largely viewed the GECF as an "ineffective" organization that is unlikely to significantly affect gas markets. Moreover, US Congressional analysts have expressed the view that the "the current structure of natural gas markets (that is, long-term contracts and pipelines connecting individual sellers to specific buyers) is not conducive to supply or price manipulation, and significant changes would need to be made to how natural gas is bought and sold before the GECF could have influence."[124]

For the moment, it thus appears that the GECF's role is merely to facilitate information sharing and dialogue among the main gas exporters. The main question, however, is whether the GECF will evolve over time into an OPEC-like cartel. Although the GECF could seek a more proactive role on market and pricing related issues in the future, its ultimate ability to operate as a cartel is probably limited by the relative ease with which other fuels can substitute for natural gas.[125] Moreover, a gas cartel makes less sense than a cartel for oil since gas exports generally require the construction of capital-intensive pipelines or LNG facilities anchored

---

[124] Michael Ratner, *Global Natural Gas: A Growing Resource*, Cong. Research Serv., Report R41543, Dec. 22, 2010, http://assets.opencrs.com/rpts/R41543_20101222.pdf.

[125] IEA WEO-2009.

by long-term contracts. As such, the main GECF member countries, such as Qatar, that have large infrastructure projects and long-term supply contracts are less likely to favor cartel behavior than other members of the GECF. In addition, the recent success of US shale gas and the possibility of global shale gas development makes it less likely that the GECF could influence price as an increasing number of countries now have the potential to become major gas producers.

12.6 LNG AS A SHIPPING AND VEHICLE FUEL

More than ever before, LNG has the opportunity to win the race to fuel the heavy-duty fleets of the 21st century as ships, trains, trucks and buses search for low-cost, cleaner-burning and abundant fuels. In some areas, such as shipping, the switch to LNG fuels is largely being driven by the enforcement of IMO (International Maritime Organization) emission standards. As the world continues to find new solutions to reduce emissions, LNG as a fuel is likely to gain even wider acceptance in the coming decades as a fuel for more and more types of vehicles.

12.6.1 LNG as a Shipping Fuel

Historically, stringent environmental and emission regulations were not applied to the global shipping industry. However, the significant increase in global trade over the past few decades has resulted in increased emissions of air pollutants and greenhouse gases from the shipping industry, which now carries almost 90 percent of global trade by weight. As a result, increased attention has been focused on the shipping industry and in an effort to join the global community in emission reduction efforts the IMO recently adopted new regulations to reduce air pollution and greenhouse gas (GHG) emissions.[126]

By way of background, the IMO regulates pollution from ships through the International Convention on the Prevention of Pollution from Ships, known as MARPOL 73/78. In 1997 a new annex (Annex VI) was added to MARPOL 73/78 to regulate airborne emissions from ships, and specifically SO_x, NO_x, ODS, and VOC.[127] MARPOL Annex VI entered into

[126] International Maritime Organization (IMO), Marine Environment Protection Committee, 58th Session, July 28, 2008 (MEPC 58/4/5), "Prevention of Air Pollution from Ships."

[127] International Maritime Organization (IMO), *Air Pollution and Greenhouse*

force on May 19, 2005, and a revised Annex VI with tighter emission limits was adopted in October 2008 and entered into force on July 1, 2010.[128]

In general, the revised Annex VI sets limits on NO_x and SO_x emissions from ship exhausts, and prohibits deliberate emissions of ozone depleting substances. The IMO emission standards are commonly referred to as Tiers I, II, and III and the requirements are divided into two categories: global requirements, and more stringent requirements applicable to ships in geographically limited Emission Control Areas (ECAs). An ECA can be designated for SO_x, and PM, or for NO_x, or for all three types of emissions from ships. The Baltic Sea and the North Sea are already enforced as ECAs. The US coastlines will be enforced as an ECA starting in 2012 and other areas are under consideration.[129]

In order to comply with the new IMO SO_x and NO_x emission reduction limits, shipowners currently have three options if they want to continue sailing in ECAs starting from 2015. They can switch to a low-sulfur fuel, install an exhaust gas scrubber, or switch to LNG as a shipping fuel.[130] While switching to a low-sulfur fuel requires only minor modifications to the ship's fuel system, the availability of low-sulfur fuel is already limited and rising demand is expected to increase its price. An exhaust gas scrubber can be installed to remove sulfur from the engine exhaust by using chemicals or saltwater but scrubbers require significant ship alterations. Moreover, scrubbers also increase power consumption, which leads to an increase in CO_2 emissions.[131]

Accordingly, the third alternative, to use LNG as a fuel, is emerging as potentially the best option to meet the new IMO regulations. Since LNG is a clean-burning fuel, no additional abatement measures are required in order to meet the ECA requirements for ships fueled with LNG.

Gas Emissions, http://www.imo.org/OurWork/Environment/PollutionPrevention/AirPollution/Pages/Default.aspx.

[128] Annex 13, Resolution MEPC.176(58), Adopted on 10 October 2008, Amendments to the ANNEX of the Protocol of 1997 to Amend the International Convention for the Prevention of Pollution from Ships, 1973, As Modified by the Protocol of 1978 Relating Thereto (Revised MARPOL Annex VI), available at http://www.imo.org/ourwork/environment/pollutionprevention/airpollution/documents/23-add-1.pdf.

[129] Det Norske Veritas (DNV), *Greener Shipping in North America*, Feb. 2011, http://www.dnv.com/resources/reports/greener_shipping_north_america.asp. DNV is a global provider of risk management services with the purpose of safeguarding life, property and the environment. DNV has done extensive work and analysis on LNG as a shipping fuel.

[130] Ibid.

[131] Ibid.

However, an LNG-fueled ship requires purpose-built or modified engines and a sophisticated system of special fuel tanks, a vaporizer, and double-insulated piping. A key challenge has been available space for cylindrical LNG fuel tanks on board ships but new hull integrated tanks are expected to resolve this issue.[132]

LNG as a shipping fuel has the added advantage of being the only option that offers any noticeable reduction in carbon dioxide (CO_2) emissions. According to the IMO, in 2007 international shipping was estimated to have contributed about 2.7 percent of the global emissions of CO_2. In recognition of the need for the industry to reduce its CO_2 emissions, the IMO has been pursuing the control of GHG emissions from international shipping for a number of years. In July 2011, the IMO adopted mandatory technical and operational energy efficiency measures, which will significantly reduce the amount of CO_2 emissions from international shipping.[133]

While the adoption of the IMO GHG measures is a significant step towards the regulation of GHG emissions, there is some indication that some states, most notably the EU, consider them insufficient and may establish even stricter standards.[134] While it is a bit early to predict how the GHG emission regulations will eventually play out, even stricter standards would bode well for LNG fuel emerging as the best of the three options. Indeed, some industry experts are already advocating for LNG since it may offer the best economics to shipowners and the best environmental effect to the public.[135]

Nonetheless, the application of LNG-fueled ships has developed more slowly than some expected, given the "superior qualities of LNG compared to other fuels,"[136] with most of the developments thus far occurring

---

[132] Ibid.

[133] IMO, *Breakthrough at IMO: Adoption of the First Ever Global and Legally Binding CO_2 Standard for an Industry Sector*, available at http://www.imo.org/OurWork/Environment/PollutionPrevention/AirPollution/Documents/GHG%20Flyer%20WEB.pdf.

[134] DNV, *CO_2 Emissions from Ships: the Latest IMO Regulatory Developments and What They Mean for Bulk Carriers*, available at http://www.dnv.com/industry/maritime/publicationsanddownloads/publications/updates/bulk/2011/3_2011/co2emissionsfromshipsthelatestimoregulatorydevelopmentsandwhattheymeanforbulkcarriers.asp.

[135] Det Norske Veritas (DNV), *Greener Shipping in North America*, Feb. 2011, http://www.dnv.com/resources/reports/greener_shipping_north_america.asp.

[136] DNV, *Greener Shipping in the Baltic Sea*, June 2010, available at http://www.dnv.com/industry/maritime/publicationsanddownloads/publications/updates/lng/2010/01_2010/balticsea.asp.

in Norway, which currently has more than 20 LNG-fueled ships.[137] While initially the focus has been on short sea shipping involving relatively small ships, such as LNG-fueled ferries, more recent attention has been focused on the international ocean-going fleet, with DNV recently demonstrating the feasibility of LNG-fueled large ships through concept studies of a container ship and a VLCC-size oil tanker.[138]

One potential limitation to the widespread use of LNG as a fuel is that it requires an LNG fuel supply infrastructure and a large fleet of LNG-fueled ships.[139] These two are mutually interdependent since shipowners will not invest in LNG-fueled ships until an LNG fuel supply infrastructure is in place, and LNG fuel suppliers will not invest in infrastructure until there is a large fleet of LNG-fueled ships.[140]

More recently, it appears that LNG as a ship fuel is gaining some momentum as more key players throughout the shipping industry are analyzing the benefits of LNG-fueled ships, either as conversions or new builds. A recent survey of the LNG-fueled marine industry indicated that there are 48 non-LNG carrier ships fueled with LNG.[141] The more recent and greater interest in LNG as a shipping fuel seems to be creating some momentum that may eventually increase its use and the speed at which LNG fuel is introduced around the world. For example, at least one expert has predicted that 500 LNG-fueled ships will be on order by 2015 and several thousand by 2020.[142]

12.6.2 LNG as a Vehicle Fuel

Natural gas as a vehicle fuel can be used in a compressed form (CNG) or liquid form (LNG) in virtually the whole spectrum of vehicles, including

---

137 Det Norske Veritas (DNV), *Greener Shipping in North America*, Feb. 2011, available at http://www.dnv.com/resources/reports/greener_shipping_north_america.asp.

138 Ibid.

139 DNV, *Greener Shipping in the Baltic Sea*, June 2010, http://www.dnv.com/industry/maritime/publicationsanddownloads/publications/updates/lng/2010/01_2010/balticsea.asp.

140 Ibid.

141 Zeus Development Corporation, *LNG-Fueled Marine Industry at 48 Ships and Growing, Zeus Conference to Examine Implications*, Dec. 13, 2011, Marketwire, available at www.marketwire.com. Additional information about the LNG-fueled shipping industry can be found at Zeus Development Corporation, http://www.zeusintel.com/WLNGF2012.

142 DNV, *LNG as Ship Fuel is Gaining Momentum*, available at http://www.dnv.com/press_area/press_releases/2011/lngasshipfuelisgainingmomentum.asp.

cars, vans, buses, trucks, trains, and even airplanes. Due to the fact that natural gas has low energy density at atmospheric pressure and temperature compared to liquid fuels, compression or liquefaction is needed to reach an acceptable vehicle range. In general, light-duty vehicles, such as passenger cars, use CNG or are converted gasoline-powered vehicles. Heavy-duty vehicles (HDVs) require more energy to run and thus tend to use LNG to maintain an acceptable range.[143]

Despite the fact that natural gas vehicle (NGV) technology has been around for a long time and is well established in some countries, worldwide the use of natural gas in the road transport sector remains negligible but has the potential to grow in the coming years.[144] According to the IEA, in 2008 there were an estimated 9.6 million NGVs on the roads, mainly in Pakistan, Argentina, Brazil, India, Iran and Italy. The majority of NGVs are passenger cars but buses account for more of the consumption.[145] In most cases, NGVs were introduced as a means of monetizing abundant local supplies, but more recently the environmental benefits of natural gas over gasoline and diesel have increased demand for natural gas in the transportation sector. This is especially the case in some large cities throughout the world that have begun to use natural gas for all public vehicles such as buses.[146]

Despite the potential environmental benefits of NGVs, expanding the NGV fleet globally has faced numerous barriers throughout the years. One barrier is that natural gas must be stored in cylinders on the vehicle, which reduces overall vehicle storage space. However, the more significant barrier is the "chicken-and-egg" problem identified in the section above regarding LNG as a shipping fuel – the absence of an existing fuel distribution network has tended to discourage the uptake of NGVs. For this reason, public buses and other fleet vehicles have tended to dominate the use of NGVs. It is possible, however, that more stringent emission standards could encourage faster deployment of NGVs, especially in countries with abundant natural gas resources and low prices.[147]

This is starting to take shape in the United States as a result of the abundance of low-priced gas resulting from the shale gas revolution in recent years. The US is a particularly interesting and relevant market

---

143 IEA Working Paper, *The Contribution of Natural Gas Vehicles to Sustainable Transport* (2010) available at http://www.iea.org/publications/free_new_Desc.asp?PUBS_ID=2304.

144 IEA WEO-2009.

145 Ibid.

146 Ibid.

147 Ibid.

Table 12.3 Typical economics of an LNG truck in the US in the absence of tax credits and subsidies

Natural gas price (US$/MBtu)	5.00
LNG costs (US$/gallon diesel equivalent)	2.03
Diesel fuel consumption (gallons/y)	20,000
Diesel price (US$/g)	3.00
Fuel costs savings (US$/y)	19,384
Incremental costs LNG truck (US$)	40,000
Payback period (y)	2.00
Savings over seven-year lifetime (US$)	95,688

Source: IEA, Neandross (2009)

to study in terms of LNG use since the approach to NGVs in the US is mostly fleet-driven with a focus on HDVs that typically use LNG. While the availability of natural gas passenger cars is limited to just a couple of models, there are a wide variety of HDV models now available with trucks that can reach a range up to 600 miles on LNG.[148]

Heavy-duty natural gas vehicles are seen in many different applications in the United States, including specialized vehicles in marine ports and airports, school buses, waste collection, transit buses and long-haul trucks. Non-road applications including railroad may also be promising. One reason for this is the fact that the economics of an LNG vehicle seem to make sense. While incremental costs and savings for an NGV truck vary, Table 12.3 summarizes some typical data as reported by the industry and illustrates that, even in the absence of tax credits and subsidies, the fuel price differential is sufficient to make the economics of NGVs potentially attractive for fleet owners.

Even though the economics look quite attractive for US fleet owners, legislation may also be on its way in the US that would provide incentives for the use of natural gas in transportation to both vehicle owners and infrastructure builders. In November 2011, the US Senate introduced the NAT GAS Act of 2011, which would provide a five-year extension of tax credit incentives for the purchase of NGVs. The bill would also encourage manufacturers to produce dedicated NGVs. The bi-partisan bill was introduced to accelerate the production and use of natural gas-fueled vehicles

[148] IEA Working Paper, *The Contribution of Natural Gas Vehicles to Sustainable Transport* (2010), available at http://www.iea.org/publications/free_new_Desc.asp?PUBS_ID=2304.

and the legislation would expand the NGV tax incentive to help offset an NGV's incremental cost, subject to caps depending upon vehicle size.[149]

According to Clean Energy,[150] a leading provider of natural gas fuel for transportation in North America, the backing by the US Congress is critical to helping the owners of the approximately eight million heavy-duty trucks in the US convert from diesel to cleaner-burning, less expensive natural gas. In some US markets, natural gas is $1.50 per gallon less expensive than diesel or gasoline, which would significantly reduce costs for vehicle and fleet owners. Moreover, natural gas could reduce "greenhouse gas emissions up to 30% in light-duty vehicles and 23% in medium to heavy-duty vehicles." Lastly, and driven primarily by the increase in shale gas development, "natural gas is a secure North American energy source with 98% of the natural gas consumed produced in the U.S. and Canada."[151]

Aside from the economics and policy support that may be needed to spur investment in natural gas transportation, other factors are also important such as having "champions" – major companies in a certain industry who take the lead will cause others to follow.[152] The rise of shale gas in the US combined with the increased availability of LNG means that natural gas producers in the US find themselves more and more looking for demand in an oversupplied market. So, obviously, any new market presenting itself can certainly count on a healthy interest from gas producers, who can help to drive NGV market development by either investing directly or actively supporting lobbying or PR/marketing campaigns.[153] This development is illustrated by a number of recent industry activities.

For example, EnCana's Natural Gas Vehicles Drive Project is "converting several fleet vehicles to natural gas, purchasing natural gas-powered Honda Civic GX vehicles for employee use and embarking on a consumer- and industry-focused education campaign about natural gas as a transportation fuel."[154] More recently, one of the largest producers of natural

---

[149] *Clean Energy Supports 2011 Natural Gas Act*, LNG World News, Nov. 21, 2011, available at http://www.lngworldnews.com/usa-clean-energy-supports-2011-nat-gas-act/.

[150] Clean Energy, http://www.cleanenergyfuels.com.

[151] *Clean Energy Supports 2011 Natural Gas Act*, LNG World News, Nov. 21, 2011,http://www.lngworldnews.com/usa-clean-energy-supports-2011-nat-gas-act/.

[152] IEA Working Paper, *The Contribution of Natural Gas Vehicles to Sustainable Transport* (2010), available at http://www.iea.org/publications/free_new_Desc.asp?PUBS_ID=2304.

[153] Ibid.

[154] Ibid., citing http://coloradoenergynews.com/2009/07/natural-gas-vehicle-program-unveiled-by-encana.

gas in the US, Chesapeake Energy, has emerged as a primary champion of natural gas vehicles with its July 2011 announcement that it will invest $1 billion over ten years in natural gas vehicle technologies. According to some, Chesapeake's announcement demonstrates that natural gas producers believe prices will remain low long-term unless they can find new users for the fuel.[155]

Chesapeake and others in the industry, including Mr T. Boone Pickens, the billionaire investor, have long argued that the US should burn more natural gas in cars and trucks because it is cleaner and cheaper than oil and is domestically produced.[156] But natural gas vehicles have made only limited inroads in the transportation sector, in large part because of the chicken-and-egg problem: "Drivers don't want to buy natural-gas vehicles until there are plenty of places where they can fill them up, but service stations don't want to invest in natural-gas infrastructure until there are more drivers who would use it."[157]

Chesapeake hopes to resolve this dilemma by funding the construction of about 150 natural gas filling stations on major highways through a $150 million investment in Clean Energy Fuels, which will help build "America's natural-gas superhighway."[158] The initial focus of the efforts to build the US natural gas superhighway will be on fueling stations for heavy-duty vehicles, which should benefit LNG as a vehicle fuel.[159]

In recent years, the type of project undertaken in the US demonstrates the focus on fleet owners and HDVs and this focus is likely to continue for the near future. For example, large transportation company JB Hunt secured $19 million in state and federal funding for 262 LNG trucks and two refueling stations, and logistics company Cal Cartage received $12

---

155 Ben Casselman, *Chesapeake will Invest in Uses for Natural Gas*, Wall Street Journal, July 12, 2011, http://online.wsj.com/article/SB10001424052702304584404576440332696732422.html.

156 Ibid.

157 Ibid.

158 Ibid. It is worth noting that environmental groups in the US have been somewhat skeptical of natural gas vehicle proposals, arguing that while converting from oil to natural gas could provide some environmental benefits, gas drilling has its own environmental risks. According to a vehicle analyst for the Natural Resources Defense Council, "Improving vehicle efficiency is the cleanest, fastest and cheapest way to cut our oil consumption and carbon pollution from transportation."

159 See Clarissa Kell-Holland, *Could LNG be a Lifeline for Truckers?* Land Line Magazine, March/April 2009, noting that trucking companies are increasingly using LNG to fuel their fleets and that more LNG infrastructure is being built.

million in funding for 132 LNG trucks (currently they have 400 LNG trucks) plus $1 million for two LNG fuel stations.[160]

In the US, more LNG refueling stations and networks have been emerging, albeit slowly, over the past few years, with station builder Clean Energy recently reporting that it currently builds one station per week.[161] In terms of LNG infrastructure there are 60 fuel stations, eight production plants plus landfill LNG sites. The "stationary" characteristic of the LNG market is illustrated by the fact that 54 percent of the market consists of transit buses, 30 percent refuse vehicles, 12 percent port applications and only 4 percent road use, although the industry expects the latter to grow substantially in the next few years as it targets the fleet of, in total, three million trucks in the United States.[162]

California has a high concentration of LNG and CNG stations, with many in and around Los Angeles, which converted its refuse trucks fleet to LNG a few years ago. To illustrate the costs involved, an LNG project for the city of San Bernardino received $1.7 million in grants, of which $1.23 million was used for the LNG refueling station and $492,000 was used to purchase 20 LNG trucks.[163]

While it remains to be seen how many fleets will convert to LNG, the initial results look promising. For example, in April 2011, the US Department of Energy (DOE) announced the National Clean Fleets Partnership, which is a public–private effort to help large companies reduce diesel and gasoline use.[164] Each of the partners involved has already shown a commitment to using clean energy technologies and reducing petroleum use. For example, Ryder announced it will use hundreds of heavy-duty LNG trucks and build out additional infrastructure, such as LNG fueling stations and two additional maintenance facilities, to support them. This project alone is expected to save 1.5 million gallons of diesel.[165]

---

[160] IEA Working Paper, *The Contribution of Natural Gas Vehicles to Sustainable Transport* (2010), http://www.iea.org/publications/free_new_Desc.asp?PUBS_ID=2304.

[161] Ibid.

[162] Ibid.

[163] Ibid.

[164] Fact Sheet, *National Clean Fleets Partnership*, April 1, 2011, available at http://energy.gov/articles/fact-sheet-national-clean-fleets-partnership.

[165] *Six New Corporations Join the National Clean Fleets Partnership*, July 7, 2011, available at http://energy.gov/articles/six-new-corporations-join-national-clean-fleets-partnership.

Conclusion: the future looks bright for LNG as a fuel for the 21st century

As the preceding chapters hopefully demonstrate, the LNG business is both exciting and challenging, with the past decade being one of the most dynamic periods in time for the LNG industry. The start of the millennium saw a sharp increase in LNG trade as major projects came online to supply the world's growing economies. While demand for all energy dropped sharply as the global economic recession took hold in late 2008 to 2009, the industry proved resilient by rebounding in 2010 with a record 21 percent growth rate![1]

In 2011, the industry again showed its strength by stepping up to supply Japan with record amounts of LNG after Japan's nuclear capacity was shut down following the March 2011 earthquake and tsunami. As this book goes to print, 2011 LNG trade data are just starting to be released with preliminary reports indicating that 2011 was another strong year of growth for the LNG industry, with LNG trade up 9.4 percent compared with 2010.[2]

Whereas the start of the 21st century saw the emergence of the US as a major LNG importer, the US is now expected to become an exporter by the end of this decade with the potential to become a major player in the LNG markets. While the shale gas revolution has thus far been limited to North America, more and more countries are assessing their resources and global shale gas development could take off over the next decade and significantly alter global gas markets. Indeed, as this book goes to print, there are an increasing number of opportunities[3] and

---

[1] GIIGNL, The LNG Industry in 2010, www.giignl.org.

[2] GIIGNL, The LNG Industry in 2011, www.giignl.org.

[3] In July 2012, ICIS Heren announced the world's first cleared LNG swap traded on July 16, 2012, with the contract settling on the ICIS East Asia Index (EAX) for physical LNG. The deal was done for September 2012 at $13.90/MMBtu. Up to this point, there has been a very small over-the-counter swaps market for LNG, with a limited pool of counterparties and no clearing services, so this swap could signify more active clearing and broking which could accelerate liquidity growth in the LNG market in the coming years. *World's First Cleared*

challenges[4] for the global gas industry and there will no doubt be many more by the time this book is published.

Regardless of the issues that may arise, however, over the past decade the LNG industry has adapted and evolved to accommodate various market forces, geopolitical realities and technological advances and there is no doubt that the industry will continue to seize the opportunities and meet the challenges of the 21st century.

LNG Swap Trades – Settles on ICIS East Asia Index, ICIS Heren, www.icis.com/energy.

[4] In May 2012, Argentina announced the seizure and nationalization of a 51 percent stake in Argentina's energy company, YPF SA, from majority stakeholder Repsol YPF SA. While Repsol has canceled LNG cargoes to Argentina, the ultimate impact on global gas markets remains unclear. Alejandro Lifschitz, *Exclusive: Repsol Cancels LNG Cargoes to Argentina*, Reuters, May 18, 2012, http://www.reuters.com/article/2012/05/18/us-repsol-argentina-idUSBRE84H0SI20120518 (noting that Argentina relies on LNG imports to meet 20–30 percent of domestic natural gas consumption and that Repsol had agreed to supply Argentina with 10 LNG cargoes in 2012). See also Zaida Espana, *IEA Says Firms May Avoid Argentina after YPF Seizure*, Reuters, May 11, 2012, http://www.reuters.com/article/2012/05/11/us-iea-repsol-idUSBRE84A0HO20120511.

APPENDICES

APPENDIX A WORLDWIDE LIQUEFACTION PLANTS

Country	Location	Status	No. of trains	Liquefaction capacity 10^6 t per year*	Liquefaction capacity bcm**	Storage tanks*	Storage capacity m^3*	Owners	Start-up year and train
ATLANTIC BASIN									
Algeria	Arzew GL 4Z	O	3	0.93	1.5	3	33,000	Sonatrach	1964
Algeria	Arzew GL 1Z	O	6	8.19	11.2	3	300,000	Sonatrach	1978
Algeria	Arzew GL 2Z	O	6	7.98	10.9	3	300,000	Sonatrach	1981
Algeria	Arzew GL 3Z (Gassi Touil)	C	1	4.7	6.4			Sonatrach	2012
Algeria	Skikda GL 1K	O	3	3.13	4.3	5	308,000	Sonatrach	1972
Algeria	Skikda	C	1	4.5	6.1			Sonatrach	2012–13
Angola LNG	Soyo	C	1	5.2	7.1		360,000	Sonangol, Chevron, ENI, Total, BP	2012

Cameroon	Kribi	A	1	3.5	4.8			Société Nationale des Hydrocarbures (NOC), GDF Suez	2016
Egypt (Segas)	Damietta	O	1	5	6.5	2	300,000	SEGAS	2005
Egypt LNG T1&2	Idku	O	2	7.2	9.8	2	280,000	Egyptian LNG (BG, Petronas, EGPC, EGAS, GDF Suez)	2005
Equatorial New Guinea	Bioko Island	O	1	3.7	4.6	2	272,000	EG LNG (Marathon, Mitsui, Sonagas, Marubeni)	2007
Equatorial New Guinea	EG LNG T2	P			6			EG LNG (Marathon, GEPetrol. Mitsui, Marubeni, E.On, Union Fenosa Gas)	2016
Libya	Marsa el Brega	O	3	0.6	1	2	96,000	Sirte Oil Co. (NOC)	1970
Libya	Mellitah	P			5.2			Eni	2016
Nigeria	Bonny Island	O	3	9.6		3	252,600	Nigeria LNG (NNPC, Shell, Total, ENI)	1999–2000

Country	Location	Status	No. of trains	Liquefaction capacity 10^6 t per year*	Liquefaction capacity bcm**	Storage tanks*	Storage capacity m^3*	Owners	Start-up year and train
ATLANTIC BASIN									
Nigeria	Bonny Island	O	2	8.1		"	"	Nigeria LNG (NNPC, Shell, Total, ENI)	2006
Nigeria	Bonny Island	O	1	4		1	84,200	Nigeria LNG (NNPC, Shell, Total, ENI)	2008
Nigeria	NLNG Train 7	P	1	8.4	10.9			NNPC, Shell, BP, Total	2013+
Nigeria	Brass LNG	P	2	10	13.6			NNPC, ENI, ConocoPhillips, Total)	2014
	Olokola LNG	P	4	20				NNPC, Chevron, Shell, BG	2013+
Norway	Hammerfest	O	1	4.3	5.6	2	250,000	Snøhvit AG (StatoilHydro, Total, GDF Suez, RWE-DEA, Hess	2007
Trinidad & Tobago	Atlantic LNG Point Fortin T1	O	1	3.3	4.5	4	520,000	Atlantic LNG (BP, BG, GDF Suez, Repsol, NGC)	1999

Trinidad & Tobago	Atlantic LNG Point Fortin T2	O	2	6.6	9	“	“	Atlantic LNG (BP, BG, Repsol)	2002 Train 2 2003 Train 3
Trinidad & Tobago	Atlantic LNG Point Fortin T3	O	1	5.2	7.1	“	“	BP, BG, Repsol, NGC	2006 Train 4
MIDDLE EAST									
Abu Dhabi	Das Island	O	3	5.6	7.9	3	240,000	ADGAS (ADNOC, BP, Total, Mitsui)	1977 Trains 1 and 2 1994 Train 3
Oman	Qalhat	O	2	7.1	9.8	2	240,000	Oman LNG (Oman govt., Shell, Total, Korea LNG Mitsubishi, Mitsui, Partex, Itochu)	2000 Trains 1 and 2
Oman	Qalhat	O	1	3.6	4.9			Qalhat LNG (Oman govt., Oman LNG, Itochu, Mitsubishi, Union Fenosa Gas, Osaka Gas)	2006

Country	Location	Status	No. of trains	Liquefaction capacity 10^6 t per year*	Liquefaction capacity bcm**	Storage tanks*	Storage capacity m^3*	Owners	Start-up year and train
MIDDLE EAST									
Qatar	Ras Laffan (Qatargas 1 - T1&T2	O	2	6.4		4	340,000	Qatargas (QP, ExxonMobil, Total, Marubeni, Mitsui)	1999
Qatar	Ras Laffan (Qatargas 1 - T3)	O	1	3.1		"	"	Ras Laffan LNG (QP, ExxonMobil, total, Marubeni, Mitsui)	1999
Qatar	Ras Laffan (Qatargas 2- T4)	O	1	7.8	10.6	8	1,160,000	QP, ExxonMobil)	2009
Qatar	Ras Laffan (Qatargas 2- T5)	O	1	7.8	10.6	"	"	Qatargas II (QP, ExxonMobil, Total)	2009
Qatar	Ras Laffan (Qatargas 3- T6)	O	1	7.8	10.6	"	"	QP, Conoco, Mitsuil	2010

Qatar	Ras Laffan (Qatargas 4- T7)	O	1	7.8	10.6	“	“		2011
Qatar	Ras Laffan (RasGas 1 - T1&T2)	O	2	6.6	9	6	840,000	QP, ExxonMobil, Kogas, Itochu, Nissho Iwai, LNG Japan	1999–2000
Qatar	Ras Laffan (RasGas 2 - T3)	O	1	4.7	6.4			QP, ExxonMobil	2004
Qatar	Ras Laffan (RasGas 2 - T4)	O	1	4.7	6.4			QP, ExxonMobil	2005
Qatar	Ras Laffan (RasGas 2 - T5)	O	1	4.7	6.4			QP, ExxonMobil	2007
Qatar	Ras Laffan (RasGas 3 - T6)	O	1	7.8	10.6			QP, ExxonMobil	2009
Qatar	Ras Laffan (RasGas 3 - T7)	O	1	7.8	10.6			QP, ExxonMobil	2010
Yemen	Bal Haf (T1&T2)	O	2	6.7	9.2	2	140,000	Yemen LNG (Total, Yemen govt., Hunt Oil, SK, Kogas, Hyundai, GASSP)	2009 Train 1 2010 Train 2

Country	Location	Status	No. of trains	Liquefaction capacity 10^6 t per year*	Liquefaction capacity bcm**	Storage tanks*	Storage capacity m^3*	Owners	Start-up year and train
PACIFIC BASIN									
Australia	Withnell Bay, WA	O	4	12.1		4	260,000	Northwest Shelf (Woodside, Shell, BHP, BP, Chevron, Mitsubishi/ Mitsui)	1989
Australia	Withnell Bay, WA	O	1	4.3		1	65,000	Woodside, Shell, BHP, BP, Chevron, Australia Japan LNG	2008
Australia	Darwin LNG Wickham Point, NT	O	1	3.4	4.5	1	188,000	Darwin LNG (ConocoPhil-lips, ENI, Santos, Inpex, TEPCo, Tokyo Gas)	2006
Australia	Pluto LNG Burrup Peninsula, WA	C	3	12.9	19.5			Pluto LNG (Woodside, Tokyo Gas, Kansai Electric)	2012 Train 1 2013 Train 2 2014 Train 3

Australia	Gorgon LNG Barrow Island	C	3	15	20.4		Gorgon LNG (Chevron, Shell, ExxonMobil, Tokyo Gas, Osaka Gas, Chubu Electric)	2014
Australia	Queensland Curtis Curtis Island, Queensland	C	2	8.5	11.6		Queensland Curtis LNG (BG, CNOOC (T1), Tokyo Gas (T2)	2014 Trains 1 and 2
Australia	Gladstone LNG Gladstone, QLD	C	2	7.8	10.6		Gladstone LNG (Santos; Petronas; Kogas; Total)	2015 Trains 1 and 2
Australia	Wheatstone LNG Ashburton North, WA	C	2	8.9	11.7		Wheatstone LNG (Chevron, Shell, Apache Julimar, KUFPEC)	2016 Trains 1 and 2
Australia	Prelude (floating) Browse Basin	P		3.6	4.9		Shell Development (Australia) Proprietary Ltd.	2016
Australia	Icthys LNG Blaydi Point, Darwin Harbor, NT	P	2	8.0+	10.9		Inpex (76%); Total (24%)	2016 Train 1 2017 Train 2

Country	Location	Status	No. of trains	Liquefaction capacity 10^6 t per year*	Liquefaction capacity bcm**	Storage tanks*	Storage capacity m^3*	Owners	Start-up year and train
PACIFIC BASIN									
Australia	Australia Pacific LNG Curtis Island, QLD	P	2	9				Australia Pacific LNG, Gladstone (ConocoPhillips, Origin Energy, Sinopec Corp.)	2015 Train 1 2016 Train 2
Brunei	Lumut	O	5	7.2	9.8	3	195,000	Brunei LNG (Brunei govt., Shell, Mitsubishi)	1973
Canada	Port of Kitimat, BC	P	1	5	6.8			Kitimat LNG (Apache Corp., EOG Resources, Encana)	2015
Indonesia	Blang Lancang Arun	O	3	4.75		4	508,000	PT Arun (Pertamina, ExxonMobil, JILC)	1978–79
Indonesia	Bontang - Badak (Badak	O	8	22.2	29.4	6	630,000	PT Badak (Pertamina,	1977–98

	A-H)							VICO, Total, JILCO	
Indonesia	Tangguh	O	2	7.6	10.3	2	340,000	Gov't Indonesia	2009
Indonesia	Masela (Abadi field; FLNG)	P	1	2.5				Inpex Holdings 90%; Energi Mega Persada (PT EMP Energi) 10%	2018 Train 1
Indonesia	Masela	P	1	2					TBA Train 2
Indonesia	Sulawesi	C	1	2				Donggi-Senoro (Mitsubishi, PT Pertamina, PT Medco Energi International)	2014
Malaysia	Bintulu MLNG 1 (Satu)	O	3	8.1		6	445,000	Malaysia LNG Sdn Bhd (Petronas, Shell, Mitsubishi)	1983
Malaysia	Bintulu MLNG 2 (Dua)	O	3	7.8		"	"	Malaysia LNG Dua (Petronas, Shell, Mitsubishi, Sarawak State Gvnt)	1995
Malaysia	Bintulu MLNG 3 (Tiga)	O	2	6.8		"	"	Malaysia LNG Tiga (Petronas, Shell, Nippon	2003

Country	Location	Status	No. of trains	Liquefaction capacity 10^6 t per year*	Liquefaction capacity bcm**	Storage tanks*	Storage capacity m^3*	Owners	Start-up year and train
PACIFIC BASIN									
								Oil, Diamond Gas, Sarawak State Gvnt)	
Papua New Guinea	Port Moresby	C	1	6.6				ExxonMobil, Oil Search, Santos, AGL, Nippon Oil, Eda Oil, local interests	2014 Train 1
Papua New Guinea	Napa Napa	P	2	4				InterOil Corp., Pacific LNG Operations Ltd., Petromin PNG Holdings Ltd.	2015 Trains 1 and 2
Papua New Guinea	Papua New Guinea (floating)	P		3				DSME, Hough LNG, Petromin LNG	2014
Peru	Peru LNG - Melchorita	O	1	4.45		2	260,000	Hunt Oil, SK Corp. Repsol, Marubeni	2010

Russia	Sakhalin 2 Sakhalin Island	O	2	9.6		2	200,000	Sakhalin Investment Co. (Gazprom, Shell, Mitsui, Mitsubishi)	2009
Russia	Shtokman LNG Teriberka, Murmansk	P	2	7.5				OAO Gazprom, Total SA, Statoil	2014+
Russia	Yamal LNG South Tambey, Yamal Peninsula	P		15				Yamal LNG (Novatek 51%, Total 20%)	2016+
US/Alaska	Kenai LNG Kenai, Alaska	O	2	1.4		3	108,000	ConocoPhillips, Marathon	1969
US	Sabine Pass Sabine, LA	P	4	18				Cheniere Energy Partners, Sabine Pass Liquefaction	2015/16 Trains 1 and 2 2017/18 Trains 3 and 4

Notes: * Capacity data from GIIGNL; ** bcm capacity from IEA. O – Operation; C – Construction; P – Proposed; A – Announced

Sources: GIIGNL World LNG 2010; IEA MTOGM 2010; O&GJ 2012 LNG World Trade Map

APPENDIX B LNG CARRIER FLEET (AS OF NOV. 11, 2011)

IMO no.	Name	LiqCub	Blt	Blt Mth	Subtype	Commercial owner
9377547	**ASEEM**	154,800	2009	11		3J/QGTC/SCI
9307176	**AL DEEBEL**	145,130	2005	10		3J/QShip
9298399	**AL THAKHIRA**	145,702	2005	10		3J/QShip
9285952	**LUSAIL**	145,000	2005	6		3J/QShip
9256200	**FUWAIRIT**	138,200	2004	1		3J/QShip
9253703	**RAAHI**	138,076	2004	12		3J/QShip/SCI
9250713	**DISHA**	136,025	2004	1		3J/QShip/SCI
9334076	**EJNAN**	145,000	2007	1		4J/Qatar Gas Transport
8013950	**WILPOWER**	125,929	1983	8		Awilco LNG
8125832	**WILGAS**	126,975	1984	7		Awilco LNG
8014409	**WILENERGY**	125,542	1983	10		Awilco LNG
9425277	**METHANE PATRICIA CAMILA**	170,683	2010	10		BG Group
9516129	**METHANE BECKI ANNE**	170,678	2010	9		BG Group
9520376	**METHANE MICKIE HARPER**	170,684	2010	12		BG Group
9412880	**METHANE JULIA LOUISE**	170,723	2010	3		BG Group
9321744	**METHANE HEATHER SALLY**	145,611	2007	6		BG Group
9321768	**METHANE ALISON VICTORIA**	145,578	2007	5		BG Group
9307188	**METHANE RITA ANDREA**	145,644	2006	3		BG Group
9321770	**METHANE NILE EAGLE**	145,598	2007	12		BG Group
9321756	**METHANE SHIRLEY ELISABETH**	145,488	2007	3		BG Group
9307190	**METHANE JANE ELIZABETH**	145,644	2006	6		BG Group
9307205	**METHANE LYDON VOLNEY**	145,644	2006	8		BG Group
9256793	**METHANE KARI ELIN**	138,267	2004	6		BG Group
7390181	**LNG AQUARIUS**	126,750	1977	6		BGT/MOL/LJ

7390167	**LNG TAURUS**	126,750	1979	8		BGT/MOL/LJ
7413232	**LNG LIBRA**	126,750	1979	4		BGT/MOL/LJ
7390208	**LNG CAPRICORN**	126,750	1978	6		BGT/MOL/LJ
7390193	**LNG ARIES**	126,750	1977	12		BGT/MOL/LJ
9333620	**BRITISH DIAMOND**	151,945	2008	10		BP
9333606	**BRITISH RUBY**	151,945	2008	6		BP
9333618	**BRITISH SAPPHIRE**	151,945	2008	9		BP
9333591	**BRITISH EMERALD**	151,945	2007	7		BP
9238040	**BRITISH INNOVATOR**	136,135	2003	2		BP
9250191	**BRITISH MERCHANT**	138,000	2003	7		BP
9238038	**BRITISH TRADER**	138,248	2002	11		BP
7347768	**BELANAK**	75,000	1975	7	FSU	Brunei Shell Tankers
9496305	**ARKAT**	147,000	2011	2		Brunei Shell Tankers
9496317	**AMALI**	147,000	2011	8		Brunei Shell Tankers
9210828	**ABADI**	136,912	2002	6		Brunei Shell Tankers
7121633	**BEBATIK**	75,056	1972	10		Brunei Shell Tankers
7347732	**BILIS**	77,731	1975	4		Brunei Shell Tankers
7359785	**BUBUK**	77,679	1975	10		Brunei Shell Tankers
9368302	**BW GDF SUEZ PARIS**	162,400	2009	8		BW Gas
9368314	**BW GDF SUEZ BRUSSELS**	162,400	2009	6		BW Gas
9230062	**BW SUEZ BOSTON**	138,059	2003	1		BW Gas
9243148	**BW SUEZ EVERETT**	138,028	2003	6		BW Gas
9256597	**BERGE ARZEW**	138,089	2004	7		BW Gas
8210209	**KOTO**	125,454	1984	1		BW Gas
9269960	**LNG LOKOJA**	148,471	2006	11		BW Gas/Marubeni
9311567	**LNG KANO**	148,565	2007	1		BW Gas/Marubeni
9311579	**LNG ONDO**	148,478	2007	9		BW Gas/Marubeni
9311581	**LNG IMO**	148,399	2008	6		BW Gas/Marubeni

IMO no.	Name	LiqCub	Blt	Blt Mth	Subtype	Commercial owner
9267003	**LNG OYO**	145,842	2005	12		BW Gas/Marubeni
9266994	**LNG ENUGU**	145,926	2005	11		BW Gas/Marubeni
9266982	**LNG RIVER ORASHI**	145,914	2004	12		BW Gas/Marubeni
9267015	**LNG BENUE**	145,952	2006	3		BW Gas/Marubeni
9275359	**MUSCAT LNG**	145,000	2004	4		Cardiff Marine
7229447	**ISABELLA**	35,491	1975	4		Chemikalien Seetransport
7328243	**ANNABELLA**	35,491	1975			Chemikalien Seetransport
9369473	**DAPENG STAR**	147,210	2009	11		China LNG Shipping
9305128	**MIN LU**	147,210	2009	8		China LNG Shipping
9308479	**DAPENG SUN**	147,210	2008	4		China LNG Shipping
9308481	**DAPENG MOON**	147,210	2008	7		China LNG Shipping
9305116	**MIN RONG**	147,210	2009	2		China LNG Shipping
9433884	**SHINJU MARU No. 2**	2,536	2008	10		Chuo Kaiun
7357452	**METHANIA**	131,235	1978	10		Distrigas
9323687	**CLEAN ENERGY**	149,700	2007	3		Dynagas
9315692	**CLEAN POWER**	149,700	2007	7		Dynagas
9317999	**CLEAN FORCE**	149,700	2007	11		Dynagas
9433717	**CASTILLO DE SANTISTEBAN**	173,600	2010	8		Elcano
9236418	**CASTILLO DE VILLALBA**	138,000	2003	10		Elcano
9064085	**LNG LERICI**	63,957	1998	3		ENI
9064073	**LNG PORTOVENERE**	63,993	1997	4		ENI
6905616	**LNG PALMARIA**	39,691	1969	10		ENI
6928632	**LNG ELBA**	39,795	1970	6		ENI
9389643	**EXPEDIENT**	150,900	2010	4	REGAS	Excelerate
9444649	**EXEMPLAR**	150,900	2010	9	REGAS	Excelerate
9381134	**EXQUISITE**	150,900	2009	10	REGAS	Excelerate
9252539	**EXCELLENCE**	138,120	2005	4	REGAS	Excelerate

9239616	**EXCELSIOR**	138,074	2005	1	REGAS	Exmar
9230050	**EXCALIBUR**	138,034	2002	10		Exmar
9361079	**EXPLORER**	150,981	2008	4	REGAS	Exmar/Excelerate
9361445	**EXPRESS**	150,900	2009	4	REGAS	Exmar/Excelerate
9322255	**EXCELERATE**	138,074	2006	10	REGAS	Exmar/Excelerate
9246621	**EXCEL**	138,134	2003	9		Exmar/Mitsui OSK
9352860	**GASLOG SAVANNAH**	154,984	2010	5		GasLog
9355604	**GASLOG SINGAPORE**	155,006	2010	7		GasLog
9306495	**PROVALYS**	154,472	2006	11		GDF Suez
7391214	**MATTHEW**	126,540	1979	6		GDF Suez
9269207	**GDF SUEZ GLOBAL ENERGY**	74,130	2004	11		GDF Suez
7390179	**LNG VIRGO**	126,750	1979	12		General Dynamics
7390143	**LNG GEMINI**	126,750	1976	9		General Dynamics
7390155	**LNG LEO**	126,750	1978	12		General Dynamics
9256614	**GOLAR WINTER**	138,250	2004	3	FSRU	Golar LNG
7373327	**GOLAR SPIRIT**	129,013	1981	9	FSRU	Golar LNG
7382744	**LNG KHANNUR**	125,003	1977	7	FSRU	Golar LNG
7361922	**GOLAR FREEZE**	125,858	1977	2	FSRU	Golar LNG
9303560	**GOLAR GRAND**	145,700	2006	1		Golar LNG
9320374	**GOLAR MARIA**	145,700	2006	6		Golar LNG
9256767	**GOLAR VIKING**	140,207	2005	1		Golar LNG
9253715	**METHANE PRINCESS**	138,000	2003	8		Golar LNG
9165011	**GOLAR MAZO**	135,225	1999	12		Golar LNG
9253105	**GOLAR ARCTIC**	138,538	2003	12		Golar LNG
7382720	**HILLI**	124,890	1975	12		Golar LNG
7382732	**GIMI**	124,872	1976	12		Golar LNG
7361934	**GANDRIA**	125,904	1977	10		Golar LNG/Bluewater Energy

IMO no.	Name	LiqCub	Blt	Blt Mth	Subtype	Commercial owner
9155078	**HANJIN MUSCAT**	138,366	1999	7		Hanjin Shipping
9176010	**HANJIN SUR**	138,333	2000	1		Hanjin Shipping
9176008	**HANJIN RAS LAFFAN**	138,214	2000	7		Hanjin Shipping
9061928	**HANJIN PYEONG TAEK**	138,366	1995	9		Hanjin Shipping
8706155	**EKAPUTRA**	136,400	1989	1		Humpuss Trans
9060534	**SURYA AKI**	19,538	1996	3		Humpuss Trans
7411961	**RAMDANE ABANE**	126,190	1981	7		Hyproc
7400704	**MOURAD DIDOUCHE**	126,130	1980	12		Hyproc
7400663	**LARBI BEN M'HIDI**	129,767	1977	6		Hyproc
7359955	**MOSTEFA BEN BOULAID**	125,260	1976	8		Hyproc
7400675	**BACHIR CHIHANI**	129,767	1979	2		Hyproc
9372999	**HYUNDAI ECOPIA**	145,000	2008	11		Hyundai Merchant Marine
9155157	**HYUNDAI COSMOPIA**	137,415	2000	1		Hyundai Merchant Marine
9155145	**HYUNDAI TECHNOPIA**	137,415	1999	7		Hyundai Merchant Marine
9179581	**HYUNDAI AQUAPIA**	137,415	2000	3		Hyundai Merchant Marine
9183269	**HYUNDAI OCEANPIA**	137,415	2000	7		Hyundai Merchant Marine
9018555	**HYUNDAI UTOPIA**	125,182	1994	6		Hyundai Merchant Marine
9075333	**HYUNDAI GREENPIA**	125,000	1996	11		Hyundai Merchant Marine
9317200	**NORTH PIONEER**	2,513	2005	11		Iino Gas Transport
9247194	**SK SUNRISE**	138,270	2003	10		Iino/Itochu
9360790	**AL ORAIQ**	210,000	2008	6		J5/Qatar Gas Transport
9360829	**UMM AL AMAD**	210,100	2008	9		J5/Qatar Gas Transport
9360817	**FRAIHA**	210,000	2008	8		J5/Qatar Gas Transport
9360805	**MURWAB**	210,100	2008	6		J5/Qatar Gas Transport
9360843	**AL THUMAMA**	216,200	2008	4		J5/Qatar Gas Transport
9360855	**AL SAHLA**	216,200	2008	6		J5/Qatar Gas Transport

9360867	**AL UTOURIYA**	216,200	2008	9		J5/Qatar Gas Transport
9338266	**AL AAMRIYA**	210,000	2008	5		J5/Qatar Gas Transport
9330745	**NEVA RIVER**	145,394	2007	12		K Line
9355379	**TANGGUH PALUNG**	154,810	2009	3		K Line/Meratus
9349007	**TANGGUH FOJA**	154,810	2008	11		K Line/Meratus
9349019	**TANGGUH JAYA**	154,967	2008	12		K Line/Meratus
9276389	**ARCTIC DISCOVERER**	142,612	2006	2		K Line/Mitsui & Co.
9275335	**ARCTIC VOYAGER**	142,759	2006	7		K Line/Mitsui & Co.
9350927	**TRINITY GLORY**	154,200	2009	3		K Line/Mitsui & Co./Shoei
9329291	**LNG EBISU**	145,000	2008	9		Kansai Electric/Mitsui
						OSK/Iino Kaiun
9351971	**PACIFIC ENLIGHTEN**	145,000	2009	3		KE/Tepco/NYK/MOL/MC
9401295	**BARCELONA KNUTSEN**	173,400	2010	4		Knutsen O.A.S. Shipping
9477593	**RIBERA DEL DUERO KNUTSEN**	173,400	2010	11		Knutsen O.A.S. Shipping
9434266	**VALENCIA KNUTSEN**	173,400	2010	8		Knutsen O.A.S. Shipping
9414632	**SEVILLA KNUTSEN**	173,400	2010	6		Knutsen O.A.S. Shipping
9338797	**SESTAO KNUTSEN**	138,000	2007	11		Knutsen O.A.S. Shipping
9326603	**IBERICA KNUTSEN**	138,000	2006	8		Knutsen O.A.S. Shipping
9246578	**CADIZ KNUTSEN**	138,826	2004	6		Knutsen O.A.S. Shipping
9236432	**BILBAO KNUTSEN**	138,000	2004	2		Knutsen O.A.S. Shipping
9275074	**PIONEER KNUTSEN**	1,100	2004	3		Knutsen O.A.S. Shipping
9373010	**K. MUGUNGWHA**	151,812	2008	10		Korea Line
9373008	**K JASMINE**	151,800	2008	5		Korea Line
9157636	**K ACACIA**	138,017	2000	1		Korea Line
9186584	**K FREESIA**	138,015	2000	6		Korea Line
7428433	**TENAGA EMPAT**	128,354	1981	3	FSU	M I S C
9331660	**SERI BALHAF**	157,720	2009	1		M I S C
9331672	**SERI BALQIS**	157,610	2009	3		M I S C

IMO no.	Name	LiqCub	Blt	Blt Mth	Subtype	Commercial owner
9331658	**SERI BIJAKSANA**	152,900	2008	4		M I S C
9331646	**SERI BEGAWAN**	152,900	2008	1		M I S C
9321665	**SERI ANGKASA**	145,130	2006	12		M I S C
9293844	**SERI AMANAH**	145,709	2006	3		M I S C
9321653	**SERI ANGGUN**	145,731	2006	11		M I S C
9261205	**PUTERI MUTIARA SATU**	137,595	2005	4		M I S C
9248502	**PUTERI FIRUS SATU**	137,489	2004	8		M I S C
9211872	**PUTERI DELIMA SATU**	137,489	2003	10		M I S C
9245031	**PUTERI ZAMRUD SATU**	137,100	2004	1		M I S C
9229647	**PUTERI NILAM SATU**	137,489	2003	9		M I S C
9213416	**PUTERI INTAN SATU**	137,489	2002	8		M I S C
9030838	**PUTERI ZAMRUD**	130,358	1996	7		M I S C
9030814	**PUTERI DELIMA**	130,405	1994	12		M I S C
9030826	**PUTERI NILAM**	130,405	1995	7		M I S C
9030840	**PUTERI FIRUS**	130,358	1997	5		M I S C
9030802	**PUTERI INTAN**	130,405	1994	7		M I S C
9331634	**SERI BAKTI**	152,944	2007	6		M I S C
9329679	**SERI AYU**	145,894	2007	10		M I S C
9293832	**SERI ALAM**	145,572	2005	9		M I S C
7428445	**TENAGA LIMA**	130,000	1981	3		M I S C
7428471	**TENAGA TIGA**	130,000	1982	1		M I S C
7428457	**TENAGA SATU**	130,000	1979	2		M I S C
7428469	**TENAGA DUA**	130,000	1981	8		M I S C
9320386	**SIMAISMA**	145,700	2006	7		Maran Gas Maritime
9331048	**MARAN GAS CORONIS**	145,700	2007	7		Maran Gas Maritime
9302499	**RASGAS ASCLEPIUS**	142,906	2005	6		Maran Gas Maritime
9324435	**AL JASSASIYA**	145,700	2007	5		Maran Gas Maritime

9308431	**UMM BAB**	142,891	2005	11		Maran Gas Maritime
9016492	**AMAN BINTULU**	18,927	1993	10		MISC/NYK
9134323	**AMAN SENDAI**	18,928	1997	5		MISC/NYK
9161510	**AMAN HAKATA**	18,942	1998	11		MISC/NYK
9274226	**ENERGY PROGRESS**	147,558	2006	11		Mitsui O.S.K.
9256602	**LNG PIONEER**	138,121	2005	7		Mitsui O.S.K.
8125868	**WAKABA MARU**	127,209	1985	4		Mitsui O.S.K.
9187356	**SURYA SATSUMA**	23,096	2000	10		Mitsui O.S.K.
9385673	**GDF SUEZ NEPTUNE**	145,130	2009	11	REGAS	Mitsui OSK/Hoegh LNG
9390680	**GDF SUEZ CAPE ANN**	145,130	2010	6	REGAS	Mitsui OSK/Hoegh LNG
9271248	**ARCTIC PRINCESS**	147,835	2005	1		Mitsui OSK/Hoegh LNG
9284192	**ARCTIC LADY**	147,208	2006	4		Mitsui OSK/Hoegh LNG
7320344	**NORMAN LADY**	87,994	1973			Mitsui OSK/Hoegh LNG
9349942	**SUN ARROWS**	19,176	2007	11		MOL/Hiroshima Gas
9361639	**BEN BADIS**	177,300	2010	11		MOL/Itochu
9360922	**ABDELKADER**	177,000	2010	2		MOL/Itochu
9275347	**LALLA FATMA N'SOUMER**	145,445	2004	10		MOL/Itochu/Sonatrach/ Hyproc
9324344	**CHEIKH BOUAMAMA**	75,558	2008	7		MOL/Itochu/Sonatrach/ Hyproc
9324332	**CHEIKH EL MOKRANI**	75,759	2007	6		MOL/Itochu/Sonatrach/ Hyproc
9085649	**AL ZUBARAH**	137,573	1996	12		MOL/NYK/K Line/Iino
9132741	**AL BIDDA**	135,279	1999	11		MOL/NYK/K Line/Iino
9086734	**AL RAYYAN**	135,358	1997	3		MOL/NYK/K Line/Iino
9085625	**AL WAJBAH**	137,308	1997	5		MOL/NYK/K Line/Iino
9085651	**BROOG**	137,529	1998	5		MOL/NYK/K Line/Iino
9085637	**DOHA**	137,262	1999	6		MOL/NYK/K Line/Iino

IMO no.	Name	LiqCub	Blt	Blt Mth	Subtype	Commercial owner
9132818	**ZEKREET**	137,482	1998	12		MOL/NYK/K Line/Iino
9086746	**AL WAKRAH**	137,568	1998	12		MOL/NYK/K Line/Iino
9132791	**AL JASRA**	135,169	2000	7		MOL/NYK/K Line/Iino
9085613	**AL KHOR**	137,354	1996	12		MOL/NYK/K Line/Iino
9338929	**GRAND MEREYA**	145,963	2008	10		MOL/Primorsk
9375721	**GDF SUEZ POINT FORTIN**	154,914	2010	2		MOL/Sumitomo/LNG Japan
9315719	**GRACE BARLERIA**	149,700	2007	9		N Y K
9322803	**LNG BORNO**	149,600	2007	9		N Y K
9322815	**LNG OGUN**	149,600	2007	6		N Y K
9074640	**AL HAMRA**	137,000	1996	12		National Gas Shipping (NGSCO)
9074626	**MUBARAZ**	135,000	1996	1		National Gas Shipping (NGSCO)
9074638	**MRAWEH**	135,000	1996	5		National Gas Shipping (NGSCO)
9074652	**UMM AL ASHTAN**	137,000	1997	5		National Gas Shipping (NGSCO)
9035852	**SHAHAMAH**	135,496	1994	10		National Gas Shipping (NGSCO)
9038452	**GHASHA**	137,514	1995	6		National Gas Shipping (NGSCO)
9035864	**ISH**	137,512	1995	11		National Gas Shipping (NGSCO)
9038440	**AL KHAZNAH**	135,496	1994	5		National Gas Shipping (NGSCO)
7708948	**LNG BONNY**	132,588	1981	12		Nigeria LNG

9241267	**LNG BAYELSA**	137,500	2003	2	Nigeria LNG
9216298	**LNG RIVERS**	137,500	2002	6	Nigeria LNG
9216303	**LNG SOKOTO**	137,425	2002	8	Nigeria LNG
9262211	**LNG ADAMAWA**	138,437	2005	6	Nigeria LNG
9262209	**LNG AKWA IBOM**	141,500	2004	11	Nigeria LNG
9262223	**LNG CROSS RIVER**	141,000	2005	9	Nigeria LNG
9262235	**LNG RIVER NIGER**	141,000	2006	6	Nigeria LNG
7360124	**LNG LAGOS**	122,255	1976	12	Nigeria LNG
7619587	**LNG EDO**	126,530	1980	5	Nigeria LNG
7702401	**LNG FINIMA**	132,588	1984	1	Nigeria LNG
7619575	**LNG ABUJA**	126,530	1980	9	Nigeria LNG
7360136	**LNG PORT HARCOURT**	122,255	1977	10	Nigeria LNG
9260603	**SHINJU MARU No. 1**	2,513	2003	7	NS United Kaiun Kaisha
9250725	**NORTHWEST SWAN**	140,500	2004	4	NWS LNG Shipping
8913174	**NORTHWEST SEAEAGLE**	125,541	1992	11	NWS LNG Shipping
9045132	**NORTHWEST STORMPETREL**	125,525	1994	12	NWS LNG Shipping
8608872	**NORTHWEST SANDERLING**	127,525	1989	4	NWS LNG Shipping
8608705	**NORTHWEST SHEARWATER**	127,500	1991	9	NWS LNG Shipping
8913150	**NORTHWEST SANDPIPER**	125,042	1993	2	NWS LNG Shipping
8608884	**NORTHWEST SNIPE**	127,747	1990	10	NWS LNG Shipping
9320075	**GASELYS**	154,472	2007	3	NYK/GDF Suez
9323675	**GRACE COSMOS**	141,000	2008	3	NYK/MBK
9315707	**GRACE ACACIA**	149,700	2007	2	NYK/MBK
9475208	**SOYO**	160,000	2011	8	NYK/MBK/Teekay
9490959	**MALANJE**	160,000	2011	9	NYK/MBK/Teekay
9490961	**LOBITO**	160,000	2011	10	NYK/MBK/Teekay
9403657	**TAITAR No. 4**	145,333	2010	10	NYK/Mitsu & Co./CPC
9403645	**TAITAR No. 2**	147,000	2009	12	NYK/Mitsu & Co./CPC

IMO no.	Name	LiqCub	Blt	Blt Mth	Subtype	Commercial owner
9403671	**TAITAR No. 3**	145,000	2010	1		NYK/Mitsu & Co./CPC
9403669	**TAITAR No. 1**	145,000	2009	9		NYK/Mitsu & Co./CPC
9043677	**DWIPUTRA**	127,386	1994	3		NYK/MOL
8014473	**SENSHU MARU**	127,167	1984	2		NYK/MOL/K Line
8702941	**LNG SWIFT**	127,580	1989	8		NYK/MOL/K Line
8110203	**ECHIGO MARU**	125,568	1983	8		NYK/MOL/K Line
8701791	**NORTHWEST SWALLOW**	127,544	1989	11		NYK/MOL/K Line
9265500	**DUKHAN**	137,661	2004	10		NYK/MOL/KL/MBK/ Qatar Shipping
9020766	**LNG VESTA**	127,547	1994	6		NYK/MOL/KL/OG/TG/ THG
9006681	**LNG FLORA**	127,705	1993	3		NYK/MOL/KL/OG/TG/ THG
9483877	**ENERGY HORIZON**	177,000	2011	9		NYK/Tokyo Gas
9405588	**ENERGY CONFIDENCE**	153,000	2009	5		NYK/Tokyo Gas
9200316	**LNG JAMAL**	133,333	2000	10		OG/NYK/MOL/KLine
9253284	**GOLAR FROST**	138,830	2004	6	FSRU	OLT
9326689	**IBRA LNG**	148,176	2006	8		Oman Shipping
9300817	**SALALAH LNG**	145,000	2005	12		Oman Shipping
9294264	**NIZWA LNG**	145,469	2005	12		Oman Shipping
9317315	**IBRI LNG**	145,173	2006	7		Oman Shipping
9210816	**SOHAR LNG**	137,248	2001	10		Oman Shipping
9341689	**LNG JUPITER**	155,999	2009	6		Osaka Gas/NYK
9341299	**LNG BARKA**	153,643	2008	12		Osaka Gas/NYK
9277620	**LNG DREAM**	145,254	2006	9		Osaka Gas/NYK
9337743	**AL HAMLA**	216,200	2008	2		OSG/QGTC
9337705	**AL GATTARA**	216,280	2007	11		OSG/QGTC

9337731	**TEMBEK**	216,200	2007	11	OSG/QGTC
9337717	**AL GHARRAFA**	216,200	2007	1	OSG/QGTC
9337975	**DUHAIL**	210,000	2008	1	Pronav/QGTC
9337963	**AL SAFLIYA**	210,000	2007	10	Pronav/QGTC
9337987	**AL GHARIYA**	210,000	2008	1	Pronav/QGTC
9337951	**AL RUWAIS**	210,000	2007	10	Pronav/QGTC
9388819	**LIJMILIYA**	261,700	2008	12	Q G T C
9372743	**AL GHUWAIRIYA**	261,700	2008	12	Q G T C
9388821	**AL SAMRIYA**	261,700	2009	2	Q G T C
9397315	**AL MAFYAR**	266,000	2009	4	Q G T C
9397298	**AL MAYEDA**	266,000	2009	3	Q G T C
9443413	**RASHEEDA**	266,276	2010	8	Q G T C
9397303	**MEKAINES**	266,000	2009	3	Q G T C
9443683	**AL DAFNA**	266,000	2009	9	Q G T C
9388833	**BU SAMRA**	266,000	2008	12	Q G T C
9337755	**MOZAH**	266,000	2008	9	Q G T C
9372731	**UMM SLAL**	266,000	2008	11	Q G T C
9443401	**AAMIRA**	266,237	2010	4	Q G T C
9418365	**SHAGRA**	266,000	2009	11	Q G T C
9431214	**ZARGA**	266,433	2010	3	Q G T C
9431123	**AL KARAANA**	210,100	2009	10	Q G T C
9360831	**AL SHEEHANIYA**	210,100	2009	2	Q G T C
9431147	**AL BAHIYA**	210,100	2010	1	Q G T C
9431111	**AL KHATTIYA**	210,100	2009	7	Q G T C
9397353	**ONAIZA**	210,100	2009	3	Q G T C
9397341	**AL SADD**	210,100	2009	3	Q G T C
9431135	**AL NUAMAN**	210,100	2009	10	Q G T C
9397286	**AL GHASHAMIYA**	266,000	2009	4	Q G T C

IMO no.	Name	LiqCub	Blt	Blt Mth	Subtype	Commercial owner
9397339	**AL REKAYYAT**	216,200	2009	6		Q G T C
9337729	**MESAIMEER**	216,200	2009	3		Q G T C
9397327	**AL KHARAITIYAT**	216,200	2009	6		Q G T C
7391197	**LNG DELTA**	124,014	1978	5		Shell
9253222	**GEMMATA**	135,269	2004	1		Shell
9236626	**GALLINA**	135,269	2002	12		Shell
9236614	**GALEA**	135,269	2002	9		Shell
7391202	**GALEOMMA**	124,014	1978	12		Shell
9319404	**TRINITY ARROW**	154,982	2008	3		Shoei
9157624	**SK SUMMIT**	135,244	1999	8		SK Shipping
9180231	**SK SPLENDOR**	135,603	2000	3		SK Shipping
9180243	**SK STELLAR**	138,540	2000	12		SK Shipping
9157739	**SK SUPREME**	135,490	2000	1		SK Shipping
9038816	**Y. K. SOVEREIGN**	127,125	1994	12		SK Shipping
9468437	**NORGAS UNIKUM**	12,000	2011	6		Skaugen I.M.
9468449	**BAHRAIN VISION**	12,000	2011	10		Skaugen I.M.
9378278	**NORGAS INNOVATION**	10,000	2010	1		Skaugen I.M.
9378292	**NORGAS INVENTION**	10,000	2011	1		Skaugen I.M.
9378280	**NORGAS CREATION**	10,000	2010	8		Skaugen I.M.
9482299	**SONANGOL ETOSHA**	160,500	2011	11		SONANGOL
9475600	**SONANGOL SAMBIZANGA**	160,500	2011	10		SONANGOL
6910702	**SCF ARCTIC**	71,651	1969			Sovcomflot
6901892	**SCF POLAR**	71,650	1969	9		Sovcomflot
9325893	**TANGGUH TOWUTI**	145,700	2008	10		Sovcomflot/NYK
9334284	**TANGGUH BATUR**	145,700	2008	12		Sovcomflot/NYK
9338955	**GRAND ANIVA**	145,000	2008	1		Sovcomflot/NYK
9332054	**GRAND ELENA**	145,580	2007	10		Sovcomflot/NYK

9383900	**STENA CRYSTAL SKY**	171,800	2011	5	Stena
9413327	**STENA CLEAR SKY**	171,800	2011	5	Stena
9315393	**STENA BLUE SKY**	145,700	2006	6	Stena
9372963	**STX KOLT**	145,700	2008	11	STX Pan Ocean
9390185	**STX FRONTIER**	153,000	2010	4	STX Pan Ocean
9361990	**TANGGUH SAGO**	155,000	2009	3	Teekay
9230048	**HISPANIA SPIRIT**	138,517	2002	8	Teekay
9333632	**TANGGUH HIRI**	155,000	2008	11	Teekay
9247364	**GALICIA SPIRIT**	140,624	2004	7	Teekay
9259276	**MADRID SPIRIT**	138,000	2004	12	Teekay
9236420	**CATALUNYA SPIRIT**	138,000	2003	8	Teekay
9001784	**ARCTIC SPIRIT**	89,089	1993	12	Teekay
9001772	**POLAR SPIRIT**	88,996	1993	6	Teekay
9342487	**MAERSK MAGELLAN**	165,500	2009	3	Teekay LNG Partners
9339260	**MAERSK ARWA**	165,500	2008	9	Teekay LNG Partners
9336737	**MAERSK METHANE**	165,500	2008	2	Teekay LNG Partners
9336749	**MAERSK MARIB**	165,500	2008	5	Teekay LNG Partners
9369899	**WOODSIDE DONALDSON**	165,500	2009	10	Teekay LNG Partners
9369904	**MAERSK MERIDIAN**	165,500	2010	1	Teekay LNG Partners
9321732	**MAERSK QATAR**	145,130	2006	4	Teekay LNG Partners
9255854	**MAERSK RAS LAFFAN**	138,270	2004	3	Teekay LNG Partners
9360893	**AL SHAMAL**	217,000	2008	6	Teekay/QGTC
9360908	**AL KHUWAIR**	217,000	2008	6	Teekay/QGTC
9360879	**AL HUWAILA**	217,000	2008	5	Teekay/QGTC
9360881	**AL KHARSAAH**	217,000	2008	6	Teekay/QGTC
9325697	**AL AREESH**	151,700	2007	1	Teekay/QGTC
9325702	**AL DAAYEN**	151,700	2007	3	Teekay/QGTC
9325685	**AL MARROUNA**	151,816	2006	9	Teekay/QGTC

IMO no.	Name	LiqCub	Blt	Blt Mth	Subtype	Commercial owner
9247962	**PACIFIC NOTUS**	137,006	2003	9		Tokyo Electric/NYK/ Mitsubishi Corp
9343106	**ALTO ACRUX**	147,798	2008	3		Tokyo Electric/NYK/ Mitsubishi Corp
9376294	**CYGNUS PASSAGE**	145,000	2009	1		Tokyo Electric/NYK/ Mitsubishi Corp
9264910	**PACIFIC EURUS**	135,000	2006	3		Tokyo Electric/NYK/ Mitsubishi Corp
9269180	**ENERGY ADVANCE**	147,624	2005	3		Tokyo Gas
9355264	**ENERGY NAVIGATOR**	147,558	2008	6		Tokyo LNG/Mitsui OSK
9245720	**ENERGY FRONTIER**	147,599	2003	9		Tokyo LNG/Mitsui OSK
9324277	**NEO ENERGY**	146,735	2007	2		Tsakos
9469235	**KAKUREI MARU**	2,512	2008	11		Tsurumi Sunmarine
7359670	**TRANSGAS**	129,323	1977	12		Unknown
9404584	**CORAL METHANE**	7,350	2009	4		Veder A.
Under Construction						
9486738	STX JINHAE 5003	160,000	2015	4		Anangel Shipping Ent.
9486740	STX JINHAE 5004	160,000	2015	4		Anangel Shipping Ent.
9627966	DAEWOO 2290	160,500	2014	2		Awilco LNG
9627954	DAEWOO 2289	160,500	2013	9		Awilco LNG
9645970	HYUNDAI 2580	155,000	2013	9		BW Gas
9640437	HYUNDAI 2571	155,000	2014	11		BW Gas
9640645	HYUNDAI 2572	155,000	2015	3		BW Gas
9636711	DAEWOO 2297	159,000	2014	3		Cardiff Marine
9636735	DAEWOO 2400	159,000	2014	7		Cardiff Marine
9636747	DAEWOO 2401	159,000	2014	9		Cardiff Marine

9636723	DAEWOO 2298	159,000	2014	5		Cardiff Marine
9610779	SAMSUNG 1942	160,000	2015	5		Chevron
9610767	SAMSUNG 1941	160,000	2015	2		Chevron
9606950	SAMSUNG 1921	154,800	2013	12		Chevron
9606948	SAMSUNG 1920	154,800	2013	10		Chevron
9583677	SHEN HAI	147,210	2012	11		China LNG Shipping
9629586	HYUNDAI 2556	155,000	2013	7		Dynagas
9637492	HYUNDAI 2558	155,000	2014	3		Dynagas
9629598	HYUNDAI 2557	155,000	2013	10		Dynagas
9637507	HYUNDAI 2565	155,000	2014	8		Dynagas
9638525	DAEWOO 2402	173,400	2014	5	FSRU	Excelerate
9480356	SAMSUNG 1839	220,000	2012	4	FLNG	FLEX LNG
9438107	SAMSUNG 1762	220,000	2014	11	FLNG	FLEX LNG
9514573	SAMSUNG 1850	220,000	2012	9	FLNG	FLEX LNG
9438092	SAMSUNG 1761	220,000	2013	9	FLNG	FLEX LNG
9638915	SAMSUNG 2044	155,000	2014	10		GasLog
9634098	SAMSUNG 2042	155,000	2014	3		GasLog
9600528	SAMSUNG 1946	155,000	2013	3		GasLog
9626285	SAMSUNG 2017	155,000	2013	7		GasLog
9634086	SAMSUNG 2041	155,000	2013	12		GasLog
9600530	SAMSUNG 1947	155,000	2013	6		GasLog
9638903	SAMSUNG 2043	155,000	2014	6		GasLog
9626273	SAMSUNG 2016	155,000	2013	5		GasLog
9637765	SAMSUNG	170,000	2013	9	REGAS	Golar LNG
9626027	SAMSUNG 2026	160,000	2014	4		Golar LNG
9635315	SAMSUNG 2047	160,000	2014	7		Golar LNG
9624926	SAMSUNG 2022	160,000	2013	7		Golar LNG
9624940	SAMSUNG 2024	160,000	2014	1		Golar LNG

IMO no.	Name	LiqCub	Blt	Blt Mth	Subtype	Commercial owner
Under Construction						
9637325	SAMSUNG 2048	160,000	2014	9		Golar LNG
9624914	SAMSUNG 2021	160,000	2013	4		Golar LNG
9624938	SAMSUNG 2023	160,000	2013	10		Golar LNG
9626039	SAMSUNG 2027	160,000	2014	8		Golar LNG
9629536	HYUNDAI ULSAN 2549	170,000	2014	2	FSRU	Hoegh L. & Co.
9629524	HYUNDAI ULSAN 2548	170,000	2013	12	FSRU	Hoegh L. & Co.
9627497	DAEWOO 2291	156,000	2014	6		Maran Gas Maritime
9633173	DAEWOO 2296	156,000	2013	12		Maran Gas Maritime
9627502	DAEWOO 2292	156,000	2015	6		Maran Gas Maritime
9633161	DAEWOO 2295	156,000	2013	10		Maran Gas Maritime
9627485	DAEWOO 2288	156,000	2013	6		Maran Gas Maritime
9633434	HYUNDAI SAMHO S625	156,000	2014	4		Maran Gas Maritime
9633422	HYUNDAI SAMHO S624	156,000	2013	12		Maran Gas Maritime
9613161	HUDONG ZHONGHUA	170,000	2016	4		Mitsui O.S.K.
9613135	HUDONG ZHONGHUA	170,000	2014	5		Mitsui O.S.K.
9613159	HUDONG ZHONGHUA	170,000	2015	9		Mitsui O.S.K.
9613147	HUDONG ZHONGHUA	170,000	2015	1		Mitsui O.S.K.
9645736	MITSUBISHI NAGASAKI	153,000	2014	4		Mitsui O.S.K.

9645748	MITSUBISHI NAGASAKI	153,000	2015	2	Mitsui O.S.K.
9540716	KAWASAKI 1665	177,000	2013	1	N Y K
9607760	MITSUBISHI NAGASAKI	145,000	2013	6	N Y K
9621077	MITSUBISHI NAGASAKI 2289	145,400	2014	2	N Y K
9491812	CUBAL	160,000	2012	1	NYK/MBK/Teekay
9468451	DINGHENG JIANGSU 2007-003	12,000	2012	2	Skaugen I.M.
9468463	DINGHENG JIANGSU 2007-004	12,000	2012	6	Skaugen I.M.
9378307	NORGAS CONCEPTION	10,000	2011	11	Skaugen I.M.
9482304	SONANGOL BENGUELA	160,500	2011	12	SONANGOL
9630004	STX JINHAE 1910	170,200	2013	8	Sovcomflot
9630028	STX JINHAE 1911	170,200	2014	3	Sovcomflot
9644421	DAEWOO	171,800	2015	2	Stena
9644419	DAEWOO	171,800	2014	2	Stena
9636785	SAMSUNG 2045	160,000	2013	12	Thenamaris
9636797	SAMSUNG 2046	160,000	2014	3	Thenamaris
9640023	SAMSUNG 2049	160,000	2014	7	Thenamaris
9617698	MEYER WERFT 665	15,000	2012	12	Veder A.

Source: *EA Gibson*

APPENDIX C WORLDWIDE LNG REGASIFICATION TERMINALS

	Location/site/ project	No. of storage tanks	Total storage capacity in cm (liq)	No. of vaporizers (excluding back-up capacity)	Nominal send-out capacity in NG bcm/y	Capacity, million tpy*	Status	Start-up year	Principal owner
France	Fos-sur-Mer (Fos Tonkin)	3	150,000	15	5.5	5.1	O	1972	Elengy (GDF Suez)
	Montoir de Bretagne	3	360,000	11	10	7.5	O	1980	Elengy (GDF Suez)
	Montoir de Bretagne expansion					1.8	C	2015	
	Fos-Cavaou	3	330,000	4	8.25	6.0	O	2009–2010	Société du Terminal Methanier Fos Cavaou (GDF Suez, Total SA)
	Antifer, La Havre					6.5	A	2015	Gaz de Normandie (Poweo 73%; CIM SNC 27%)
	Dunkerque					5.8	C	2015	EDF 65%; Fluxys 25%; Total 10%

Spain	Barcelona	6	540,000	13	17.08	12.5	O	1968, 2009	Enagas
	Huelva	4	460,000	9	11.83	8.6	O	1988; 2008	Enagas
	Cartagena	4	437,000	9	11.8	8.6	O	1989, 2009	Enagas
	Bilbao	2	300,000	4	7	5.9	O	2003	BBG (Enagas, EVE, RREFF)
	Mugardos (El Ferrol)	2	300,000	3	3.6	2.6	O	2007	Reganosa Group, Union Fenosa, Endesa, Sonatrach, local companies
	Sagunto	3	450,000	5	8.76	6.3	O	2007, 2009	Union Fenosa, Iberdrola, Oman Oil, Osaka Gas
	El Musel, Gijon					5.1	C	2013	Enagas
	Palos de la Frontera (Andalusia)					3.0	A	2015	Energas
Italy	Panigaglia	2	100,000	4	3.32	2.5	O	1969	GNL Italia
	Rovigo (Atlantic LNG)			5	8	5.8	O	2009	Adriatic LNG (ExxonMobil, QP, Edison)

	Location/site/ project	No. of storage tanks	Total storage capacity in cm (liq)	No. of vaporizers (excluding back-up capacity)	Nominal send-out capacity in NG bcm/y	Capacity, million tpy*	Status	Start-up year	Principal owner
	Livorno (FRSU)					2.7–3.4	C	2011+	OLT Offshore LNG Toscana SPA (Endesa, IRIDE, ASA, OLT-E, Golar)
	Porto Empedocle					5.8	A	2016	Nuove Energie (Enel 90%; Siderurgica Investimenti Group 10%)
Belgium	Zeebrugge	4	380,000	11	9	6.6	O	1987	GDF Suez, Publigaz, Fluxys
Turkey	Marmara Ereglisi	3	255,000	7	6.2	4.8	O	1992	Botas
	Aliaga/Izmir	2	280,000	5	6	4.4	O	2006	Egegaz
Greece	Revithoussa	2	130,000	6	5	2.3	O	2000	Depa S.A.
Portugal	Sines	2	240,000	5	5.2	4.00	O	2004	Ren Atlantico
	Sines Expansion Project	1	150,000					2009–2012	
United Kingdom	Isle of Grain	8	1,000,000	14	19.5	14.8	O	2005	National Grid

	Teesside (GasPort)	1	138,000			3.0	O	2007	Excelerate Energy
	Milford Haven (South Hook LNG)	5	775,000	15	21	15.6	O	2009	South Hook LNG (QP, ExxonMobil, Total)
	Milford Haven (Dragon LNG)	2	320,000	6	6	6.6	O	2009	Dragon LNG (BG, Petronas, 4Gas)
	Walney Island, FLNG					5.8	A	2013	Port Meridan Energy (Hoegh LNG)
Netherlands*	Gate LNG					8.8	O	2011	Gasunie, Vopak, Dong Energy, Essent, OMV Gas International, E.ON Ruhrgas
Poland*	Świnoujście					1.0	C	2014	PGNiG
Canada	St. John, NB (Canaport LNG)	3	160,000	8	10	7.5	O	2009	Canaport LNG (Irving Oil, Repsol)
United States	Everett, Mass.	2	155,000	4	6.9	6.1	O	1971	GDF Suez LNG
	Elba Island, Ga.	5	535,000	11	16.3	11.4	O	1978–2010	Southern LNG (El Paso) (restarted 2001,

Location/site/ project	No. of storage tanks	Total storage capacity in cm (liq)	No. of vaporizers (excluding back-up capacity)	Nominal send-out capacity in NG bcm/y	Capacity, million tpy*	Status	Start-up year	Principal owner
								expanded 2006&2010)
Lake Charles, La.	4	425,000	14	24.3	14.0	O	1982	Trunkline (Southern Union)
Cove Point, Md.	5	380,000	10	10.74	13.8	O	1978–2003	Dominion Cove Point (restarted 2003)
Gulf Gateway, offshore Louisiana	1	150,000		4.6	3.3	Clsd	2005	Excelerate Energy (facility retired 2011)
Northeast Gateway, offshore Mass.	1	150,000		4.6	3.0	O	2008	Excelerate Energy
Freeport, Tex. (Freeport LNG)	2	330,694	7	18	11.4	O	2008	Freeport LNG Development (Michael Smith, ConocoPhil-lips)
Hackberry, La. (Cameron LNG)	3	480,000	10	15.5	11.4	O	2009	Cameron LNG (Sempra)

	Sabine Pass, La (Cheniere).	3	480,000	16	27	30.5	O	2008	Cheniere Energy
	Sabine Pass, Tex (Golden Pass)	5	775,000		9.8	15.6	O	2010	Golden Pass (ExxonMobil, QP, ConocoPhil-lips)
	Neptune LNG, offshore Mass.	2	290,000		3.9	3.8	O	2010	GDF Suez LNG
	Pascagoula, Miss. (Gulf LNG)					5.0	O	2011	Gulf LNG Energy (GE Energy Finan. Ser., Sonangol, El Paso)
Dominican Republic	Andres (Punta Caucedo)	1	160,000	2	2.32	1.8	O	2003	AE SAndres Corp.
Mexico	Altamira, Tamulipas	2	300,000	5	7.8	3.8	O	2006	Shell (50%), Total (25%), Mitsui (25%)
	Enseñada, Baja California (Energia Costa Azul)	2	320,000	6	10.33	7.5	O	2008	Energía Costa Azul (Sempra)
	Enseñada, Baja California expansion					7.5	A	TBD	

	Location/site/ project	No. of storage tanks	Total storage capacity in cm (liq)	No. of vaporizers (excluding back-up capacity)	Nominal send-out capacity in NG bcm/y	Capacity, million tpy*	Status	Start-up year	Principal owner
	Manzanillo, Colima					3.8	O	2011	Terminal de GNL (Samsung, Korea Gas, Mitsui)
Puerto Rico	Penuelas, Puerto Rico	1	160,000	2	3.75	2.7	O	2000	Gas Natural and International Power- Mitsui
Argentina	Bahia Blanca (GasPort)			6	3	3	O	2008	Repsol, YPF
	Escobar (GasPort)					3.8	O	2011	Enarsa, YPF
	Cuatreros, Bahia Blanca (GasPort)					3	O	2012	Enarsa, YPF
	Buenos Aires-Montevideo (FLNG)					2.7–4.5	C	2013	Enarsa, ANCAP, UTE
	San Antonio Oeste, Rio Negro (FLNG)					5	A	2014	Enarsa
Brazil	Pecem (FRSU)	1	129,000	2	2.5	1.5	O	2009	Petrobras

	Guanabara Bay (FRSU)	1	138,000	2	5	3.5	O	2009	Petrobras
	Bay of All Saints, Bahia (FRSU)					3.7	A	2013	Petrobras
Chile	Quintero	3	344,000	3	3.65	2.5	O	2009	GNL Quintero S.A.
	Mejillones (FSU)	1	154,500	3	2	1.5	O	2010	GNLM
Dubai	Jebel Ali (FSRU)	1	125,850		3	3.72	O	2010	Dubai Supply Authority (charter from Golar LNG)
Kuwait	Mina Al Ahmadi (GasPort)	1	150,000		7.07	0.01	O	2009	Kuwait national Petroleum Co. (KNPC)
China	Dapend, Shenzhen	3	480,000	7	4.9	6.7	O	2006	Dapeng LNG: CNOOC, BP
	Fujian	2	320,000		3.7	2.6	O	2008	Fujain LNG (CNOOC 60%, Fujian NV & Dev. Corp. 40%)
	Shanghai, Mengtougou	3	120,000		0.2		O	2008	Shanghai Gas Group
	Shanghai, Yangshan	3	495,000		4.1	3.0	O	2009	Shanghai LNG (CNOOC 45%,

	Location/site/ project	No. of storage tanks	Total storage capacity in cm (liq)	No. of vaporizers (excluding back-up capacity)	Nominal send-out capacity in NG bcm/y	Capacity, million tpy*	Status	Start-up year	Principal owner
	(Ximentang Isle)								Shenenergy Group 55%)
	Shanghai expansion					3.0	A*	TBD	
	Dalian					3.5	O	2011	Kunlun Energy (75%), Dalian Port (20%), Dalian Investment (5%)
	Zhejiang Ningbo					3.5	C	2012	CNOOC, Zhejiang Energy Group, Ningbo Power Development
	Jiangsu Rudong					3.5	O	2011	Kunlun Energy (55%), Pacific Oil and Gas (35%), Jiangsu Guoxin Investment Group (10%)

	Shandong Quindao					3.0	C	2013	Sinopec, China Huaneng Group
	Zhuhai Jinwan LNG, Guangdong					3.5	C	2013	CNOOC Gas & Power Group
	Zhuhai Jinwan LNG expansion					3.5	A*	2015	
	Tianjin (FSRU)					2.2	A	2013	CNOOC Gas & Power Group
	Tianjin					6.0	A	2015	CNOOC Gas & Power Group
	Yuedong LNG, Jieyang, Guangdong					2.0	C	2013	CNOOC
	Shenzhen city, Guangdong					2.5	A*	2014	PetroChina
	Heiyangang, Hainan Yangpu Economic Dev. Zn.					2.0	A*	2015–16	CNOOC Hainan Natural Gas (CNOOC 65%; Hainan Development Holdings 35%)
	*Approved by NDRC								
India	Dahej	4	592,000	19	12.5	11.65	O	2004, 2009	Petronet LNG

	Location/site/project	No. of storage tanks	Total storage capacity in cm (liq)	No. of vaporizers (excluding back-up capacity)	Nominal send-out capacity in NG bcm/y	Capacity, million tpy*	Status	Start-up year	Principal owner
	Hazira	2	320,000	5	3.4	3.60	O	2005	Hazira LNG Private Ltd (Shel 74%, Tota 26%)[1]
	Hazira Expansion					1.40	A	2013	
	Dabhol					5.50	O	2011	Ratnagiri LNG (NTPC, GAIL, Maharashtra State Electricity Board)
	Kochi, Kerala					2.50	C	2012	Petronet LNG
	Kochi, Kerala, expansion					2.50	A	2012–13	
Japan	Nigishi	14	1,180,000	14	15	12.10	O	1969	Tokyo Gas
	Senboku I, Osaka	4	180,000	5	2.94	2.50	O	1972	Osaka Gas
	Senboku II, Osaka	18	1,585,000	15	15.7	12.90	O	1977	Osaka Gas
	Sodegaura, Chiba	35	2,660,000	36	41.6	29.30	O	1973	Tokyo Gas

Chita Kyoda Joint Terminal, Aichi	4	300,000	14	9.89	7.6	O	1978	Toho Gas, Chubu Electric
Tobata, Kitakyushu City	8	480,000	9	10.28	6.8	O	1977	Kita Kyushu LNG
Himeji LNG	7	520,000	8	11	8.5	O	1979	Kansai Electric, Osaka Gas
Chita LNG	7	640,000	11	15.7	11.5	O	1983	Chubu Electric, Toho Gas
Higashi Ohgishima, Kawasaki City	9	540,000	9	18	15.5	O	1984	Tokyo Electric
Niigata, Higata Higashi Port (Nihonkai LNG)	8	720,000	14	11.6	9.0	O	1984	Nihonkai LNG Co. Ltd., Tohoku Electric
Himeji	8	740,000	6	6.4	5.0	O	1984	Osaka Gas, Kansai Electric
Futtsu, Chiba	10	1,110,000	13	26	20.1	O	1985	TEPCO
Yokkaichi LNG Centre	4	320,000	8	9.2	7.1	O	1988	Chubu Electric
Yanai	6	480,000	5	3.1	2.4	O	1990	Chugoku Electric
Oita, Oita City	5	460,000	6	6.27	4.9	O	1990	Oita LNG
Yokkaichi Works, Mie	2	160,000	4	2	0.7	O	1991	Toho Gas, Chubu Electric
Fukuoka	2	70,000	7	1.1	0.9	O	1993	Saibu Gas

Location/site/ project	No. of storage tanks	Total storage capacity in cm (liq)	No. of vaporizers (excluding back-up capacity)	Nominal send-out capacity in NG bcm/y	Capacity, million tpy*	Status	Start-up year	Principal owner
Omuta satellite, Fukuoka Pref.*					0.5	O	2011	Saibu Gas
Hatsukaichi, Hiroshima	2	170,000	4	1.15	0.6	O	1996	Hiroshima Gas
Kagoshima, Southern Kyushu	2	86,000	3	0.3	0.2	O	1996	Nippon Gas
Kawagoe, Mie	4	480,000	4	7.1	5.5	O	1997	Chubu Electric
Sodeshi, Shizuoka	3	337,200	8	3.9	0.9	O	1996	Shimizu LNG
Ohgishima, Yokohama	3	600,000	10	12.4	6.0 (1998)	O	1998, 2009	Tokyo Gas
Shin-Minato, Sendai City	1	80,000	3	0.38	0.3	O	1997	Sendai Gas
Chita Midorihama, Aichi	2	400,000	7	9.2	5.4	O	2001	Toho Gas, Chubu Electric
Nagasaki	1	35,000	3	0.2	0.1	O	2003	Saibu Gas

Mizushima, Okayama	1	160,000	3	1.3	0.6	O	2006	Mzushima LNG
Sakai, Osaka	3	420,000	6	8.7	2.1	O	2006	Kansai Electric
Sakaide, Shikoku	1	180,000	3	1.64	1.3	O	2010	Shikoku Elec. Power; Cosmo Oil; Shikoku Gas
Okinawa				0.3		O	2012	Okinawa Electric
Ishikari, Hokkaido					TBA	C	2012–13	Hokkaido Gas Co., Ltd
Naoetsu, Joetsu City, Nigata					1.0	C	2014	Inpex Holdings
Hachinohe, Hokkaido					TBA	C	2015	JX Nippon Oil & Energy
Hitachi, Ibaraki					TBA	A	2015	Tokyo Gas
Kita Kyushu City, Fukuoka Pref.					TBA	A	2014	Hibiki (Saibu Gas; Kyushu Electric)
Shin-Sendai					TBA	A	2016	Tohoku
Toyama Shinko, Toyama					TBA	A	2018–2019	Hokuriku Electric Power

	Location/site/ project	No. of storage tanks	Total storage capacity in cm (liq)	No. of vaporizers (excluding back-up capacity)	Nominal send-out capacity in NG bcm/y	Capacity, million tpy*	Status	Start-up year	Principal owner
	Joetsu					TBA	A	TBD	Chubu Electric, Tohoku Electric
S. Korea	Pyeong-Taek	14	1,560,000	31	40.28	31.20	O	1986	Korea Gas Corp.
	Inchon	20	2,680,000	33	40.99	32.80	O	1996	Korea Gas Corp.
	Tong-Yeong	12	1,680,000	12	20.72	15.00	O	2002	Korea Gas Corp.
	Gwangyang	3	365,000	2	2.3	1.80	O	2005	Posco
	Samcheok					18.35	C	2013	Korea Gas Corp.
Taiwan	Yung-An	6	690,000	16	23	17.9	O	1990	CPC Corp.
	Taichung	3	480,000	6	9	3.0	O	2009	CPC Corp.
Thailand*	Map Ta Phut, Rayong Prov.					5.0	O	2011	PTT LNG Co. Ltd., Egco Group, Electricity Generating Auth.
Singapore*	Jurong Island, Singapore					3.50	C	2013	Singapore LNG (Energy Market Authority)

Indonesia*	Jakarta Bay, West Java FRSU	3.80	C	2012	Nusantara Regas (Pertamina, 60%; Perusahaan Gas Negara 40%)
Malaysia*	Mukim Sungai Udang, Melaka (jetty regas)	3.80	C	2012	Petronas
Pakistan*	Karachi (offshore)	3.5	A	2013	Sui Southern Gas Co.

Note: Status: O = Operating; C = Under construction; A = Announced.

Sources: GIIGNL, *The LNG Industry* 2010; *Oil&Gas Journal* 2012 LNG World Trade Map*

APPENDIX D COMPANIES WITH MAJOR HOLDINGS IN LNG AND OTHER NOTABLE PROJECTS BY COMPANY

Company	Lead role in LNG export and other major export projects	Participation in LNG export and other major projects	LNG regasification
Shell	Brunei, Malaysia, Nigeria LNG, Oman LNG, Sakhalin II, Russia, Shell Australia LNG Prelude, Australia Pearl GTL, Qatar	North West Shelf, Australia; Qatargas 4 (Train 7); Gorgon, Australia; Greater Sunrise OK LNG, Nigeria	Altamira, Costa Azul, Mexico; Hazira, India; Broadwater, Cove Point, Elba Island, US; Fos Faster, France; Ionio LNG, Italy
Total	Yemen LNG; Shtokman (Technical), Russia; Pars LNG, Iran (suspended)	Bontang, Indonesia Brunei, Abu Dhabi; Oman; Snohvit, Norway; Nigeria LNG; Angola LNG; Brass LNG, Nigeria; Barnett Shale (Chesapeake)	Fos Cavaou, Dunkirk France; Altamira, Mexico; Hazira, India; Sabine Pass, US; South Hook, UK; Krk Island, Croatia
BP	Tangguh, Indonesia	Abu Dhabi; North West Shelf; Bontang, Indonesia	Cove Point, US; Isle of Grain, UK; Guangdong Dapeng, China
ExxonMobil	Arun, Indonesia; PNG LNG, Papua New Guinea	Qatargas 1&2, RasGas, Gorgon, Australia XTO (shale)	South Hook, UK; Golden Pass, US; Rovigo, Italy
Chevron	Angola LNG; Gorgon, Wheatstone, Australia	North West Shelf, Australia; OK LNG, Nigeria	Sabine Pass, US
ENI	Libya (?)	Segas, Egypt Qalhat, Oman Darwin LNG, Australia Brass LNG, Nigeria	Panigaglia, Italy; Spain via Union Fenosa Gas; Cameron, US

ConocoPhillips	Kenai, Alaska; Darwin, Australia; Australia Pacific LNG	Qatargas 3 (T6) Brass LNG, Nigeria Greater Sunrise	Freeport, Golden Pass, US
Marathon	Equatorial Guinea LNG	Kenai, Alaska	Elba Island, Georgia (US)
BG Group	Egyptian LNG; Queensland Curtis LNG, Australia	Atlantic LNG, Trinidad; Equatorial Guinea LNG; OK LNG, Nigeria; Sabine Pass, US	Lake Charles, US; Dragon, Wales; Brindisi, Italy
Repsol	Peru LNG	Atlantic LNG, Trinidad	Canaport, Canada
GDF Suez	Bonaparte LNG, Australia	Egyptian LNG T1; Snohvit, Norway; Trinidad T1	France; UK; Belgium; India; US; Chile; Canada; Italy; Pakistan
Woodside	North West Shelf, Pluto, Browse LNG, Australia; Greater Sunrise		
StatoilHydro	Snohvit, Norway	Egyptian LNG; Gladstone LNG, Australia	Dragon LNG, UK; Port Dickson, Malaysia
Inpex	Ichthys, Australia; Masela, Indonesia	Bontang, Indonesia	Naoetsu
Cheniere	Sabine Pass LNG, US (proposed)	Sabine Pass LNG, US (proposed)	Sabine Pass LNG, US

Sources: IEA MTOGM 2010; Company data

APPENDIX E NORTH AMERICAN LNG IMPORT/EXPORT TERMINALS – PROPOSED/POTENTIAL (AS OF JULY 17, 2012)

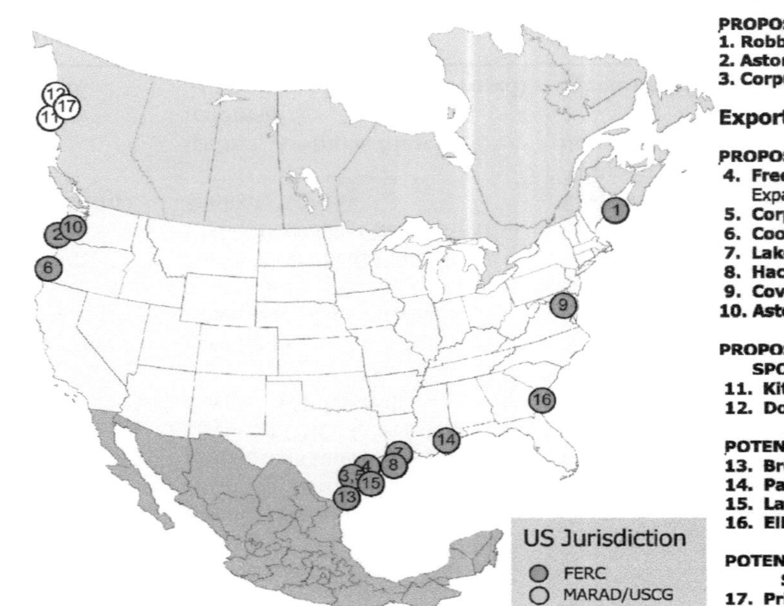

Import Terminal

PROPOSED TO FERC

1. **Robbinston, ME:** 0.5 Bcfd (Kestrel Energy - Downeast LNG)
2. **Astoria, OR:** 1.5 Bcfd (Oregon LNG)
3. **Corpus Christi, TX:** 0.4 Bcfd (Cheniere – Corpus Christi LNG)

Export Terminal

PROPOSED TO FERC

4. **Freeport, TX:** 1.8 Bcfd (Freeport LNG Dev/Freeport LNG Expansion/FLNG Liquefaction)
5. **Corpus Christi, TX:** 1.8 Bcfd (Cheniere – Corpus Christi LNG)
6. **Coos Bay, OR:** 0.9 Bcfd (Jordan Cove Energy Project)
7. **Lake Charles, LA:** 2.4 Bcfd (Southern Union - Trunkline LNG)
8. **Hackberry, LA:** 1.7 Bcfd (Sempra – Cameron LNG)
9. **Cove Point, MD:** 0.75 Bcfd (Dominion – Cove Point LNG)
10. **Astoria, OR:** 1.30 Bcfd (Oregon LNG)

PROPOSED CANADIAN SITES IDENTIFIED BY PROJECT SPONSORS

11. **Kitimat, BC:** 0.7 Bcfd (Apache Canada Ltd.)
12. **Douglas Island, BC:** 0.25 Bcfd (BC LNG Export Cooperative)

POTENTIAL U.S. SITES IDENTIFIED BY PROJECT SPONSORS

13. **Brownsville, TX:** 2.8 Bcfd (Gulf Coast LNG Export)
14. **Pascagoula, MS:** 1.5 Bcfd (Gulf LNG Liquefaction)
15. **Lavaca Bay, TX:** 1.38 Bcfd (Excelerate Liquefaction)
16. **Elba Island, GA:** 0.5 Bcfd (Southern LNG Company)

POTENTIAL CANADIAN SITES IDENTIFIED BY PROJECT SPONSORS

17. **Prince Rupert Island, BC:** 1.0 Bcfd (Shell Canada)

Source: US DOE/FERC, http://www.ferc.gov/industries/gas/indus-act/lng/LNG-proposed-potential.pdf

APPENDIX F APPLICATIONS RECEIVED BY DOE/FE TO EXPORT DOMESTICALLY PRODUCED LNG FROM THE LOWER-48 STATES (AS OF JULY 16, 2012)

Company	Quantity[a]	FTA applications[b] (Docket number)	Non-FTA applications[c] (Docket number)
Sabine Pass Liquefaction, LLC	2.2 billion cubic feet per day (Bcf/d)[d]	Approved (10-85-LNG)	Approved (10-111-LNG)
Freeport LNG Expansion, L.P. and FLNG Liquefaction, LLC	1.4 Bcf/d[d]	Approved (10-160-LNG)	Under DOE Review (10-161-LNG)
Lake Charles Exports, LLC[e]	2.0 Bcf/d[d]	Approved (11-59-LNG)	Under DOE Review (11-59-LNG)
Carib Energy (USA) LLC	0.03 Bcf/d:FTA 0.01 Bcf/d: non FTA[f]	Approved (11-71-LNG)	Under DOE Review (11-141-LNG)
Dominion Cove Point LNG, LP	1.0 Bcf/d[d]	Approved (11-115-LNG)	Under DOE Review (11-128-LNG)
Jordan Cove Energy Project, L.P.	1.2 Bcf/d: FTA 0.8 Bcf/d: non-FTA[g]	Approved (11-127-LNG)	Under DOE Review (12-32-LNG)
Cameron LNG, LLC	1.7 Bcf/d[d]	Approved (11-145-LNG)	Under DOE Review (11-162-LNG)
Freeport LNG Expansion, L.P. and FLNG Liquefaction, LLC[h]	1.4 Bcf/d[d]	Approved (12-06-LNG)	Under DOE Review (11-161-LNG)
Gulf Coast LNG Export, LLC[i]	2.8 Bcf/d	Under DOE Review (12-05-LNG)	
Cambridge Energy, LLC	0.27 Bcf/d[j]	Pending Approval (12-18-LNG)	n/a
Gulf LNG Liquefaction Company, LLC	1.5 Bcf/d	Approved (12-47-LNG)	n/a
LNG Development Company, LLC (d/b/a Oregon LNG)	1.25 Bcf/d: FTA 1.3 Bcf/d: non-FTA	Approved (12-48-LNG)	Under DOE Review (12-77-LNG)
SB Power Solutions Inc.	0.07 Bcf/d	Approved (12-50-LNG)	n/a

Company	Quantity[a]	FTA applications[b] (Docket number)	Non-FTA applications[c] (Docket number)
Southern LNG Company, L.L.C.	0.5 Bcf/d	Approved (12-54-LNG)	n/a
Excelerate Liquefaction Solution I, LLC	1.38 Bcf/d	Pending Approval (12-61-LNG)	n/a
Total of all applications received		18.70 Bcf/d	14.61 Bcf/d

Notes:

(a) Actual applications were in the equivalent annual quantities.

(b) FTA – Applications to export to free trade agreement (FTA) countries. The Natural Gas Act, as amended, has deemed FTA exports to be in the public interest and applications shall be authorized without modification or delay.

(c) Non-FTA applications require DOE to post a notice of application in the Federal Register for comments, protests and motions to intervene, and to evaluate the application to make a public interest consistency determination.

(d) Requested approval of this quantity in both the FTA and non-FTA export applications. Total facility is limited to this quantity (i.e. FTA and non-FTA volumes are not additive at a facility).

(e) Lake Charles Exports, LLC submitted one application seeking separate authorizations to export LNG to FTA countries and another authorization to export to non-FTA countries. The proposed facility has a capacity of 2.0 Bcf/d, which is the volume requested in both the FTA and non-FTA authorizations. [From the October 12, 2012 update, available at http://fossil.energy.gov/programs/gasregulation/reports/Long-Term-LNG-Export-10-16-12.pdf.]

(f) Carib Energy (USA) LLC requested authority to export the equivalent of 11.53 Bcf per year of natural gas to FTA countries and 3.44 Bcf per year to non-FTA countries.

(g) Jordan Cove Energy Project, L.P. requested authority to export the equivalent of 1.2 Bcf/d of natural gas to FTA countries and 0.8 Bcf/d to non-FTA countries.

(h) DOE/FE received a new application (11-161-LNG) by FLEX to export an additional 1.4 Bcf/d of LNG from new trains to be located at the Freeport LNG Terminal, to non-FTA countries, and a separate application (12-06-LNG) to export this same 1.4 Bcf/d of LNG to FTA countries (received January 12, 2012). This 1.4 Bcf/d is in addition to the 1.4 Bcf/d FLEX requested in dockets (10-160-LNG and 10-161-LNG).

(i) An application was submitted by Gulf Coast on January 10, 2012, seeking one authorization to export LNG to any country not prohibited by US law or policy.

(j) Cambridge Energy, LLC requested authority to export up to 2 million metric tons of LNG annually, equivalent to an average of 0.27 Bcf/d of natural gas to FTA countries in Central and South America, the Caribbean, and the Asia/Pacific region.

Source: Office of Oil and Gas Global Security and Supply, Office of Fossil Energy, US Department of Energy, http://fossil.energy.gov/programs/gasregulation/Long_Term_LNG_Export_Concise_07-16-12.2.pdf

Glossary

Annual contract quantity (ACQ): The quantity delivered in each contract year, as specified in a gas or LNG sales contract.

Annual delivery program (ADP): The annual program for the delivery and receipt of LNG cargoes. The ADP is usually agreed on between sellers and buyers before the beginning of each contract year.

Associated natural gas: Gas found mixed with oil in underground reservoirs. Associated natural gas comes out of solution as a by-product of oil production.

Baseload: The volume of gas designed to be delivered by a system that does not fluctuate significantly.

Boil-off or Boil-off gas: The gas produced when LNG in a storage vessel evaporates as heat leaks into the liquid. This vaporized gas can be recovered and used as energy. The evaporation process keeps the liquid cool and is called auto-refrigeration.

Burner tip: The ultimate point of consumption for natural gas.

Cargo containment system: The method of storing LNG in a ship.

Charterparty: A contractual agreement between a shipowner and a cargo owner, whereby a ship is chartered (hired) for a single voyage or over a period of time.

CIF (cost, insurance, and freight) contract: An LNG sales and purchase agreement in which the buyer of the LNG takes ownership either when the LNG is loaded onto the vessel or during the voyage to the receiving terminal. Payment is made at the time of the transfer of ownership: however, the seller remains responsible for the transportation and the transportation cost.

Combined-cycle gas turbine (CCGT): The combination of gas turbines with a waste heat recovery steam generator and a steam turbine in an electric power generation plant.

Compressed natural gas (CNG): Natural gas compressed under high pressures, usually between 3,000 and 3,600 psi. CNG may be used as a vehicle fuel.

Condensates: Hydrocarbon liquids, existing as vapor in a natural gas reservoir that condense to liquids as their temperature and pressure decrease during production. Natural gas condensates consist mainly of pentanes

(C_5H_{12}) and heavier components. They are usually blended with crude oil for refining.

Daily average sendout: The total volume of natural gas delivered over a defined period of time divided by the total number of days in the period.

Debottlenecking: The process of increasing the capacity of a plant by making relatively minor modifications to individual systems.

Deliverability: The amount of natural gas that a well, field, pipeline, or distribution system can supply in a given period (typically 24 hours).

Delivery at Place (DAP): Term used when the seller is providing the LNG shipping to the buyer's facility.

Delivery at Terminal (DAT): Term used when the seller is providing the LNG shipping to the buyer's facility.

Delivery Ex-ship (DES) contract: An LNG sales and purchase agreement in which title to the LNG is transferred to the buyer when the LNG is unloaded at the receiving terminal and payment is made. The term DES has been replaced by DAP as per Incoterms® 2010.

Deregulation: The process of decreasing or eliminating government regulatory control over industries and allowing competitive forces to drive the market.

Downstream: Oil or gas operations that are closer to the end user (see upstream).

Draft: The vertical distance from the waterline to the lowest point of the ship, or stated another way, the least depth of water needed for the ship to float.

Dry gas: Natural gas that doesn't contain liquid hydrocarbons.

Engineering, procurement, and construction (EPC) contract: The EPC contract defines the terms under which the detailed design, procurement, construction, and commissioning of an LNG export or import facility will be conducted.

Exclusion zone: An area surrounding an LNG plant in which an operator legally controls all activities. The exclusion zone creates a buffer in the event of an LNG incident. It provides thermal-radiation protection from fires and flammable vapor-dispersion protection from un-ignited vapor clouds.

Feedstock gas (feedgas): Dry natural gas used as raw material for LNG, petrochemical, and gas-to-liquids plants.

FERC: The Federal Energy Regulatory Commission (United States). FERC regulates interstate transmission of gas and electricity as well as aspects of LNG terminal siting.

Flaring: Burning off unwanted natural gas under controlled conditions.

Force majeure: French for "greater force". Force majeure is a common clause in contracts that essentially releases one or both parties from liability or obligation when an extraordinary event beyond the control of the parties, such as war, strike, riot, crime, act of God (e.g., flood, earthquake, or volcano), prevents one or both parties from wholly or partially performing their obligations under the contract.

Free on board (FOB) contract: An LNG sales and purchase agreement in which the buyer takes title to the LNG on loading at the liquefaction plant and is responsible for transporting it to the receiving terminal.

Front-end engineering design (FEED): The engineering phase that defines the design parameters of a project in sufficient detail to permit an accurate assessment of the project's costs prior to the final investment decision and to provide the basis for the competitive bidding of an EPC contract.

Gas reserves: Gas deposits that geologists and petroleum engineers know or strongly believe can be recovered given today's prices and drilling technology. The term "reserves" is distinct from the term "resources", which includes all the deposits of gas that are still in the ground.

Gas-to-liquids (GTL): A process that converts natural gas into high-value liquid fuels similar to diesel, distillate or heating oil.

Greenfield plant: A brand-new facility built on a parcel of land which has not previously been used for that purpose.

Heads of agreement (HoA): A document outlining the main issues to be included in a subsequent contract or agreement (e.g., the sale and purchase of LNG). It serves as a guideline for both parties negotiating the final documents. HoAs may be binding or non-binding.

Heating value (HV): The amount of heat produced from the complete combustion of a unit quantity of fuel. The gross HV occurs when all the products of combustion are cooled to standard conditions and the latent heat of the water vapor formed is reclaimed. The net HV is the gross value minus the latent heat of the vaporization of the water.

Henry Hub: A point on the U.S. natural gas pipeline system in Erath, Louisiana. Henry Hub is owned by Sabine Pipe Line LLC and interconnects with nine interstate and four intrastate pipelines. It is the standard delivery point for NYMEX natural gas future contracts in the United States and is the point at which the benchmark natural gas price is set for the Gulf Coast.

Hub: A physical location where multiple pipelines interconnect and where buyers and sellers can make transactions.

Hydrocarbon: An organic chemical compound of hydrogen and carbon in gaseous, liquid, or solid phase.

Interchangeability: The ability to substitute one gaseous fuel for another in a combustion application without materially changing operational safety or performance and without materially increasing air pollutant emissions.

Liquefied natural gas (LNG): Natural gas, consisting mainly of methane (CH_4) that has been liquefied by cooling to −260°F (−161°C) at atmospheric pressure. LNG occupies approximately 1/600th of the space required for the vapor state at standard conditions.

Liquefied petroleum gases (LPG): Hydrocarbons that are gases at normal temperatures and pressures but readily turn into liquids under moderate pressure at normal temperatures, including propane, propylene, butane, isobutane, butylenes, or mixtures in any ratio.

LNG chain: The components of an LNG project that link the natural gas in the ground to the ultimate consumer. The main links are natural gas production and processing, liquefaction, shipping, regasification, and distribution.

LNG tanker: A tanker ship designed to transport LNG. An LNG tanker must have insulation and cryogenic facilities to keep LNG cool enough to stay in liquid form.

Manufactured gas: Gas produced from oil, coal, or coke.

Natural gas liquids (NGL): Liquid hydrocarbons extracted from natural gas in the field, typically ethane, propane, butane, pentane, and natural gasoline.

Natural gas processing: The treatment of field gas for use as pipeline-quality gas or as a feedstock for a liquefaction plant. Processing includes removing liquids, solids, vapors, and impurities.

Net back price: The price that the producer of natural gas receives at a defined point. The net back price is derived from the market price for the gas minus delivery charges to the market.

Nonassociated gas: Gas located in a reservoir that does not contain oil.

Off-peak: The period of a day, week, month, or year when demand is at its lowest.

Offshore LNG terminal: An LNG terminal that is located at an offshore location rather than on land.

Open access: A pipeline transportation or LNG terminaling service that is available to all shippers on a non-discriminatory basis, subject to capacity availability.

Peakshaving: The supplementing of the normal supply of gas during periods of peak demand with gas taken from LNG storage facilities, or produced from a mixture of propane and air. These facilities are usually located near load centers (market areas).

Peakshaving plant: A plant that liquefies gas from a pipeline during off-peak periods and stores it as LNG for future use, usually for local markets in cold climates.

Project financing: A common method of financing the construction of industrial infrastructure whereby the value of the plant and part or all of its anticipated revenues are pledged as collateral to secure financing from lenders, who have limited or no rights to seek debt repayment (recourse) from the owners. Also known as non-recourse financing.

Receiving terminal: A large facility capable of receiving and storing the cargo from an LNG ship and vaporizing it for delivery to a pipeline system.

Regasification: Converting LNG back to its gaseous state by the application of heat. Also known as vaporization.

Royalties: A payment to an owner, often a government, for the right to produce oil and gas resources, usually calculated as a percentage of the revenues obtained at the point of production.

Sale and purchase agreement (SPA): A contract between a seller and a buyer for the sale and purchase of a certain quantity of natural gas or LNG for delivery during a specified period at a specified price.

Sendout: The volume of gas delivered by a plant or system in a specified period of time.

Spot sales: The sale of natural gas on a short-term basis.

Stranded gas: Gas that is not close to the market and requires significant infrastructure to deliver it to the market.

Take or pay: A commitment by the buyer of natural gas or LNG to pay for a minimum quantity in a defined period regardless of whether delivery is accepted.

Time charter: A contract for a specified period of time or a particular voyage, in which the shipowner provides the vessel and crew while the charterer supplies the cargo.

Tolling rates: Pricing that charges a customer a fixed rate per volume for use of a facility.

Upstream: Oil or gas operations that are closer to production.

Vaporization: See regasification.

Weathering: The change in LNG composition over time as a result of the preferential evaporation of nitrogen and methane in the boil-off of stored LNG.

Wet gas: Natural gas that produces a liquid condensate when it is brought to the surface.

Source: Adapted from Michael D. Tusiani and Gordon Shearer, LNG A NONTECHNICAL GUIDE, (PennWell 2007); Bob Shively, John Ferrare, and Belinda Petty, UNDERSTANDING TODAY'S GLOBAL LNG BUSINESS (Energy Dynamics Publications 2010).

Index